SECOND EDITION

Numerical Methods for
Partial Differential Equations

SECOND EDITION

Numerical Methods for Partial Differential Equations

WILLIAM F. AMES

University of Iowa
Iowa City, Iowa

School of Mathematics
Georgia Institute of Technology
Atlanta, Georgia

ACADEMIC PRESS

New York   San Francisco   1977

A Subsidiary of Harcourt Brace Jovanovich, Publishers

THOMAS NELSON & SONS

London   Lagos   Melbourne   Toronto

ACADEMIC PRESS, INC.
111 Fifth Avenue, New York, New York 10003

PUBLISHED IN GREAT BRITAIN IN 1977 BY THOMAS NELSON & SONS LIMITED.

THOMAS NELSON AND SONS LTD.
Lincoln Way   Windmill Road   Sunbury-on-Thames   Middlesex   TW16 7HP
P.O. Box 73146   Nairobi   Kenya

THOMAS NELSON (AUSTRALIA) LTD.
19-39 Jeffcott Street   West Melbourne   Victoria 3003

THOMAS NELSON AND SONS (CANADA) LTD.
81 Curlew Drive   Don Mills   Ontario

THOMAS NELSON (NIGERIA) LTD.
8 Ilupeju Bypass   PMB 1303 Ikeja   Lagos

Library of Congress Cataloging in Publication Data

Ames, William F
    Numerical methods for partial differential equations.
    Second edition
    (Computer science and applied mathematics)
    Includes bibliographical references and indexes.
    1.   Differential equations, Partial—Numerical
solutions.   I.   Title.
QA374.A46   1977   515'.353   77-5786
Academic Press: ISBN 0–12–056760–1
Thomas Nelson and Sons Ltd.: ISBN 0 17 771086 1

To the women in my life

Theresa
Della
Mary
Karen
Susan
Pamela

Contents

4 Hyperbolic equations

5 Special topics

6 Weighted residuals and finite elements

Preface to second edition

Since the publication of the first edition, research in and applications of numerical analysis have expanded rapidly. The past few years have witnessed the maturation of numerical fluid mechanics and finite element techniques. Numerical fluid mechanics is addressed in substance in this second edition. I have also added material in several other areas of promise, including hopscotch and other explicit–implicit methods, Monte Carlo techniques, lines, the fast Fourier transform, and fractional steps methods. A new sixth chapter introduces the general concepts of weighted residuals, with emphasis on orthogonal collocation and the Bubnov–Galerkin method. In turn, the latter procedure is used to introduce the finite element concepts.

The spirit of the first edition was to be as self-contained as possible, to present many applications illustrating the theory, and to supply a substantial number of recent references to supplement the text material. This spirit has been retained—there are 38 more problems and 138 additional references. Also, a substantial number of additional applications have been included and references to others appended.

I wish to extend my special thanks to Ms. Mildred Buckalew for the preparation of an outstanding manuscript on the typewriter.

Georgia Institute of Technology

Preface to first edition

That part of numerical analysis which has been most changed by the ongoing revolution in numerical methods is probably the solution of partial differential equations. The equations from the technological world are often very complicated. Usually, they have variable coefficients, nonlinearities, irregular boundaries, and occur in coupled systems of differing types (say, parabolic and hyperbolic). The 'curse of dimensionality' is ever present – problems with two or three space variables, and time, are within our computational grasp.

Early development of calculational algorithms was based more upon the extension of methods for hand computation, empiricism, and intuition than on mathematical analyses. With increasing education and the subsequent development of the professional numerical analyst, the pattern is changing. New, useful methods are evolving which come closer to full utilization of the inherent powers of high-speed, large-memory computing machines. Many significant and powerful methods await discovery both for problems which are computable with existing techniques and those which are not.

Unfortunately, as in other portions of mathematics, the abstract and the applications have tended to diverge. A new field of pure mathematics has been generated and while it has produced some results of value to users, the complexities of real problems have yet to be significantly covered by the presently available theorems. Nevertheless, guidelines are now available for the person wishing to obtain the numerical solution to a practical problem.

The present volume constitutes an attempt to introduce to upper-level engineering and science undergraduate and graduate students the concepts of modern numerical analyses as they apply to partial differential equations. The book, while sprinkled liberally with practical problems and their solutions, also strives to point out the pitfalls – e.g., overstability, consistency requirements, and the danger of extrapolation to nonlinear problems methods which have proven useful on linear problems. The mathematics is by no means ignored, but its development to a keen-edge is not the major goal of this work.

The diligent student will find 248 problems of varying difficulty to test his mettle. Additionally, over 400 references provide a guide to the research and practical problems of today. With this text as a bridge, the applied student should find the professional numerical analysis journals more understandable.

I wish to extend special thanks to Mrs. Gary Strong and Mrs. Steven

Dukeshier for the typing of a difficult manuscript and Mr. Jasbir Arora for preparation of the ink drawings. Lastly, the excellent cooperation and patience of Dr. Alan Jeffrey and my publishers have made the efforts of the past two years bearable.

1
Fundamentals

1-0 Introduction

Numerical calculation is commonplace today in fields where it was virtually unknown before 1950. The high-speed computing machine has made possible the solution of scientific and engineering problems of great complexity. This capability has, in turn, stimulated research in numerical analysis since effective utilization of such devices depends strongly upon the continual advance of research in relevant areas of mathematical analysis. One measure of the growth is the upsurge of books devoted to the subject in the years after 1953. A second measure is the development, during the same period, of at least six research journals whose primary concern is numerical analysis. The major research journals are *SIAM Journal of Numerical Analysis, Mathematics of Computation, Numerische Mathematik, Journal of Computational Physics, Computer Journal,* and *ACM Journal.*†

Finite difference approximations for derivatives were already in use by Euler [1]‡ in 1768. The simplest finite difference procedure for dealing with the problem $dx/dt = f(x, t)$, $x(0) = a$ is obtained by replacing $(dx/dt)_{n-1}$ with the crude approximation $(x_n - x_{n-1})/\Delta t$. This leads to the recurrence relation $x_0 = a$, $x_n = x_{n-1} + \Delta t f(x_{n-1}, t_{n-1})$ for $n > 0$. This procedure is known as Euler's method. Thus we see that for one-dimensional systems the finite difference approach has been deeply ingrained in computational algorithms for quite some time.

For two-dimensional systems the first computational application of finite difference methods was probably carried out by Runge [2] in 1908. He studied the numerical solution of the Poisson equation $u_{xx} + u_{yy} = $ constant. At approximately the same time Richardson [3], in England, was carrying on similar research. His 1910 paper was the earliest work on the application of iterative methods to the solution of continuous equilibrium problems by finite differences. In 1918 Liebmann [4], in considering the finite difference approximation to Laplace's equation, suggested an improved method of iteration. Today the name of Liebmann is associated with any method of iteration by single steps in which a fixed calculation sequence is followed.

The study of errors in finite difference calculations is still an area of prime research interest. Early mathematical convergence proofs were carried out by LeRoux [5], Phillips, and Wiener [6], and Courant, Friedrichs, and Lewy [7].

---

† SIAM is the common abbreviation for Society for Industrial and Applied Mathematics. ACM is the abbreviation for the Association for Computing Machinery.
‡ Numbers in brackets refer to the references at the end of each chapter.

Some consider the celebrated 1928 paper of Courant, Friedrichs, and Lewy as the birthdate of the modern theory of numerical methods for partial differential equations. The algebraic solution of finite difference approximations is best accomplished by some iteration procedure. Various schemes have been proposed to accelerate the convergence of the iteration. A summary of those that were available in 1950, and which are adaptable to automatic programming, is given by Frankel [8]. Other methods require considerable judgment on the part of the computer and are therefore better suited to hand computation. Higgins [9] gives an extensive bibliography of such techniques. In the latter category the method of *relaxation* has received the most complete treatment. Relaxation was the most popular method in the decade of the thirties. Two books by Southwell [10, 11] describe the process and detail many examples. The *successive over-relaxation method*, extensively used on modern computers, is an outgrowth of this highly successful hand computation procedure.

Let us now consider some of the early technical applications. The pioneering paper of Richardson [3] discussed the approximate solution by finite differences of differential equations describing stresses in a masonry dam. Equilibrium and eigenvalue problems were successfully handled. Binder [12] and Schmidt [13] applied finite difference methods to obtain solutions of the diffusion equation. The classical explicit recurrence relation

$$u_{i,j+1} = ru_{i-1,j} + (1 - 2r)u_{i,j} + ru_{i+1,j}, \qquad r = \Delta t/(\Delta x)^2$$

for the diffusion equation $u_t = u_{xx}$ was given by Schmidt [13] in 1924.

For any given continuous system there are a multiplicity of discrete models which are usually comparable in terms of their relative truncation errors. Early approximations were second order—that is, $O(h^2)$†—and these still play an important role today. Higher order procedures were promoted by Collatz [14, 15] and Fox [16]. The relative economy of computation and accuracy of second-order processes utilizing a small interval size, compared with higher order procedures using larger interval sizes, has been discussed in the papers of Southwell [17] and Fox [18].

It is quite possible to formulate a discrete model in an apparently natural way which, upon computation, produces only garbage.‡ This is especially true in propagation problems—that is, problems described by parabolic and hyperbolic equations. An excellent historical example is provided by Richardson's pioneering paper [3], in which his suggested method for the conduction equation, describing the cooling of a rod, was found to be completely unstable by O'Brien, Hyman, and Kaplan [19]. Another example concerns the trans-

---

† The notation $O(h^2)$ is read '(term of) order h^2' and can be interpreted to mean 'when h is small enough the term behaves essentially like a constant times h^2'. Later we make this concept mathematically precise.

‡ Misuse of computational algorithms has been described as GIGO—Garbage In and Garbage Out.

verse vibration of a beam. In 1936 Collatz [20] proposed a 'natural' finite difference procedure for the beam equation $u_{tt} + u_{xxxx} = 0$, but fifteen years later [21] the algorithm was found to be computationally unstable.

Nevertheless, the analyst usually strives to use methods dictated by the problem under consideration—these we call *natural methods*. Thus, a natural coordinate system may be toroidal (see Moon and Spencer [22]) instead of cartesian. Certain classes of equations have natural numerical methods which may be distinct from the finite difference methods. Typical of these are the *method of lines* for propagation problems and the *method of characteristics* for hyperbolic systems. Characteristics also provide a convenient way to classify partial differential equations.

1-1 Classification of physical problems

The majority of the problems of physics and engineering fall naturally into one of three *physical* categories: *equilibrium problems, eigenvalue problems,* and *propagation problems*.

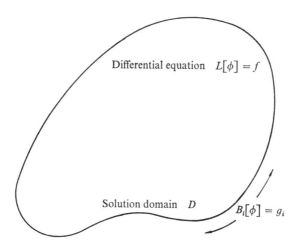

Differential equation $L[\phi] = f$

Solution domain D

$B_i[\phi] = g_i$

Fig. 1-1   Representation of the general equilibrium problem

Equilibrium problems are problems of steady state in which the equilibrium configuration ϕ in a domain D is to be determined by solving the differential equation

$$L[\phi] = f \tag{1-1}$$

within D, subject to certain boundary conditions

$$B_i[\phi] = g_i \tag{1-2}$$

on the boundary of D. Very often, but not always, the integration domain D is closed and bounded. In Fig. 1-1 we illustrate the general equilibrium problem. In mathematical terminology such problems are known as *boundary value problems.* Typical physical examples include steady viscous flow, steady temperature distributions, equilibrium stresses in elastic structures, and steady voltage distributions. Despite the apparent diversity of the physics we shall shortly see that the governing equations for equilibrium problems are *elliptic.*†

Eigenvalue problems may be thought of as extensions of equilibrium problems wherein critical values of certain parameters are to be determined in addition to the corresponding steady-state configurations. Mathematically the problem is to find one or more constants (λ), and the corresponding functions (ϕ), such that the differential equation

$$L[\phi] = \lambda M[\phi] \qquad (1\text{-}3)$$

is satisfied within D and the boundary conditions

$$B_i[\phi] = \lambda E_i[\phi] \qquad (1\text{-}4)$$

hold on the boundary of D. Typical physical examples include buckling and stability of structures, resonance in electric circuits and acoustics, natural frequency problems in vibrations, and so on. The operators L and M are of elliptic type.

Propagation problems are initial value problems that have an unsteady state or transient nature. One wishes to predict the subsequent behavior of a system given the initial state. This is to be done by solving the differential equation

$$L[\phi] = f \qquad (1\text{-}5)$$

within the domain D when the initial state is prescribed as

$$I_i[\phi] = h_i \qquad (1\text{-}6)$$

and subject to prescribed conditions

$$B_i[\phi] = g_i \qquad (1\text{-}7)$$

on the (open) boundaries. The integration domain D is open. In Fig. 1-2 we illustrate the general propagation problem. In mathematical parlance such problems are known as *initial boundary value problems.*‡ Typical physical examples include the propagation of pressure waves in a fluid, propagation of stresses and displacements in elastic systems, propagation of heat, and the development of self-excited vibrations. The physical diversity obscures the fact that the governing equations for propagation problems are *parabolic or hyperbolic.*

The distinction between equilibrium and propagation problems was well

---

† The original mathematical formulation of an equilibrium problem will generate an elliptic equation or system. Later mathematical approximations may change the type. A typical example is the boundary layer approximation of the equations of fluid mechanics. Those elliptic equations are approximated by the parabolic equations of the boundary layer. Yet the problem is still one of equilibrium.

‡ Sometimes only the terminology initial value problem is utilized.

stated by Richardson [23] when he described the first as *jury* problems and the second as *marching* problems. In equilibrium problems the entire solution is passed on by a jury requiring satisfaction of all the boundary conditions and all the internal requirements. In propagation problems the solution marches out from the initial state guided and modified in transit by the side boundary conditions.

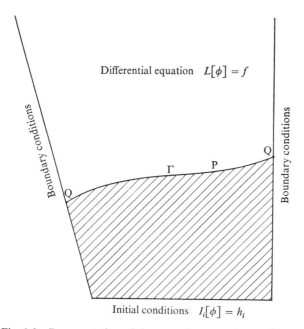

Fig. 1-2 Representation of the general propagation problem

1-2 Classification of equations

The previous physical classification emphasized the distinctive features of basically two classes of problems. These distinctions strongly suggest that the governing equations are quite different in character. From this we infer that the numerical methods for both problems must also have some basic differences. Classification of the equations is best accomplished by developing the concept of *characteristics*.

Let the coefficients $a_1, a_2, \ldots, f_1, f_2$ be functions of $x, y, u,$ and v and consider the simultaneous first-order quasilinear system†

$$a_1 u_x + b_1 u_y + c_1 v_x + d_1 v_y = f_1$$
$$a_2 u_x + b_2 u_y + c_2 v_x + d_2 v_y = f_2 \ddagger$$

(1-8)

† A quasilinear system of equations is one in which the highest order derivatives occur linearly.

‡ We shall often use the notation u_x to represent $\partial u/\partial x$.

This set of equations is sufficiently general to represent many of the problems encountered in engineering where the mathematical model is second order.

Suppose that the solution for u and v is known from the initial state to some curve Γ.† At any boundary point P of this curve, we know the continuously differentiable values of u and v and the directional derivatives of u and v in directions *below* the curve (see Fig. 1-2).

We now seek the answer to the question: 'Is the behavior of the solution just above P uniquely determined by the information below and on the curve?' Stated alternatively: 'Are these data sufficient to determine the directional derivatives at P in directions that lie above the curve Γ?' By way of reducing this question, suppose that θ (an angle with the horizontal) specifies a direction along which σ measures distance. If u_x and u_y are known at P, then the directional derivative

$$u_\sigma|_\theta = u_x \cos \theta + u_y \sin \theta = u_x \frac{dx}{d\sigma} + u_y \frac{dy}{d\sigma} \tag{1-9}$$

is also known, so we restate the question in the simpler form: 'Under what conditions are the derivatives u_x, u_y, v_x, and v_y uniquely determined at P by values of u and v on Γ?' At P we have four relations, Eqns (1-8) and

$$du = u_\sigma \, d\sigma = u_x \, dx + u_y \, dy$$
$$dv = v_\sigma \, d\sigma = v_x \, dx + v_y \, dy \tag{1-10}$$

whose matrix form is

$$\begin{bmatrix} a_1 & b_1 & c_1 & d_1 \\ a_2 & b_2 & c_2 & d_2 \\ dx & dy & 0 & 0 \\ 0 & 0 & dx & dy \end{bmatrix} \begin{bmatrix} u_x \\ u_y \\ v_x \\ v_y \end{bmatrix} = \begin{bmatrix} f_1 \\ f_2 \\ du \\ dv \end{bmatrix} \tag{1-11}$$

With u and v known at P the coefficient functions $a_1, a_2, \ldots, f_1, f_2$ are known. With the direction of Γ known, dx and dy are known; and if u and v are known along Γ, du and dv are also known. Thus, the four equations [Eqns (1-11)] for the four partial derivatives have known coefficients. A unique solution for u_x, u_y, v_x, and v_y exists if the determinant of the 4×4 matrix in Eqns (1-11) is *not zero*. If the determinant is not zero, then the directional derivatives have the same value above and below Γ.

The exceptional case, when the determinant is zero, implies that a multiplicity of solutions are possible. Thus, the system of Eqns (1-11) does not determine the partial derivatives uniquely. Consequently, discontinuities in

---

† We restrict this discussion to a finite domain in which discontinuities do not occur. Later developments consider the degeneration of smooth solutions into discontinuous ones. Additional information is available in Jeffrey and Taniuti [24] and Ames [25].

the partial derivatives may occur as we cross Γ. Upon equating to zero the determinant of the matrix in Eqns (1-11) we find the *characteristic equation*

$$(a_1c_2 - a_2c_1)(dy)^2 - (a_1d_2 - a_2d_1 + b_1c_2 - b_2c_1)\,dxdy$$
$$+ (b_1d_2 - b_2d_1)(dx)^2 = 0 \quad (1\text{-}12)$$

which is a quadratic equation in dy/dx. If the curve Γ (Fig. 1-2) at P has a slope such that Eqn (1-12) is satisfied, then the derivatives u_x, u_y, v_x, and v_y are not uniquely determined by the values of u and v on Γ. The directions specified by Eqn (1-12) are called *characteristic directions*; they may be real and distinct, real and identical, or not real according to whether the discriminant

$$(a_1d_2 - a_2d_1 + b_1c_2 - b_2c_1)^2 - 4(a_1c_2 - a_2c_1)(b_1d_2 - b_2d_1) \quad (1\text{-}13)$$

is positive, zero, or negative. This is also the criterion for classifying Eqns (1-8) as hyperbolic, parabolic, or elliptic. They are *hyperbolic* if Eqn (1-13) is positive—that is, has two *real* characteristic directions; *parabolic* if Eqn (1-13) is zero; and *elliptic* if there are no real characteristic directions.

Next consider the quasilinear second-order equation

$$au_{xx} + bu_{xy} + cu_{yy} = f \quad (1\text{-}14)$$

where a, b, c, and f are functions of x, y, u, u_x, and u_y. The classification of Eqn (1-14) can be examined by reduction to a system of first-order equations† or by independent treatment. Taking the latter course we ask the conditions under which a knowledge of u, u_x, and u_y on Γ (see Fig. 1-2) serve to determine u_{xx}, u_{xy},‡ and u_{yy} uniquely so that Eqn (1-14) is satisfied. If these derivatives exist we must have

$$d(u_x) = u_{xx}\,dx + u_{xy}\,dy$$
$$d(u_y) = u_{xy}\,dx + u_{yy}\,dy \quad (1\text{-}15)$$

---

† Transformation of Eqn (1-14) into a system of first-order equations is *not unique*. This 'nonuniqueness' is easily demonstrated. Substitutions (i) $w = u_x$, $v = u_y$, and (ii) $w = u_x$, $v = u_x + u_y$ both reduce Eqn (1-14) to two first-order equations.
For (i) we find the system
$$aw_x + bw_y + cv_y = f$$
$$w_y - v_x = 0$$
and for (ii) we have
$$aw_x + (b - c)w_y + cv_y = f$$
$$w_y - v_x - w_x = 0$$

Some forms may be more convenient than others during computation. An example of this, from a paper by Swope and Ames [26], will be discussed in Chapter 4.

‡ Throughout, unless otherwise specified, we shall assume that the continuity condition, under which $u_{xy} = u_{yx}$, is satisfied.

Eqns (1-15), together with Eqn (1-14), has the matrix form

$$\begin{bmatrix} a & b & c \\ dx & dy & 0 \\ 0 & dx & dy \end{bmatrix} \begin{bmatrix} u_{xx} \\ u_{xy} \\ u_{yy} \end{bmatrix} = \begin{bmatrix} f \\ d(u_x) \\ d(u_y) \end{bmatrix} \tag{1-16}$$

Thus the solution for u_{xx}, u_{xy}, and u_{yy} exists, and it is unique unless the determinant of the coefficient matrix vanishes, that is

$$a(dy)^2 - b\,dy\,dx + c(dx)^2 = 0. \tag{1-17}$$

Accordingly, the characteristic equation for the second-order quasilinear equation is (1-17). Equation (1-14) is hyperbolic if $b^2 - 4ac > 0$, parabolic if $b^2 - 4ac = 0$, and elliptic if $b^2 - 4ac < 0$. Since a, b, and c are functions of x, y, u, u_x, and u_y, an equation may change its type from region to region.

In the hyperbolic case there are two real characteristic curves. Since the higher order derivatives are indeterminate along these curves they provide paths for the propagation of discontinuities. Indeed, shock waves and other disturbances do propagate into media along characteristics.

The characteristic directions for the linear wave equation

$$u_{xx} - \alpha^2 u_{yy} = 0 \quad (\alpha \text{ constant}) \tag{1-18}$$

are $\quad (dy)^2 - \alpha^2(dx)^2 = 0$

or $\quad\quad\quad y \pm \alpha x = \beta. \tag{1-19}$

These are obviously straight lines.

A more complicated example is furnished by the nozzle problem. The governing equations of steady two-dimensional irrotational isentropic flow of a gas are (see, for example, Shapiro [27]):

$$uu_x + vu_y + \rho^{-1}p_x = 0$$

$$uv_x + vv_y + \rho^{-1}p_y = 0$$

$$(\rho u)_x + (\rho v)_y = 0 \tag{1-20}$$

$$v_x - u_y = 0$$

$$p\rho^{-\gamma} = \text{constant}, \quad \frac{dp}{d\rho} = c^2$$

where u and v are velocity components, p is pressure, ρ is density, c is the velocity of sound, and γ is the ratio of specific heats (for air $\gamma = 1.4$).

By multiplying the first of Eqns (1-20) by ρu, the second by ρv, using $dp = c^2\,d\rho$, and adding the two resulting equations we find that Eqns (1-20) are equivalent to the following pair of first-order equations for u and v,

$$(u^2 - c^2)u_x + (uv)u_y + (uv)v_x + (v^2 - c^2)v_y = 0$$

$$-u_y + v_x = 0 \tag{1-21}$$

where $5c^2 = 6c^{*2} - (u^2 + v^2)$ and the quantity c^* is a reference sound velocity chosen as the sound velocity when the flow velocity $[(u^2 + v^2)^{1/2}]$ is equal to c. This problem can be put in dimensionless form by setting

$$u' = u/c^*, \qquad v' = v/c^*, \qquad c' = c/c^*, \qquad x' = x/l, \qquad y' = y/l$$

where l is one-half the nozzle width. Inserting these values in Eqns (1-21), and dropping the primes, the dimensionless equations are

$$(u^2 - c^2)u_x + (uv)u_y + (uv)v_x + (v^2 - c^2)v_y = 0$$
$$-u_y + v_x = 0$$

$$(1-22)$$

with $c^2 = 1.2 - 0.2(u^2 + v^2)$.

The characteristic directions are obtained from the particular form of Eqns (1-11)

$$\begin{bmatrix} (u^2 - c^2) & uv & uv & (v^2 - c^2) \\ 0 & -1 & 1 & 0 \\ dx & dy & 0 & 0 \\ 0 & 0 & dx & dy \end{bmatrix} \begin{bmatrix} u_x \\ u_y \\ v_x \\ v_y \end{bmatrix} = \begin{bmatrix} 0 \\ 0 \\ du \\ dv \end{bmatrix}$$

as

$$\left.\frac{dy}{dx}\right|_\alpha = \frac{uv + c[u^2 + v^2 - c^2]^{1/2}}{u^2 - c^2} \qquad (1\text{-}23a)$$

$$\left.\frac{dy}{dx}\right|_\beta = \frac{uv - c[u^2 + v^2 - c^2]^{1/2}}{u^2 - c^2} \qquad (1\text{-}23b)$$

where α and β are labels used to distinguish the two directions. When the flow is *subsonic*, $u^2 + v^2 < c^2$, the characteristics are complex, and Eqns (1-22) are therefore elliptic; when the flow is transonic, $u^2 + v^2 = c^2$ and Eqns (1-22) are parabolic; and, when $u^2 + v^2 > c^2$, the flow is supersonic and Eqns (1-22) are hyperbolic.

In Chapter 4 the characteristics will be utilized in developing a numerical method for hyperbolic systems.

PROBLEMS

1-1 In dimensionless form the *threadline* equation from Swope and Ames [26] is

$$y_{tt} + \alpha y_{xt} + \tfrac{1}{4}(\alpha^2 - 4)y_{xx} = 0 \qquad (1\text{-}24)$$

where $\alpha = 2v/c$. Find the characteristics of this equation and classify it.

1-2 The one-dimensional isentropic flow of a perfect gas is governed by the equations of momentum, continuity, and gas law which are, respectively,

$$u_t + uu_x + \rho^{-1}p_x = 0 \qquad (1\text{-}25a)$$

$$\rho_t + \rho u_x + u\rho_x = 0 \qquad (1\text{-}25b)$$

$$p\rho^{-\gamma} = \alpha = \text{constant}, \qquad c^2 = dp/d\rho \qquad (1\text{-}25c)$$

where x is displacement, t is time, $u(x, t)$, $p(x, t)$, and $\rho(x, t)$ are the velocity, pressure, and density, respectively. c is the velocity of sound in the gas and γ, the ratio of specific heats, is constant.

(a) Eliminate the pressure and write the two equations for u and ρ.

(b) Find the characteristics of the system and classify it.

1-3 The nonlinear longitudinal oscillations of a string have been modeled by Zabusky [28] as

$$y_{tt} = [1 + \epsilon y_x]^\alpha y_{xx}. \tag{1-26}$$

Find the characteristics and classify.

1-4 Nearly uniform transonic flow of a real gas has been examined by Tomotika and Tamada [29] with the equation

$$w_{\psi\psi} = k[w^2]_{\phi\phi}, \quad k > 0. \tag{1-27}$$

Find the characteristics and classify.

1-5 Show that the introduction of a velocity potential ϕ (defined by $\partial\phi/\partial x = u$, $\partial\phi/\partial y = -v$) allows Eqns (1-22) to be transformed to a single second-order equation.

1-6 During compression of a plastic bar the conditions of force equilibrium give rise to the equations (Hill [30])

$$(2k)^{-1}p_x + (\cos 2\psi)\psi_x + (\sin 2\psi)\psi_y = 0$$
$$(2k)^{-1}p_y + (\sin 2\psi)\psi_x - (\cos 2\psi)\psi_y = 0 \tag{1-28}$$

where ψ and p are the dependent variables. Find the characteristics.

1-7 Reduce the pair of equations in Problem 1-6 to a single second-order equation.

1-8 In Eqns (1-8) suppose that $f_1 = f_2 = 0$ and that the coefficient functions $a_1, b_1, \ldots, d_2$ are functions only of the dependent variables u and v. Show that this is *reducible* to a linear system by interchanging the roles of the dependent and independent variables.

1-3 Asymptotics

We often wish to evaluate a certain number defined in a particular way, but because of the large number of operations required, direct evaluation is not feasible. In such cases an alternative method to obtain information—or, to develop a useful approximation—would be highly desirable. A situation like this is considered to belong to asymptotics. Asymptotics is certainly not a new field, but it is only recently that special courses and books have been devoted to the subject. Some of the more useful volumes are those of Erdelyi [31], Jeffreys [32], and de Bruijn [33].

Usually this new method gives improved accuracy when the number of operations involved in the definition increases. Thus, a typical (and one of the oldest) asymptotic results is Stirling's formula

$$\lim_{n \to \infty} \frac{n!}{e^{-n} n^n \sqrt{(2\pi n)}} = 1. \tag{1-29}$$

For each n, $n!$ is evaluated without any theoretical problems. However, for larger n, the number of necessary operations increases. But Eqn (1-29) suggests, as a reasonable approximation to $n!$, the quantity

$$e^{-n} n^n \sqrt{(2\pi n)} \tag{1-30}$$

and for larger n, the relative error decreases.

Since Eqn (1-29) is a limit expression it has, as it stands, little value for numerical purposes. We can draw no conclusion from Eqn (1-29) about $n!$ for any selected value of n. It is a statement about an infinite number of values of n, and no statement about any special value of n can be inferred.

Equation (1-29) will be written, for further discussion, as

$$\lim_{n \to \infty} F(n) = 1 \tag{1-31}$$

which is an *information suppressing* notation. In fact Eqn (1-31) expresses the statement:

'for each $\epsilon > 0$ there exists $N(\epsilon)$ such that $|F(n) - 1| < \epsilon$ whenever $n > N(\epsilon)$' (1-32)

When developing a proof that $\lim_{n \to \infty} F(n) = 1$, one usually produces information of the form of Eqn (1-32) with explicit construction of $N(\epsilon)$. This knowledge of $N(\epsilon)$ actually means numerical information about F, but when notation (1-31) is employed this information is suppressed. The knowledge of a function $N(\epsilon)$ with the property of (1-32) is replaced by the knowledge of the existence of such a function. One of the reasons for the great success of analysis is that a notation has been found which is still useful even though much information is suppressed. The existence of functions $N(\epsilon)$ is easier to handle than the functions themselves.

We shall find a weaker form of suppression of information most useful. Bachmann and Landau [see ref. 34] introduced the O (big 'oh') notation which does not suppress a function but only a number; that is, it replaces the *knowledge of a number* with certain properties by the *knowledge that such a number exists*! Clearly the O notation suppresses much less information than the limit notation and yet, as we shall see, it is easy to handle.

Let S be any set and f, g, ϕ be real or complex functions defined on S. Then the notation

$$f(s) = O[\phi(s)], \quad s \text{ in } S \tag{1-33}$$

means that a positive number A exists, independent of s, such that

$$|f(s)| \leq A |\phi(s)| \quad \text{for all } s \text{ in } S. \tag{1-34}$$

Thus, the O symbol means 'something that is, in absolute value, at most a constant multiple of the absolute value of'. If $\phi(s) \neq 0$ for all s in S, then Eqn (1-34) means that $f(s)/\phi(s)$ is bounded in S.

Some obvious examples are

$$\sin x = O(x) \quad (-\infty < x < \infty), \qquad e^x - 1 = O(x) \quad (-1 < x < 1).$$

In the latter example we are obviously interested in small values of x—in fact it is the fault of the large values of x that renders the statement $e^x - 1 = O(x)$ $(-\infty < x < \infty)$ untrue. Thus, a finite interval is indicated.

There are cases where one has difficulty in determining a suitable interval. To eliminate these nonessential minor inconveniences a modified O notation is introduced—of course, some information is still suppressed. We shall explain it for the case where $x \to \infty$ and note that obvious modifications lead to similar notation for other examples.

The notation

$$f(x) = O[\phi(x)], \quad (x \to \infty) \tag{1-35}$$

is to be interpreted to mean that a real number c exists such that

$$f(x) = O[\phi(x)], \quad c < x < \infty \tag{1-36}$$

that is, real numbers c and K exist such that

$$|f(x)| \leq K |\phi(x)| \quad \text{for all } x, \quad c < x < \infty$$

Some examples are

$$e^{-x} = O(x^{-m}) \quad (x \to \infty), \quad m \text{ a positive integer};$$
$$\sin x = O(x) \quad (x \to 0)$$

If $f(x)$ and $\phi(x)$ are continuous in $0 \leq x < \infty$ and $\phi(x) \neq 0$ in this interval, then we can replace Eqn (1-35) by the relation

$$f(x) = O(\phi(x)) \quad (0 \leq x < \infty)$$

that is, a O formula of the type given by Eqn (1-33). Of course this replacement is possible because f/ϕ is continuous and therefore bounded on $0 \leq x \leq c$.

Care must be exercised in using these and other asymptotic notations. The symbol $O(\phi(s))$ was not defined, only the meanings of several complete formulae are given. The isolated expression $O(\phi(x))$ cannot be defined in such a way that Eqn (1-33) remains equivalent to Eqn (1-34). Clearly $f(x) = O(\phi(x))$ implies $3f(x) = O(\phi(x))$. If $O(\phi(x))$ were to denote anything, then $f(x) = O(\phi(x)) = 3f(x)$, whereupon $f(x) = 3f(x)$! Where is the difficulty? The trouble lies in the abuse of the 'equals' sign. This sign is an equivalence relation, implying symmetry. But there is no symmetry here. For example, $O(x) = O(x^2)$ $(x \to \infty)$ is correct, but $O(x^2) = O(x)$ $(x \to \infty)$ is not.

Proper algebraic use of these notations must bear the latter warning in mind. A few examples will illustrate their proper usage.

The relation

$$O(x) + O(x^3) = O(x) \quad (x \to 0) \tag{1-37a}$$

means that if $f(x) = O(x)$ $(x \to 0)$, $g(x) = O(x^3)$ $(x \to 0)$, then

$$f(x) + g(x) = O(x) \quad (x \to 0) \tag{1-37b}$$

Similarly,

$$O(x^2) + O(x^4) = O(x^4) \quad (x \to \infty)$$

$$\exp[O(1)] = O(1) \quad (-\infty < x < \infty)$$

$$\exp[O(x^2)] = \exp[O(x^4)] \quad (x \to \infty)$$

Notation that is analogous to

$$G(x) = 1 + x + x^2 + O(x^3) \quad (x \to 0) \tag{1-38}$$

will often be used, subsequently, in this volume. This means that there exists a function f such that $G(x) = 1 + x + x^2 + f(x)$ and $f(x) = O(x^3)$ $(x \to 0)$.

A useful common interpretation of all of these formulae can be stated in the following manner. *Any relation involving the O notation is to be interpreted as a class of functions.* If the range $0 < x < a$ is considered, then $O(1) + O(x^3)$ represents the class of all functions having the form $f(x) + g(x)$, where $f(x) = O(1)$ $(0 < x < a)$, and $g(x) = O(x^3)$ $(0 < x < a)$.

A second information suppressing symbol (little 'oh') will also be used. This notation

$$f(x) = o[\phi(x)] \quad (x \to \infty) \tag{1-39}$$

is to be interpreted to mean that $f(x)/\phi(x)$ tends to zero as $x \to \infty$. Since this suppresses a great deal more information than the corresponding O notation it is a much stronger assertion and therefore less important than O relations. Equation (1-39) implies Eqn (1-35) since convergence implies boundedness from some point onward.

Equation (1-39) is to be read as 'something that tends to zero, multiplied by'. This is similar to the convention used in defining the O symbol. Examples include

$$\cos x = 1 + o(x) \quad (x \to 0)$$

$$n! = e^{-n} n^n \sqrt{(2\pi n)}(1 + o(1)) \quad (n \to \infty)$$

$$e^x = 1 + x + o(x) \quad (x \to 0) \tag{1-40}$$

$$o[f(x)g(x)] = o(f(x))O(g(x)) \quad (x \to 0)$$

$$o[f(x)g(x)] = f(x)o(g(x)) \quad (x \to 0)$$

PROBLEMS

1-9 Using the definitions show that

$$\sin x = O(x) \quad (-\infty < x < \infty); \quad 2x + x\sin x = O(x) \quad (x \to \infty).$$

1-10 Give an example of the class of functions described by

$$O(x^{-1}) + O(\ln x) \quad (x \to \infty).$$

1-11 Write a O notation for xe^x as $x \to 0$. What is a corresponding 'little oh' notation?

1-12 Establish the validity of Eqns (1-40).

Continuous domain in two dimensions
(a)

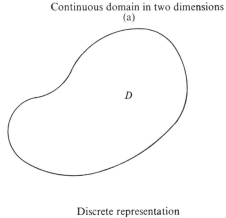

Discrete representation
(b)

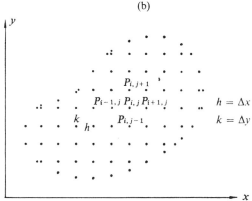

Fig. 1-3 Discrete approximation of a continuous two-dimensional domain

1-4 Discrete methods

The ultimate goal of discrete methods is the reduction of continuous systems to 'equivalent' discrete (lumped parameter) systems which are suitable for high-speed computer solution. One is initially deceived by the seeming elementary nature of the techniques. Their usage is certainly widespread but alas they are often misapplied and abused. A little knowledge is dangerous since these approximations raise many serious and difficult mathematical questions of adequacy, accuracy, convergence, and stability.

The basic approximation involves the replacement of a continuous domain D by a pattern, network, or mesh of discrete points within D, such as that shown in Fig. 1-3 for two dimensions.† Instead of developing a solution defined everywhere in D, only approximations are obtained at the isolated points labeled $P_{i,j}$. Intermediate values, integrals, derivatives, or other operator values may be obtained from this discrete solution by interpolatory techniques.

Discretization of the governing equations and boundary conditions of the continuous problem may be accomplished *physically*, but is more often carried out *mathematically*. The specialist sometimes finds the physical approach useful in motivating further analyses. In such a modus operandi the discrete (physical) model is given the lumped physical characteristics of the continuous system. For example, a heat conducting slab could be replaced by a network of heat conducting rods. The governing equations are then developed by direct application of the physical laws to the discrete system.

On the other hand, in the *mathematical* approach the continuous formulation is transformed to a‡ discrete formulation by replacing derivatives by, say, finite difference approximations. When the continuous problem formulation is already available this procedure is simpler and more flexible. We shall restrict our attention to the mathematical approach in what follows. Development of discrete approximations can proceed by several avenues, notably finite difference methods, variational methods, and the method of lines. In this chapter we shall confine ourselves to the first of these.

1-5 Finite differences and computational molecules §

Partial derivatives can be approximated by finite differences in many ways. All such approximations introduce errors, called *truncation errors*, whose presence will be signified by employing the asymptotic O notation. Several simple approximations will be developed here, leaving refinements for later chapters.

Let the problem under consideration be the two-dimensional boundary value problem

$$Lu = f, \qquad u = u(x, y) \tag{1-41}$$

in a domain D (see Fig. 1-3) subject to certain boundary conditions on the boundary of D. Let the points $P_{i,j}$ form a discrete approximation for D with uniform spacing $h = \Delta x$, $k = \Delta y$. A simple approximation for $\partial u/\partial x|_{i,j}$ will now be developed, where the notation $u_{i,j} = u(ih, jk)$ will ultimately be employed for the exact solution and $U_{i,j}$ for the discrete approximation.

---

† A variety of patterns is possible and sometimes preferable to that shown in Fig. 1-3.
‡ Note that more than one discrete formulation is possible.
§ This terminology, due to Bickley [35], is useful for visualizing the course of the computation. Other designations are stencil, star, and lozenge (see Milne [36] or Forsythe and Wasow [37]).

Development of the Taylor series for $u(x + \Delta x, y)$ about (x, y) gives

$$u(x + \Delta x, y) = u(x, y) + \Delta x \frac{\partial u}{\partial x}(x, y) + \frac{(\Delta x)^2}{2!} \frac{\partial^2 u}{\partial x^2}(x, y)$$

$$+ \frac{(\Delta x)^3}{3!} \frac{\partial^3 u}{\partial x^3}(x, y) + O[(\Delta x)^4] \qquad (1\text{-}42)$$

which, upon division by Δx, results in the relation

$$\frac{\partial u}{\partial x}(x, y) = [u(x + \Delta x, y) - u(x, y)]/\Delta x + O(\Delta x). \qquad (1\text{-}43)$$

The *forward difference* of Eqn (1-43) provides the simple *first-order*† approximation

$$\frac{\partial u}{\partial x} \approx [u(x + \Delta x, y) - u(x, y)]/\Delta x‡ \qquad (1\text{-}44)$$

for $\partial u/\partial x$ evaluated at (x, y). In the double subscript notation Eqn (1-43) would be written

$$\left.\frac{\partial u}{\partial x}\right|_{i,j} = \frac{1}{h}[u_{i+1,j} - u_{i,j}] + O(h)§ \qquad (1\text{-}45)$$

—a notation we shall commonly employ. The quantity $O(h)$ represents the asymptotic notation for the *truncation error* of this approximation.

As an alternative to the forward difference approximation of Eqn (1-45) a *backward difference* is obtained in a similar fashion. The Taylor series for $u(x - \Delta x, y)$ about (x, y) is

$$u(x - \Delta x, y) = u(x, y) - \frac{\partial u}{\partial x}(\Delta x) + \frac{1}{2!}\frac{\partial^2 u}{\partial x^2}(\Delta x)^2$$

$$- \frac{1}{3!}\frac{\partial^3 u}{\partial x^3}(\Delta x)^3 + O[(\Delta x)^4] \qquad (1\text{-}46)$$

where all derivatives are evaluated at (x, y). Upon division by Δx we find the relation

$$\left.\frac{\partial u}{\partial x}\right|_{i,j} = \frac{1}{h}[u_{i,j} - u_{i-1,j}] + O(h) \qquad (1\text{-}47)$$

which, upon suppression of the truncation error, yields a *backward difference* approximation, also of first order in truncation error.

Perhaps the easiest way to visualize the development of a higher order

---

† If $e(x)$ is an approximation to $E(x)$ we say it is of order n, with respect to some quantity Δx, if n is the largest possible positive real number such that $|E - e| = O[(\Delta x)^n]$ as $\Delta x \to 0$.

‡ We use $\approx$ as the symbol representing approximation.

§ Unless otherwise stated, $O(h^n)$ as $h \to 0$ is intended.

approximation to $\partial u/\partial x$ is to subtract Eqn (1-46) from Eqn (1-42). The result, where all derivatives are evaluated at (x, y), is

$$u(x + \Delta x, y) - u(x - \Delta x, y) = 2\Delta x \frac{\partial u}{\partial x} + \frac{(\Delta x)^3}{3} \frac{\partial^3 u}{\partial x^3} + O[(\Delta x)^5]\dagger \quad (1\text{-}48)$$

and upon division by $2\Delta x$ generates the *second-order* approximation

$$\left. \frac{\partial u}{\partial x} \right|_{i, j} = \frac{u_{i+1, j} - u_{i-1, j}}{2\Delta x} + O[(\Delta x)^2]. \quad (1\text{-}49)$$

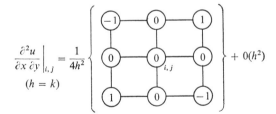

Fig. 1-4   Simple computational molecules for partial derivatives

While it is true that Eqn (1-49) has a second-order truncation error this does not mean that its application will always give rise to a more useful numerical technique than Eqn (1-45). Indeed, in Chapter 2 we shall show that *an always unstable method will result if Eqn (1-49) is applied in a naive way to the diffusion equation* $u_t = u_{xx}$!

Elementary approximations for second partial derivatives are obtainable

† This occurs since during the subtraction process all the even-order terms cancel.

from the Taylor series of Eqns (1-42) and (1-46). For example, on addition of those two equations we find

$$\frac{1}{(\Delta x)^2} \{u(x + \Delta x, y) - 2u(x, y) + u(x - \Delta x, y)\} = \frac{\partial^2 u}{\partial x^2} + O[(\Delta x)^2]. \quad (1\text{-}50)$$

In the index notation we would write

$$\frac{\partial^2 u}{\partial x^2}\bigg|_{i,j} = \frac{u_{i+1,j} - 2u_{i,j} + u_{i-1,j}}{h^2} + O(h^2). \quad (1\text{-}51)$$

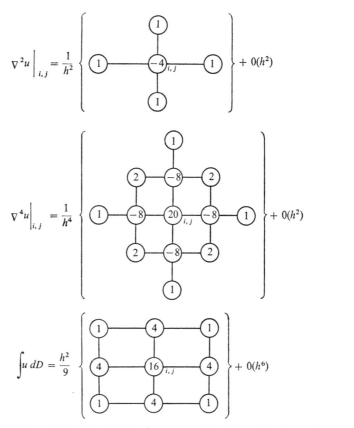

Fig. 1-5   Computational molecules for common two dimensional operators

Equations (1-49), (1-51), and a corresponding result for $\partial^2 u/\partial x \partial y$ may be pictorially represented by the computational molecules of Fig. 1-4. The numbers, in the various positions, represent the multipliers that are to be applied to the values of u at these stations.

Other common two-dimensional operators include the Laplacian ($\nabla^2 u$), the biharmonic ($\nabla^4 u$), and the integral (two-dimensional Simpson's rule). Simple approximations for these are developed in a manner analogous to the above discussion. The corresponding computation molecules are given in Fig. 1-5. For example, consider the Laplacian $\nabla^2 u = u_{xx} + u_{yy}$ with equal spacing $(h = k)$ by applying Eqn (1-51) and the corresponding relation for u_{yy}. Thus

$$\nabla^2 u \mid_{i,j} = u_{xx}\mid_{i,j} + u_{yy}\mid_{i,j}$$

$$= \frac{1}{h^2} \{u_{i+1,j} - 2u_{i,j} + u_{i-1,j} + u_{i,j+1} - 2u_{i,j} + u_{i,j-1}\} + O(h^2)$$

$$= \frac{1}{h^2} \{u_{i+1,j} + u_{i-1,j} + u_{i,j+1} + u_{i,j-1} - 4u_{i,j}\} + O(h^2). \quad (1\text{-}52)$$

In the next section we shall describe methods for developing higher order approximations.

PROBLEMS

1-13 Using two-dimensional Taylor series develop the computational molecule for u_{xy} shown in Fig. 1-4.

1-14 Using two-dimensional Taylor series with $h = k$, verify that the truncation error for $\nabla^2 u\mid_{i,j}$ is

$$-\frac{h^2}{12}\left[\frac{\partial^4 u}{\partial x^4} + \frac{\partial^4 u}{\partial y^4}\right]_{i,j} - \frac{h^4}{360}\left[\frac{\partial^6 u}{\partial x^6} + \frac{\partial^6 u}{\partial y^6}\right]_{i,j} - \cdots \quad (1\text{-}53)$$

1-15 Using Taylor series with $h = k$, develop the discrete relation

$$\nabla^2 u\mid_{i,j} = \frac{1}{2h^2}\{u_{i+1,j-1} + u_{i+1,j+1} + u_{i-1,j-1} + u_{i-1,j+1} - 4u_{i,j}\}$$

$$-\frac{h^2}{12}\left[\frac{\partial^4 u}{\partial x^4} + 6\frac{\partial^4 u}{\partial x^2 \partial y^2} + \frac{\partial^4 u}{\partial y^4}\right]_{i,j} + O(h^4) \quad (1\text{-}54)$$

and show Eqn (1-54) in computational molecule form.

1-16 Add the computational molecules of Fig. 1-5 and Problem 1-15 to obtain yet a third molecule for $\nabla^2 u\mid_{i,j}$. Note that while those of Fig. 1-5 and Problem 1-15 are *five-point* molecules, this involves *nine points*.

1-6 Finite difference operators

The following notations for various difference and related operators is now standard. We shall think of them as being applied to a function y of one independent variable x ($y = f(x)$) over a constant interval size $h = x_{n+1} - x_n$, and we denote by y_n that value of f at x_n—that is, $y_n = f(x_n)$. Now define the following operators:

Forward difference: $\quad \Delta y_n = y_{n+1} - y_n$ $\qquad\qquad$ (1-55a)

Backward difference: $\quad \nabla y_n = y_n - y_{n-1}$ $\qquad\qquad$ (1-55b)

Central difference: $\delta y_n = y_{n+1/2} - y_{n-1/2}$ (1-55c)

Averaging: $\mu y_n = \frac{1}{2}[y_{n+1/2} + y_{n-1/2}]$ (1-55d)

Shift: $E y_n = y_{n+1}$ (1-55e)

Integral: $J y = \int_x^{x+h} y(t)\, dt$ (1-55f)

Differential: $D y = dy/dx$ (1-55g)

For most purposes the operators of Eqns (1-55) can be manipulated according to the laws of elementary algebra. The general theory and exceptions to these laws do not concern us here, but can be found in Hildebrand [38] or Milne-Thompson [39]. Formal manipulation of these operators will provide a convenient vehicle to produce finite difference approximations for derivatives. Before proceeding note that all finite difference formulae are based upon polynomial approximation—that is, they give *exact results when operating upon a polynomial of the proper degree*. In all other cases the formulae are approximations and are usually expressed in series form. Since only a finite number of terms can of necessity be used, the truncation error is of concern.

From the definitions of Eqns (1-55) one can immediately infer the obvious formal relations

$$\Delta = E - 1,\dagger \qquad DJ = JD = \Delta, \qquad \nabla = 1 - E^{-1}$$

$$\mu = \frac{1}{2}(E^{1/2} + E^{-1/2}), \qquad \delta = E^{1/2} - E^{-1/2} \qquad (1\text{-}56)$$

$$\mu^2 = 1 + \frac{1}{4}\delta^2, \qquad 1 = \mu\mu^{-1} = \mu(1 + \frac{1}{4}\delta^2)^{-1/2}$$

$$= \mu(1 - \frac{1}{8}\delta^2 + \frac{3}{128}\delta^4 + \cdots)$$

which will be useful in the sequel.

To proceed, the operator D must be related to the other operators. For this purpose the Taylor series

$$f(x + h) = f(x) + \frac{h}{1!}f'(x) + \frac{h^2}{2!}f''(x) + \cdots \qquad (1\text{-}57)$$

can be expressed in the operational form

$$Ef(x) = \left[1 + \frac{hD}{1!} + \frac{h^2 D^2}{2!} + \cdots\right]f(x) = e^{hD}f(x). \qquad (1\text{-}58)$$

Thus we deduce the interesting relation between the operator E and D as

$$E = e^{hD}. \qquad (1\text{-}59)$$

We interpret this statement and, more generally, equality between operators to mean that the operators E and $\sum_{n=0}^{N} (h^n D^n/n!)$ yield identical results when applied to any polynomial of degree N for any N.

† The symbol 1 is used to denote the identity operator, that is $1 y_n = y_n$. By E^{-p} and Δ^p we shall mean y_{n-p} and $\Delta(\Delta(\ldots)$ p times.

From Eqns (1-59) and (1-56) the relations

$$hD = \log E = \log (1 + \Delta) = -\log (1 - \nabla)$$

$$= 2 \sinh^{-1} \delta/2 = 2 \log [(1 + \tfrac{1}{4}\delta^2)^{1/2} + \tfrac{1}{2}\delta] \qquad (1\text{-}60)$$

are obtained by formal operations. Upon expansion of $\log (1 + \Delta)$ we obtain the forward difference relation for the first derivative at the relevant point

$$\left.\frac{dy}{dx}\right|_i = \frac{1}{h}\left[\Delta - \frac{1}{2}\Delta^2 + \frac{1}{3}\Delta^3 - \cdots\right]y_i. \qquad (1\text{-}61)$$

Forward difference approximations for higher order derivatives follow from Eqn (1-60) since

$$\left.\frac{d^k y}{dx^k}\right|_i = D^k y|_i = \frac{1}{h^k}[\log (1 + \Delta)]^k y_i$$

$$= \frac{1}{h^k}\left[\Delta^k - \frac{k}{2}\Delta^{k+1} + \frac{k(3k + 5)}{24}\Delta^{k+2}\right.$$

$$\left. - \frac{k(k + 2)(k + 3)}{48}\Delta^{k+3} + \cdots\right]y_i. \qquad (1\text{-}62)$$

Thus for the second derivative we obtain the forward difference relation

$$h^2 y_i'' = \left[\Delta^2 - \Delta^3 + \frac{11}{12}\Delta^4 - \frac{5}{6}\Delta^5 + \cdots\right]y_i$$

$$= (y_{i+1} - 2y_i + y_{i-1})$$

$$- (y_{i+2} - 3y_{i+1} + 3y_i - y_{i-1}) + \frac{11}{12}\Delta^4 y_i + \cdots \qquad (1\text{-}63)$$

From $E = 1 + \Delta$ we can write

$$E^p y_i = y(x_i + ph) = (1 + \Delta)^p y_i = \left(\sum_{j=0}^{p}\binom{p}{j}\Delta^j\right)y_i$$

$$= y_i + p(\Delta y_i) + \binom{p}{2}\Delta^2 y_i\dagger + \cdots \qquad (1\text{-}64)$$

where $\binom{p}{j} = p!/j!(p - j)!$. This is a finite difference formula for the calculation of $y(x_i + ph)$ in the form of an infinite series which terminates if p is an integer.

Sometimes the forward differences at the previous point $(i - 1)$ are necessary rather than at the relevant point (i). In Eqn (1-61) we can set $y_i = E y_{i-1}$ and obtain

$$\left.\frac{dy}{dx}\right|_i = \frac{1}{h}[\log (1 + \Delta)]y_i = \frac{1}{h}[\log (1 + \Delta)](1 + \Delta)y_{i-1}$$

$$= \frac{1}{h}\left[\Delta + \frac{1}{2}\Delta^2 - \frac{1}{6}\Delta^3 + \cdots\right]y_{i-1} \qquad (1\text{-}65)$$

$\dagger \; \Delta^p y_i = \sum_{j=0}^{p}(-1)^j\binom{p}{j}y_{i+p-j}$ is a convenient computational form.

which uses the forward differences at the previous point. The corresponding relation for the second derivative is

$$h^2 y_i'' = \left[\Delta^2 - \frac{1}{12} \Delta^4 + \frac{1}{12} \Delta^6 - \cdots \right] y_{i-1}. \tag{1-66}$$

In a completely similar way, the corresponding *backward* difference formulae are obtained in the form

$$\frac{d^k y}{dx^k}\bigg|_i = -\frac{1}{h^k} [\log (1 - \nabla)]^k y_i$$

$$= \frac{1}{h^k} \left(\nabla + \frac{1}{2} \nabla^2 + \frac{1}{3} \nabla^3 + \cdots \right)^k y_i$$

$$= \frac{1}{h^k} \left[\nabla^k + \frac{k}{2} \nabla^{k+1} + \frac{k(3k+5)}{24} \nabla^{k+2} \right.$$

$$\left. + \frac{k(k+2)(k+3)}{48} \nabla^{k+3} + \cdots \right] y_i. \tag{1-67}$$

Valuable central difference formulae can also be obtained from the central difference form of Eqn (1-60), $D = (2/h) \sinh^{-1} \delta/2$. Since the right-hand member is an odd function of δ its expansion in powers of δ would involve odd central differences. This can be avoided by multiplying the right-hand side by μ and dividing by its equivalent $[1 + \delta^2/4]^{1/2}$; thus,

$$\frac{dy}{dx}\bigg|_i = \frac{2\mu/h \, \sinh^{-1} \delta/2}{[1 + \delta^2/4]^{1/2}} y_i$$

$$= \frac{\mu}{h} \left\{ \delta - \frac{1^2}{3!} \delta^3 + \frac{1^2 \cdot 2^2}{5!} \delta^5 - \cdots \right\} y_i \tag{1-68}$$

—a mean odd central difference. Higher derivatives of even order $2n$ are obtained by using D^{2n}, where

$$D^{2n} = \left[\frac{2}{h} \sinh^{-1} \frac{\delta}{2} \right]^{2n}$$

$$= \frac{1}{h^{2n}} \left[\delta - \frac{1^2}{2^2 \cdot 3!} \delta^3 + \frac{1^2 \cdot 3^2}{2^4 \cdot 5!} \delta^5 - \frac{1^2 \cdot 3^2 \cdot 5^2}{2^6 \cdot 7!} \delta^7 + \cdots \right]^{2n}. \tag{1-69}$$

Thus,

$$\frac{d^2 y}{dx^2}\bigg|_i = \frac{1}{h^2} \left[\delta^2 - \frac{1}{12} \delta^4 + \frac{1}{90} \delta^6 - \frac{1}{560} \delta^8 + \cdots \right] y_i. \tag{1-70}$$

Higher derivatives of order $2n + 1$ are obtained by multiplying the operator in Eqn (1-68) by D^{2n} of Eqn (1-69).

With central differences we have also simple and useful formulae for the derivatives at a half-way point from the relations

$$\frac{d^{2n}y}{dx^{2n}}\bigg|_{i+1/2} = \frac{\mu}{\sqrt{(1 + \delta^2/4)}} \left[\frac{2}{h} \sinh^{-1}\frac{\delta}{2}\right]^{2n} y_{i+1/2} \qquad (1\text{-}71)$$

$$\frac{d^{2n+1}y}{dx^{2n+1}}\bigg|_{i+1/2} = \left[\frac{2}{h} \sinh^{-1}\frac{\delta}{2}\right]^{2n+1} y_{i+1/2} \qquad (1\text{-}72)$$

Thus,

$$y'_{i+1/2} = \frac{1}{h}\left[\delta - \frac{1}{24}\delta^3 + \frac{3}{640}\delta^5 - \frac{7}{7168}\delta^7 + \cdots\right]y_{i+1/2}$$

an ordinary central difference, and

$$y''_{i+1/2} = \frac{\mu}{h^2}\left[\delta^2 - \frac{5}{24}\delta^4 + \frac{259}{5760}\delta^6 - \frac{3229}{322560}\delta^8 + \cdots\right]y_{i+1/2}$$

We shall appeal to various of these formulae when required.

PROBLEMS

1-17 Establish the relations

$$\nabla = E^{-1}\Delta, \qquad \Delta\nabla = \nabla\Delta = \Delta - \nabla = \delta^2$$

$$\mu\delta = \tfrac{1}{2}(\Delta + \nabla), \qquad \mu^2 = 1 + \tfrac{1}{4}\delta^2.$$

1-18 Establish Eqn (1-66).

1-19 Set $n = 0$ and obtain a relation for calculating $y_{i+1/2}$. Neglecting central differences of order higher than 2 find a way of calculating $y_{i+1/2}$ in terms of integral tabular points.

1-20 Neglecting differences of order higher than 4 determine an approximation for $\partial^2 u/\partial x^2$ from Eqns (1-63), (1-66), (1-67), and (1-70).

1-21 Neglecting differences of order higher than 4 determine a finite difference approximation for $u_{xx} + u_{yy}$ from Eqn (1-70).

1-7 Errors

As a vehicle for discussing the origin of errors we shall use the dimensionless Laplace equation in the square $0 \le x \le 1, 0 \le y \le 1$,

$$u_{xx} + u_{yy} = 0 \qquad (1\text{-}73)$$

together with the (mixed) boundary conditions

$$u(0, y) = f(y), \quad u(x, 0) = g(x), \quad u_x(1, y) = h(y), \quad u_y(x, 1) = k(x). \qquad (1\text{-}74)$$

A discrete network $P_{i,j} = (ih, jk)$ is formed in the square where $h = \Delta x = 1/N$, $k = \Delta y = 1/M$. Equation (1-73) will be approximated using the five-point computational molecule of Fig. 1-5 in the discrete form

$$u_{i,j} = \frac{h^2 k^2}{2(h^2 + k^2)} \left[\frac{1}{h^2} (u_{i+1,j} + u_{i-1,j}) + \frac{1}{k^2} (u_{i,j+1} + u_{i,j-1}) \right] + O(h^2) + O(k^2). \quad (1-75)$$

Normally there will be a large number of algebraic equations similar to Eqn (1-75) to be solved. This is usually accomplished by methods of iteration which will be discussed subsequently.

Strictly speaking, the error $(O(h^2) + O(k^2))$ indicated in Eqn (1-75) is the *truncation error* of the finite difference equation and not of the solution. This point is emphasized here because, as in the analytic methods, the boundary conditions are essential to the proper solution. These, if they are not of Dirichlet type $(u = f)$†, must be approximated by finite differences thereby introducing an additional or boundary truncation error. Thus, in our example problem, both $u_x(1, y)$ and $u_y(x, 1)$ must be approximated. The error in the solution, due to replacement of the continuous problem by the discrete model, will be called the *discretization error* after Wasow [40]. A useful rule of thumb, to be made more precise later, states that the order of the overall discretization error is the *smallest order* of all those approximations used unless they are somehow related.‡

When the discrete equations are not solved exactly an additional error is introduced. This error, called *round-off error*, is present in iterative solutions (machine or manual) since the iteration is only continued until no change takes place up to a certain number of digits. If the iteration is continued without decimal place limitation, any subsequent changes are considered to be round-off errors.

The interval size h (and k) affects the discretization error and round-off error in the opposite sense. The first decreases as h decreases, while the second generally increases. *It is for this reason that one cannot generally assert that decreasing the mesh size always increases the accuracy.*

Reasonably rigorous expressions for the truncation errors associated with the finite difference formulae of Section 1-6 are obtainable by methods discussed, for example, by Hildebrand [38, p. 64]. As a practical matter we can usually find the first neglected term in the Taylor series, as indicated in Section 1-5, and base our expected accuracy on an estimate of this quantity. If only an asymptotic order estimate is required, then the first neglected difference,

---

† It is sometimes useful to approximate these conditions by averaging, or by some other procedure.

‡ In a discrete method for the diffusion equation we shall relate Δt and $(\Delta x)^2$ so that the overall error is second order even though the approximation for u_t is first order (see Chapter 2).

say $\Delta^k u_i$, when replaced by $h^k D^k u(\zeta)$, where ζ is some point in the range limited by the pivotal points, will specify that estimate. Thus, if differences of order 6 and higher are neglected in Eqn (1-66), then $h^2 y''$ is approximated to $O(h^6)$ or y'' to $O(h^4)$. Of course, this estimate suppresses the coefficient of the error term which is always smaller with central differences than with sloping differences.

The contribution of the round-off error is more difficult to ascertain because of its random nature. Probability methods must be employed, thereby requiring an assumption concerning the probability density function of the error. Thus we cannot guarantee that an error is less than, say, ϵ but can only estimate the probability that this is the case.

In a wide class of situations the use of the Gaussian (normal) distribution is justified (see, for example, Feller [41]). The details must be substantially omitted here. Suppose that the errors are symmetrically distributed about a zero mean with the probability of the occurrence of an error between x and $x + dx$ being

$$p(x)\,dx = \frac{1}{\sqrt{(2\pi)}\sigma}\, e^{-x^2/2\sigma^2}\,dx \tag{1-76}$$

where σ is a constant parameter to be estimated. $p(x)$ is called the *probability density function of the distribution*. The probability that an error will not exceed x algebraically is given by the *normal distribution function*

$$G(x) = \int_{-\infty}^{x} p(t)\,dt = \frac{1}{\sqrt{(2\pi)}\sigma} \int_{-\infty}^{x} e^{-t^2/2\sigma^2}\,dt. \tag{1-77}$$

The numerical coefficient, $1/\sqrt{(2\pi)}\sigma$, has been determined in accordance with the requirement of unit probability that any error lies somewhere in $(-\infty, \infty)$.

Thus,

$$G(\infty) = 1 = \int_{-\infty}^{\infty} p(t)\,dt. \tag{1-78}$$

The probability $P(x)$ that an error chosen at random lies on the interval $-|x|$ to $|x|$ is given by

$$P(x) = G(|x|) - G(-|x|) = \int_{-|x|}^{|x|} p(t)\,dt = 2\int_{0}^{|x|} p(t)\,dt$$

or

$$P(x) = \sqrt{\frac{2}{\pi}}\,\sigma^{-1} \int_{0}^{|x|} e^{-t^2/2\sigma^2}\,dt. \tag{1-79}$$

The probability $Q(x)$ that the error exceeds $|x|$ is $Q(x) = 1 - P(x)$. By an elementary transformation of the integral, Eqn (1-79) can also be written as

$$P(x) = \frac{2}{\sqrt{\pi}} \int_{0}^{|x|/\sqrt{2}\sigma} e^{-s^2}\,ds = \mathrm{erf}\left(\frac{|x|}{\sqrt{(2)}\sigma}\right) \tag{1-80}$$

where 'erf' designates the *error function*.

The quantity σ is called the *standard deviation* of the distribution (σ^2 is the *variance*) while $1/[\sqrt{(2\pi)}\sigma]$ is called the *modulus of precision*. The points of inflection of the curve representing $p(x)$ lie at a distance σ on each side of the maximum of $p(x)$, which occurs at zero. The modulus of precision is a measure of the steepness of the frequency curve near its maximum.

Let ϵ be a random variable. The *expected value*† of any function $f(\epsilon)$, relative to the assumed distribution, is denoted by

$$E[f(\epsilon)] = \int_{-\infty}^{\infty} p(\epsilon)f(\epsilon)\,d\epsilon$$

$$= \frac{1}{\sqrt{(2\pi)}\sigma} \int_{-\infty}^{\infty} e^{-\epsilon^2/2\sigma^2} f(\epsilon)\,d\epsilon \tag{1-81}$$

if this integral exists. Thus

$$E(\epsilon) = 0$$
$$E(|\epsilon|) = \sqrt{(2/\pi)}\sigma \tag{1-82}$$
$$E(\epsilon^2) = \sigma^2$$

and all higher 'moments' can be expressed as functions of σ.

The assumed normal distribution is determined by choosing σ equal to the square root of the mean of the squared errors of the true distribution—that is,

$$\sigma = \epsilon_{\text{r.m.s.}} \tag{1-83}$$

Of course $\epsilon_{\text{r.m.s.}}$ can only be estimated from a sample of, say, the deviations of n measurements from their mean value (zero here), thus

$$\epsilon_{\text{r.m.s.}}^2 \approx \frac{1}{n-1} \sum_{i=1}^{n} \epsilon_i^2. \tag{1-84}$$

Once σ is approximated, Eqn (1-79) can be used to estimate the probability that the magnitude of a random error will not exceed a certain specified quantity. A few values of $P(\epsilon)$ are given below:

$\epsilon/\epsilon_{\text{r.m.s.}}$	0.674	0.842	1.036	1.282	1.645	2.576
$P(\epsilon)$	0.500	0.600	0.700	0.800	0.900	0.990

Thus 80 per cent of the errors will be less than $1.282\epsilon_{\text{r.m.s.}}$, and only 1 per cent will exceed $2.576\epsilon_{\text{r.m.s.}}$ *provided the distribution is sufficiently close to Gaussian.*

If the distribution of values of u is unaffected by the value of v, and vice versa, u and v are said to be (mutually) independent. Suppose that $\epsilon = u + v$,

† The term *mean* value is sometimes used here but expected value is the present terminology.

where u and v are independent and both have zero mean, then the expected value of ϵ^2 is the sum of the expected values of u^2, $2uv$, and v^2. Since u and v are independent, $E(2uv) = 2E(u)\,E(v) = 0$, from Eqns (1-82). Consequently, Eqn (1-84) implies that

$$\epsilon_{\text{r.m.s.}} = [u^2_{\text{r.m.s.}} + v^2_{\text{r.m.s.}}]^{1/2} \tag{1-85}$$

and more generally the r.m.s. value of the sum of n independent random variables u_i, $i = 1, \ldots, n$ *each with zero mean, is the square root of the sum of the squares of the r.m.s. values of the component errors.* That is, if $\epsilon = \sum_{i=1}^{n} u_i$, then $\epsilon^2_{\text{r.m.s.}} = \sum_{i=1}^{n} (u_{i\,\text{r.m.s.}})^2$.

Two further theorems will be helpful: *If u and v are independent normally distributed random variables with standard deviations* σ_u *and* σ_v, *then* $\epsilon = u + v$ *is normally distributed, with variance* $\sigma^2 = \sigma_u^2 + \sigma_v^2$ *and* $\sigma_{u+v} = (u + v)_{\text{r.m.s.}}$.

Let the approximation equation [Eqn (1-84)] for $\epsilon_{\text{r.m.s.}}$ be calculated for a very large number of sets of samples, each containing n errors chosen at random from the same distribution. If the mean of all these estimates is selected as the 'best' approximation to $\epsilon_{\text{r.m.s.}}$ then the deviations of the various estimates from this best one is normally distributed with standard deviation $\sigma \approx \epsilon_{\text{r.m.s.}}/\sqrt{(2n)}$ when n is sufficiently large (see, for example, Johnson and Leone [42], p. 139).

The error, ϵ, which arises from rounding a number to m decimal places by the *even rule*† is never larger in magnitude than one half unit in the place of the mth digit of the rounded number. Thus the probability density function has the constant value

$$p = \begin{cases} \dfrac{1}{2|\epsilon|_{\max}} & \text{when } |\epsilon| < |\epsilon|_{\max} = 5 \times 10^{-m-1} \\[2mm] 0 & \text{elsewhere} \end{cases} \tag{1-86}$$

which is poorly approximated by any normal distribution. However, the cumulative distribution function corresponding to errors which are exactly or approximately linear combinations of many such errors will be approximated by a normal distribution. Consequently, the error analysis may be confidently based upon the result of treating the individual errors as though they were normally distributed.

If x is a random variable taking all values between $-\frac{1}{2}$ and $\frac{1}{2}$ with equal probability, the r.m.s. value of x is

$$\int_{-1/2}^{1/2} x^2 \, dx = \tfrac{1}{6}\sqrt{3} \approx 0.2887. \tag{1-87}$$

---

† The conventional process of rounding a number to m decimal places consists of replacing that number by an m digit approximation with minimum error. If two roundings are possible the one for which the mth digit is even is selected.

If ϵ is round-off due to rounding in the mth decimal place, all values between $-\frac{1}{2} \times 10^{-m}$ and $\frac{1}{2} \times 10^{-m}$ are equally likely. Consequently,

$$\epsilon_{r.m.s.} \approx 0.2887 \times 10^{-m}. \tag{1-88}$$

We now appeal to the last general theorem of this section. Suppose n numbers are each rounded to m decimal places. The error in the sum of the results is approximately normally distributed with an r.m.s. value of

$$0.2887 \times \sqrt{(2n)} \times 10^{-m}, \quad \text{for large } n. \tag{1-89}$$

Let $n = 200$ numbers be added. The error in the sum of the results will be less than 6 units in the mth place. From the table following Eqn (1-84) the probability of an error of 17 units is less than 0.1 and the odds are 99 to 1 that the error will not exceed 27 units. Nevertheless, an error of 1000 units in the mth place is possible but highly unlikely.

Clearly error analysis is one of the prime considerations in the development and application of any numerical method. While it is an extensively cultivated field the tools available are often inadequate, especially in nonlinear problems. As we proceed error computation will be stressed.

PROBLEMS

1-22   Calculate $E(\epsilon^3)$ and thus show that the third moment can be expressed as a function of σ.

1-23   Let $f(x)$ and $g(x)$ be the probability density functions of ϵ_1 and ϵ_2, respectively, where ϵ_1 and ϵ_2 are independent random variables. Show that the distribution function of $\epsilon_1 + \epsilon_2$ is

$$\iint\limits_{s+t<x} f(s)g(t)\, ds\, dt = \int_{-\infty}^{x} \left[\int_{-\infty}^{\infty} f(u-t)g(t)\, dt \right] du$$

and hence the probability density function of $x = \epsilon_1 + \epsilon_2$ is

$$p(x) = \int_{-\infty}^{\infty} f(x-t)g(t)\, dt.$$

1-24   Suppose that the probability density function analogous to Eqn (1-76) is $p(x) = ae^{-|x|}, -\infty < x < \infty$. Find the cumulative distribution function and the value of a so that there will be unit probability that any error will lie somewhere on $-\infty < x < \infty$. Find $P(x)$ exactly.

1-25   Repeat Problem 1-24 for $p(x) = a/(1 + x), x \geq 0$.

1-8   Stability and convergence

The presence of round-off error or any other computational error may lead to numerical instability.† A qualitative description of instability will set the

---

† One must carefully distinguish between a *physical* instability, such as a shock or jump, and *numerical* instability

stage for subsequent discussion. To this end we consider the simple ordinary differential equation

$$\frac{dy}{dx} = y - x \tag{1-90}$$

whose general solution is

$$y = Ae^x + (x + 1) \tag{1-91}$$

where A is the integration constant. If the initial condition $y(0) = 1$ is specified, then $A = 0$ so that the exact solution is a simple linear term growing slowly with increasing x when compared to an exponential term. In an approximate solution of Eqn (1-90) the exponential term is likely to be introduced as a result of round-off errors. If at $x = \zeta$ this component is η it will be ηe^5 at $x = \zeta + 5$. Thus the resulting error will soon obliterate the desired solution and lead to entirely spurious results. Any numerical scheme which allows the growth of error, eventually 'swamping' the true solution, is unstable. These numerical phenomena must be avoided by restrictive action, such as limiting the interval size or adopting an alternative method.

A more mathematical description of stability is necessary. The general question has been considered by a number of researchers and excellent summaries are found in Forsythe and Wasow [37, pp. 29 *et seq.*] and Richtmyer [43]. The definition given here is the one in general use. Let $U(x, t)$ be the solution of a given difference approximation, solvable step by step in the t direction. The effect of a mistake or, more likely, round-off error in machine computation may replace $U(x_0, t_0)$ by $U(x_0, t_0) + \epsilon$ at the grid point (x_0, t_0). If the solution procedure is continued with the value $U(x_0, t_0) + \epsilon$ without new errors being introduced, and if at subsequent points the value $U^*(x, t)$ is obtained, then we denote by $U^*(x, t) - U(x, t)$ the *departure* of the solution resulting from the error ϵ at (x_0, t_0). When errors are committed or introduced at more than one point, cumulative departures result which are not additive except in linear problems. If δ is the maximum absolute error—that is, $|\epsilon(x, t)| < \delta$—and h the interval size, then a procedure is said to be *pointwise stable* if the cumulative departure tends to zero as $\delta \to 0$ and does not increase faster than some power of h^{-1} as $h \to 0$.

When the corresponding continuous solution remains bounded (see Crandall [44]) a finite difference process within the semi-infinite strip $0 < x < 1$, $t > 0$ is called *stepwise unstable* if for a fixed network and fixed homogeneous boundary conditions there exist initial disturbances for which the finite difference solutions $U_{i,j}$ become unbounded as $j \to \infty$. We shall further explore this idea in Chapter 2 where it will be applied to parabolic equations.

A different definition of instability is given by John [45]. Instead of the fixed network of the previous paragraph a fixed interval $0 < t < T$ is considered with a sequence of finite difference solutions for successively finer networks. If, as $h \to 0$, the finite difference solutions at $t = T$ can become unbounded, the process is called *unstable*.

A second fundamental concept, that of *convergence*, is often related to stability. To introduce this idea we utilize the partial differential equation

$$L(u) = 0 \text{ in a region } D, \quad u = g \text{ on } \Gamma \qquad (1\text{-}92)$$

where Γ is the boundary of D. Associated with Eqn (1-92) is a finite difference process whose grid, no matter the configuration, depends upon certain parameters, usually the interval sizes. Let us suppose that there is only one, say h, and write the finite difference problem as

$$L_h(U) = 0 \text{ in } D, \quad U = g_h \text{ on } \Gamma.\dagger \qquad (1\text{-}93)$$

We say the *finite difference scheme converges if $U(P)$ converges to the solution* $u(P)$, with the same boundary values, as $h \to 0$.

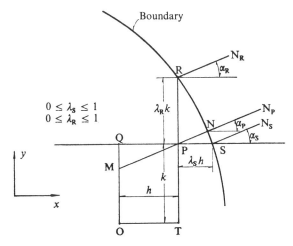

Fig. 1-6   Geometry of an irregular point P near the boundary

1-9   Irregular boundaries

Many physical problems have irregular boundaries. Upon discretization interior mesh points adjacent to the boundary are not the mesh distances h and k from the boundary. In such cases various procedures may be used. An accurate method requires the governing difference equation to apply at every internal point. At a point such as P in Fig. 1-6 we must modify the regular computational molecule by taking into account both the boundary conditions and the nonstandard spacings $\lambda_R k$ and $\lambda_S h$. This method was used by Mikeladze [46] although it was also used earlier in manual relaxation procedures by Shortley and Weller [47].

---

† This is clearly not the most general situation; it is possible for D and Γ to depend upon h, e.g. where a curvilinear boundary is replaced by straight line segments.

A second, simpler, and less accurate method used by Gerschgorin [48] and Collatz [49] approximates the partial differential equation with the *standard* finite difference molecule only at those points where it can be fitted without changes. There remains an outer ring of mesh points, which may be interior or exterior, at which the u values are not required to satisfy the governing equation. Instead, these values are determined by the requirement that they satisfy finite difference approximations to the boundary conditions—the uneven spacings are included here.

To illustrate the construction of the irregular molecule we consider the situation in Fig. 1-6 and obtain approximations for u_{xx}, u_y, u_{xx}, and u_{yy} simultaneously. Assume this to be a Dirichlet problem—that is, u is known on the boundary. Hence, u_R and u_S are given. If P is assumed to be the origin, then by expanding $u(x, y)$ about P we get, in a Taylor series,

$$u(x, y) = u_P + x\left(\frac{\partial u}{\partial x}\right)_P + y\left(\frac{\partial u}{\partial y}\right)_P + \left(\frac{\partial^2 u}{\partial x^2}\right)_P x^2/2$$

$$+ \left(\frac{\partial^2 u}{\partial x\, \partial y}\right)_P xy + \left(\frac{\partial^2 u}{\partial y^2}\right)_P y^2/2 + \cdots \quad (1\text{-}94)$$

Thinking again of P as $(0, 0)$ the points R, S, T, and Q can be written as

$$(0, \lambda_R k), \quad (\lambda_S h, 0), \quad (0, -k), \quad \text{and} \quad (-h, 0). \quad (1\text{-}95)$$

Substituting the four points of Eqn (1-95) into Eqn (1-94) we obtain the four equations [neglecting terms of $O(h^3)$ and $O(k^3)$]

$$\begin{bmatrix} 0 & \lambda_R k & 0 & \frac{1}{2}(\lambda_R k)^2 \\ \lambda_S h & 0 & \frac{1}{2}(\lambda_S h)^2 & 0 \\ 0 & -k & 0 & \frac{1}{2}k^2 \\ -h & 0 & \frac{1}{2}h^2 & 0 \end{bmatrix} \begin{bmatrix} (u_x)_P \\ (u_y)_P \\ (u_{xx})_P \\ (u_{yy})_P \end{bmatrix} = \begin{bmatrix} u_R - u_P \\ u_S - u_P \\ u_T - u_P \\ u_Q - u_P \end{bmatrix}$$

whose solutions are

$$(u_x)_P = h^{-1}\left[\frac{1}{\lambda_S(1 + \lambda_S)}\, u_S - \frac{\lambda_S}{(\lambda_S + 1)}\, u_Q - \frac{1 - \lambda_S}{\lambda_S}\, u_P\right] + O(h^2)$$

$$(1\text{-}96)$$

$$(u_{xx})_P = 2h^{-2}\left[\frac{1}{\lambda_S(\lambda_S + 1)}\, u_S + \frac{1}{(\lambda_S + 1)}\, u_Q - \frac{1}{\lambda_S}\, u_P\right] + O(h) \quad (1\text{-}97)$$

$$(u_y)_P = k^{-1}\left[\frac{1}{\lambda_R(1 + \lambda_R)}\, u_R - \frac{\lambda_R}{(\lambda_R + 1)}\, u_T - \frac{1 - \lambda_R}{\lambda_R}\, u_P\right] + O(k^2)$$

$$(1\text{-}98)$$

$$(u_{yy})_P = 2k^{-2}\left[\frac{1}{\lambda_R(\lambda_R + 1)}\, u_R + \frac{1}{(\lambda_R + 1)}\, u_T - \frac{1}{\lambda_R}\, u_P\right] + O(k) \quad (1\text{-}99)$$

Of course extensions can be obtained by the same expansion if points Q and T are also unevenly spaced.

The procedure for treating those boundary conditions involving the normal derivative $\partial u/\partial n$ hinge on the expression for the directional derivative

$$\left.\frac{\partial u}{\partial n}\right|_\alpha = \frac{\partial u}{\partial x}\cos\alpha + \frac{\partial u}{\partial y}\sin\alpha \qquad (1\text{-}100)$$

where the angle α (with the x-axis) specifies the direction. In this case we know $\partial u/\partial n$ along N_R, N_P, and N_S, then by differentiation of Eqn (1-94) we have

$$u_x = (u_x)_P + x(u_{xx})_P + y(u_{xy})_P + \cdots \qquad (1\text{-}101a)$$

$$u_y = (u_y)_P + x(u_{xy})_P + y(u_{yy})_P + \cdots \qquad (1\text{-}101b)$$

Using Eqns (1-101) we can express the normal derivatives at R, N, and S in terms of the first and second derivatives of u at P. Thus for R we find

$$\left(\frac{\partial u}{\partial n}\right)_R = [(u_x)_P + \lambda_R k(u_{xy})_P]\cos\alpha_R + [(u_y)_P + \lambda_R k(u_{yy})_P]\sin\alpha_R \qquad (1\text{-}102)$$

and related expressions at N and S. Adding to these three equations the two Taylor series for u_Q and u_T gives five equations for the derivatives u_x, u_y, u_{xy}, u_{xx}, and u_{yy} at P.

The procedure of Gerschgorin will also be discussed in the nomenclature of Fig. 1-6. The regular interior approximation is used at points Q and T. However, at P a finite difference approximation to the boundary condition is employed. Suppose that u is specified on the boundary so that u_S is known. By linear interpolation from S to Q we obtain

$$u_P = \frac{\lambda_S}{1 + \lambda_S}u_Q + \frac{1}{1 + \lambda_S}u_S \qquad (1\text{-}103)$$

as the expression for u at P. Of course, we could equally well have interpolated in the other direction from R to T and obtained

$$u_P = \frac{\lambda_R}{1 + \lambda_R}u_T + \frac{1}{1 + \lambda_R}u_R. \qquad (1\text{-}104)$$

Finally, the average of these two interpolations can also be used. In some cases higher order interpolation may be justified to reduce the boundary error. Various interpolation schemes are given, for example, in Hildebrand [38], one of which is employed in an example of boundary layer flow in Chapter 2.

In the case where $\partial u/\partial n$ is specified as the boundary condition we know $\partial u/\partial n|_N$ which may be approximated by $(u_P - u_M)/\overline{PM}$. The value u_M is obtained by, say, linear interpolation between Q and 0, and the various lengths can be expressed in terms of h and α_P. Thus

$$u_P = u_Q(1 - \tan\alpha_P) + u_0\tan\alpha_P + h\left.\frac{\partial u}{\partial n}\right|_N\sec\alpha_P. \qquad (1\text{-}105)$$

Other methods of dealing with boundary condition are given by Shaw [50], Allen [51], Batschelet [52], Fox [53], Forsythe and Wasow [37], and Viswanathan [54]. Walsh and Young [55, theorem 4.1] demonstrated that in the selection of boundary values U_R for R on C it is not necessarily best to select the given (continuous) value at this point since this gives undue weight to one point. Rather, it is sometimes better to select U_R as an average of values for points on C in the vicinity of R. In certain cases this 'trick' can decrease the error by an order of magnitude.

PROBLEMS

1-26   Develop an expression for $(\partial u/\partial n|_S$, $(\partial u/\partial n|_N$ analogous to Eqn (1-102) and write out the five equations for u_x, u_y, u_{xx}, u_{xy}, and u_{yy} at P of Fig. 1-6.

1-27   Use quadratic interpolation in the method of Gerschgorin and therefore replace Eqns (1-103) and (1-104) by more accurate expressions.

1-28   Let $\Delta x = h = k = \Delta y$ and find an expression for $u_{xx} + u_{yy}$ at the point P by applying the method of Mikeladze to the Dirichlet problem. What is the truncation error?

1-29   Let $\partial u/\partial x = 0$ be the normal derivative at $x = 1$. Show that by introducing a 'false' boundary at $1 + h$ the boundary condition can be simulated to $O(h^2)$ by the condition

$$u(1 - h, y) = u(1 + h, y) \tag{1-106}$$

for each relevant y, where $\Delta x = h$. If a backward difference is applied the method is first order. For a simple boundary value problem in a square the use of centered differences has been shown by Giese [56] to yield a smaller discretization error than would result from one-sided differences.

1-30   Will the method of Mikeladze apply to equations containing the term u_{xy}? Why?

1-10   Choice of discrete network

In digital computing two considerations favor a regular spacing of the mesh points or nodes $P_{i,j}$. For irregular nets the form of the approximating difference equations changes from zone to zone creating bothersome programming details. Second, for maximum speed, computers require simplicity of net structure. Thus on two counts regular networks are preferable. The only regular polygons which can completely fill the plane are rectangles, triangles, and hexagons. Consequently, there are only three finite difference networks of corresponding regularity. These are shown in Fig. 1-7. The hexagonal cell shape is rather complicated and often difficult to fit into problem boundaries. Therefore, the only networks in common use are the rectangular (square) and triangular nets. The triangular nets are most often chosen to be composed of equilateral triangles.

The preceding paragraph is only intended to provide a general orientation. Curvilinear coordinate systems and the application of characteristics to hyperbolic problems may dictate the use of curvilinear nets. A net composed of

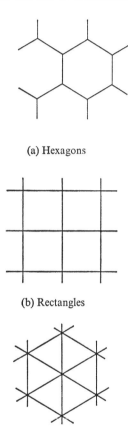

(a) Hexagons

(b) Rectangles

(c) Equilateral triangles

Fig. 1-7   Some types of finite difference nets

arcs of characteristics is indicated in Fig. 1-8. Problems involving curvilinear nets will be discussed, where appropriate, in succeeding chapters.

1-11   Dimensionless forms

An extremely useful organizational tool of the scientist and engineer, that of *dimensionless analysis*, is also strongly recommended prior to numerical computation. When the mathematical formulation is cast in dimensionless form all unnecessary symbols and units are removed and the problem appears in a very general but simple form. Several typical examples are given below. For further information such books as Langhaar [57], Bridgman [58], and Birkhoff [59] should be consulted.

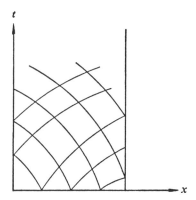

Fig. 1-8   Curvilinear net of characteristic arcs

As a first example consider the problem of heat conduction in a homogeneous rod of length L which is insulated along its length and at the end $x = L$. All temperature changes result from heat transfer at the ends and by heat conduction along the rod. Let distance along the rod be denoted by x, time by t, and temperature by $T(x, t)$. The particular propagation problem under examination will be that in which the rod, initially at a uniform high temperature T_1, is cooled by bringing it into a medium whose temperature is $T_0 \ll T_1$. The mathematical problem (see, for example, Carslaw and Jaeger [60]) consists of solving

$$\frac{\partial^2 T}{\partial x^2} = \frac{\rho c_p}{k} \frac{\partial T}{\partial t} \tag{1-107}$$

subject to the initial condition

$$T(x, 0) = T_1 \quad \text{for all } x, \quad 0 < x < L \tag{1-108}$$

and boundary conditions

$$T(0, t) = T_0 \tag{1-109}$$

$$\frac{\partial T}{\partial x}(L, t) = 0 \quad \Bigg\} \, t > 0 \tag{1-110}$$

Here ρ is the mass per unit volume, c_p is the specific heat, and k is the thermal conductivity. The physical formulation and the corresponding open-ended physical integration domain is shown in Fig. 1-9.

A problem can be brought to dimensionless form in more than one way.

A useful starting point is to first introduce the most obvious dimensionless factors while keeping in mind that the final dimensionless problem should be as free of parameters as possible with simple initial and boundary conditions.

In the case at hand let us set

$$x' = x/L \quad \text{and} \quad \psi = \frac{T - T_0}{T_1 - T_0}. \tag{1-111}$$

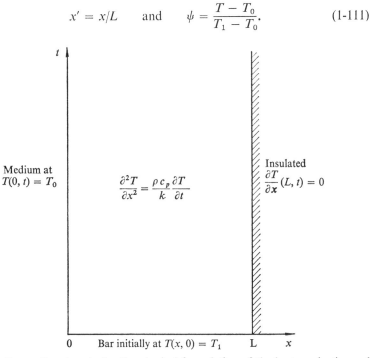

Fig. 1-9 Integration domain for the physical formulation of the heat-conducting rod cooled at one end

The use of x/L is an obvious transformation. The temperature transformation is not only dimensionless but also simplifies the initial condition to $\psi(x', 0) = 1$, $0 < x' < 1$ and the boundary condition, Eqn (1-109), to $\psi(0, t) = 0$. When Eqn (1-111) is applied to Eqn (1-107) there results

$$\frac{\partial^2 \psi}{(\partial x')^2} = \frac{\rho c_p}{k} L^2 \frac{\partial \psi}{\partial t}. \tag{1-112}$$

Of course, the left-hand side is dimensionless thereby suggesting that a dimensionless time variable should be

$$t' = \frac{t}{(\rho c_p/k)L^2} \tag{1-113}$$

which agrees with the physical dimensionality.

Upon dropping the primes from the dimensionless variables the dimension-less form of our problem is

$$\frac{\partial^2 \psi}{\partial x^2} = \frac{\partial \psi}{\partial t} \tag{1-114}$$

subject to the initial condition

$$\psi(x, 0) = 1, \quad 0 < x < 1 \tag{1-115}$$

and boundary conditions

$$\psi(0, t) = 0 \tag{1-116}$$

$$\frac{\partial \psi}{\partial x}(1, t) = 0 \left.\right\} t > 0 \tag{1-117}$$

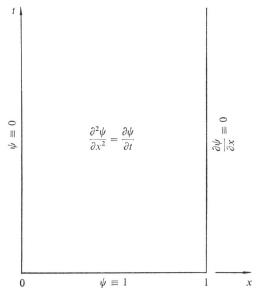

Fig. 1-10 Integration domain for the dimensionless formulation of the heat-conducting rod

These are shown on a diagram of the dimensionless integration domain in Fig. 1-10.

Ames *et al.* [61] have researched the problem of longitudinal wave propagation along a moving string of length L, intial cross-sectional area A_0, and initial mass per unit length m_0. The governing equations are

$$v \frac{\partial v}{\partial x} + \frac{\partial v}{\partial t} = m^{-1} \frac{\partial T}{\partial x} \tag{1-118}$$

$$\frac{\partial m}{\partial t} + \frac{\partial (mv)}{\partial x} = 0 \tag{1-119}$$

and

$$m(T + EA_0) = EA_0 m_0. \tag{1-120}$$

Upon setting

$$x' = \frac{x}{L}, \qquad v' = \frac{v}{C_0}, \qquad C_0^2 = \frac{EA_0}{m_0}$$

$$t' = \frac{t}{L/C_0}, \qquad T' = \frac{T}{EA_0}, \qquad m' = \frac{m}{m_0} \tag{1-121}$$

the dimensionless equations, upon dropping the primes, become

$$v\frac{\partial v}{\partial x} + \frac{\partial v}{\partial t} = m^{-1}\frac{\partial T}{\partial x} \tag{1-122}$$

$$\frac{\partial m}{\partial t} + \frac{\partial}{\partial x}(mv) = 0 \tag{1-123}$$

$$m(T + 1) = 1. \tag{1-124}$$

PROBLEMS

1-31 The transverse vibration of a flexible string of length L, stretched with large tension T and fixed at both ends, is governed to first approximation by the wave equation

$$\frac{\partial^2 y}{\partial x^2} = \frac{\rho}{T}\frac{\partial^2 y}{\partial t^2} \tag{1-125}$$

where x denotes distance along the string, t is time, $y(x, t)$ is transverse displacement from the equilibrium position, and ρ is the uniform mass per unit length. The initial conditions are

$$y(x, 0) = \frac{w_0}{2T}(L - x)x, \qquad \frac{\partial y}{\partial t} = 0, \quad 0 < x < L \tag{1-126}$$

and the boundary conditions are

$$y(0, t) = y(L, t) = 0, \quad t > 0. \tag{1-127}$$

Obtain a dimensionless formulation and draw the corresponding integration domain.

1-32 Consider the steady boundary layer equations, without a pressure gradient, for flow over a flat plate of length L,

$$u\frac{\partial u}{\partial x} + v\frac{\partial u}{\partial y} = v\frac{\partial^2 u}{\partial y^2} \tag{1-128a}$$

$$\frac{\partial u}{\partial x} + \frac{\partial v}{\partial y} = 0. \tag{1-128b}$$

(a) Define a 'stream' function ψ such that $\psi_y = u, \psi_x = -v$ thereby satisfying Eqn (1-128b) identically.

(b) Transform Eqn (1-128a) into (x, ψ) coordinates (von Mises transformation).

Ans: $u_x = v(uu_\psi)_\psi$†

† Note that $\partial u/\partial x$ of Eqns (1-128) is taken at constant y while u_x here is at constant ψ. The two are not equal!

(c) With the initial condition $u(0, \psi) = f(\psi)$ and boundary conditions

$$\frac{\partial u}{\partial \psi}(x, 0) = 0, \qquad u(x, \infty) = U$$

develop a dimensionless formulation.

REFERENCES

1. Euler, L. *Institutiones Calculi Integralis.* St. Petersburg, 1768. (See *Leonhardi Euleri Opera Omnia*, Ser. I, vol. XI, p. 424. Teubner Verlag, Leipzig, 1913.
2. Runge, C. *Z. Math. Phys.*, **56**, 225, 1908.
3. Richardson, L. F. *Trans. R. Soc.*, **A210**, 307, 1910.
4. Liebmann, H. *Sber. bayer. Akad. Wiss. Münch.*, **3**, 385, 1918.
5. LeRoux, J. *J. Math. pures appl.*, **10**, 189, 1914.
6. Phillips, H. B. and Wiener, N. *J. Math. Phys.*, **2**, 105, 1923.
7. Courant, R., Friedrichs, K., and Lewy, H. *Math. Ann.*, **100**, 32, 1928.
8. Frankel, S. P. *Mathl. Tabl. natn. Res. Coun., Wash.*, **4**, 65, 1950.
9. Higgins, T. J. In *Numerical Methods of Analysis in Engineering* (L. E. Grinter, ed.), pp. 169–198. Macmillan, London and New York, 1949.
10. Southwell, R. V. *Relaxation Methods in Engineering Science.* Oxford University Press, London and New York, 1940.
11. Southwell, R. V. *Relaxation Methods in Theoretical Physics.* Oxford University Press, London and New York, 1946.
12. Binder, L. *Über Aussere Wärmleitung und Erwärmung Elektrischer Maschinen*, Dissertation, Technische Hochschule, München. W. Knapp Verlag, Halle, Germany, 1911.
13. Schmidt, E. *A. Föppl Festschrift*, p. 179. Springer Verlag OHG, Berlin, 1924.
14. Collatz, L. *Schr. math. Semin. Inst. angew. Math. Univ. Berl.*, **3**, 1, 1935.
15. Collatz, L. *Eigenwertaufgaben mit Technischen Andwendungen*, p. 346. *Akad.* Verlag, m.b.H., Leipzig, 1949.
16. Fox, L. *Proc. R. Soc.*, **A190**, 31, 1947.
17. Southwell, R. V. In *Numerical Methods of Analysis in Engineering* (L. E. Grinter, ed.), pp. 66–74. Macmillan, London and New York, 1949.
18. Fox, L. *Mathl. Tabl. natn. Res. Coun., Wash.*, **7**, 14, 1953.
19. O'Brien, G. G., Hyman, M. A., and Kaplan, S. *J. Math. Phys.*, **29**, 223, 1951.
20. Collatz, L. *Z. angew. Math. Mech.*, **16**, 239, 1936.
21. Collatz, L. *Z. angew. Math. Mech.*, **31**, 392, 1951.
22. Moon, P. and Spencer, D. E. *Field Theory Handbook.* Springer, Berlin, 1961.
23. Richardson, L. F. *Mathl. Gaz.*, **12**, 415, 1925.
24. Jeffrey, A. and Taniuti, T. *Nonlinear Wave Propagation with Applications to Physics and Magnetohydrodynamics.* Academic Press, New York, 1964.
25. Ames, W. F. *Nonlinear Partial Differential Equations in Engineering.* Academic Press, New York, 1965.
26. Swope, R. D. and Ames, W. F. *J. Franklin Inst.*, **275**, 36, 1963.
27. Shapiro, A. H. *The Dynamics and Thermodynamics of Compressible Fluid Flow*, vol. 1, chap. 9. Ronald Press, New York, 1954.
28. Zabusky, N. J. *J. Math. Phys.*, **3**, 1028, 1962.
29. Tomotika, S. and Tamada, K. *Q. appl. Math.*, **1**, 381, 1949.
30. Hill, R. *Plasticity*, p. 226. Oxford University Press, London and New York, 1950.

31. Erdelyi, A. *Asymptotic Expansions*. Dover, New York, 1956.
32. Jeffreys, H. *Asymptotic Approximations*. Oxford University Press, London and New York, 1962.
33. de Bruijn, N. G. *Asymptotic Methods in Analysis*. North-Holland, Amsterdam, 1958.
34. Landau, E. *Vorlesungen über Zahlentheorie*, vol. 2. Leipzig, 1927.
35. Bickley, W. G. *Q. Jl. Mech. appl. Math.*, **1**, 35, 1948.
36. Milne, W. E. *Numerical Solution of Differential Equations*. John Wiley, New York, 1953.
37. Forsythe, G. E. and Wasow, W. R. *Finite Difference Methods for Partial Differential Equations*. Wiley, New York, 1960.
38. Hildebrand, F. B. *Introduction to Numerical Analysis*. McGraw-Hill, New York, 1956.
39. Milne-Thompson, L. M. *Calculus of Finite Differences*. Macmillan, London, 1933.
40. Wasow, W. R. *Z. angew. Math. Phys.*, **6**, 81, 1955.
41. Feller, W. *An Introduction to Probability Theory and Its Applications*, vol. I. Wiley, New York, 1950.
42. Johnson, N. L. and Leone, F. C. *Statistics and Experimental Design*, vol. I. Wiley, New York, 1964.
43. Richtmyer, R. D. *Difference Methods for Initial Value Problems*, 2nd edit. Wiley (Interscience), New York, 1967.
44. Crandall, S. H. *J. Math. Phys.*, **32**, 80, 1953.
45. John, F. *Communs. pure appl. Math.*, **5**, 155, 1952.
46. Mikeladze, Sh. *Izv. Akad. Nauk SSSR*, **5**, 57, 1941.
47. Shortley, G. H. and Weller, R. *J. appl. Phys.*, **9**, 334, 1938.
48. Gerschgorin, S. *Z. angew. Math. Mech.*, **10**, 373, 1930.
49. Collatz, L. *Z. angew. Math. Mech.*, **13**, 55, 1933.
50. Shaw, F. S. *An Introduction to Relaxation Methods*. Dover, New York, 1950.
51. Allen, D. N. de G. *Relaxation Methods*. McGraw-Hill, New York, 1954.
52. Batschelet, E. *Z. angew. Math. Phys.*, **3**, 165, 1952.
53. Fox, L. (ed.). *Numerical Solutions of Ordinary and Partial Differential Equations*. Macmillan (Pergamon), New York, 1962.
54. Viswanathan, R. V. *Mathl. Tabl. natn. Res. Coun., Wash.*, **11**, 67, 1957.
55. Walsh, J. L. and Young, D. *J. Math. Phys.*, **33**, 80, 1954.
56. Giese, J. H. *J. Math. Phys.*, **37**, 169, 1958.
57. Langhaar, H. L. *Dimensional Analysis and Theory of Models*. Wiley, New York, 1951.
58. Bridgman, P. W. *Dimensional Analyses*, 2nd edit. Yale University Press, New Haven, Connecticut, 1931.
59. Birkhoff, G. *Hydrodynamics*. Princeton University Press, Princeton, N.J., 1960.
60. Carslaw, H. S. and Jaeger, J. C. *Conduction of Heat in Solids*, 2nd edit. Oxford University Press, London and New York, 1959.
61. Ames, W. F., Lee, S. Y., and Zaiser, J. N. *Int. J. Nonlinear Mech.*, **3**, 449, 1968.

2
Parabolic equations

2-0 Introduction

Quantitative physical laws are idealizations of reality. As knowledge grows
we observe that a given physical situation can be idealized mathematically
in a number of different ways. It is therefore important to characterize those
reasonable ideal formulations. Hadamard [1] (see, for example, Courant and
Hilbert [2]) examines this problem, asserting that a physical problem is *well
posed* if its solution *exists*, is *unique*, and depends *continuously on the auxiliary
data*. These criteria are now recognized to be for general orientation only
and may not be literally true. Indeed the improperly posed problems are
receiving increased attention (see, for example, Lavrentiev [3]).

The above criteria for a well-posed problem are physically reasonable in
most cases. Existence and uniqueness are an affirmation of the *principle of
determinism* without which experiments could not be repeated with the
expectation of consistent data. The continuous dependence criteria is an
expression of the stability of the solution—that is, a small change in any of
the problem's auxiliary data should produce only a correspondingly small
change in the solution.

Problems considered herein are usually assumed to be properly posed from
the outset. Existence and uniqueness are usually insured under physically
reasonable assumptions such as those of piecewise differentiability of coefficients and initial data. The important but difficult questions of existence and
uniqueness cannot be discussed in detail, but typical theorems will be stated
and applied where necessary. We must emphasize that some guide to existence
and uniqueness is advisable, if not absolutely necessary, before attempting a
numerical solution.

Parabolic partial differential equations arising in scientific and engineering
problems are often of the form

$$u_t = L(u) \tag{2-1}$$

where $L(u)$ is a second-order elliptic partial differential operator which may be
linear or nonlinear. Diffusion in an isotropic medium, heat conduction in an
isotropic medium, fluid flow through porous media, boundary layer flow over
a flat plate, persistence of solar prominences, and the wake growth behind a
submerged object have been modeled by the parabolic equation

$$u_t = \text{div} [f \text{ grad } u]\dagger$$
$$= \nabla \cdot [f \nabla u] \tag{2-2}$$

† The vector differential operators divergence (div, $\nabla \cdot$) and gradient (grad, ∇) are
given in Moon and Spencer [4] for many orthogonal coordinate systems.

where f may be constant, a function of the space coordinates, or a function of u or ∇u or both.

Physical problems that have been modeled by Eqn (2-2) are widespread. Some are discussed in Crank [5], Carslaw and Jaeger [6], Fox [7], and Ames [8]—the latter being devoted to nonlinear problems. Examples drawn from these and other sources will be discussed in the sequel.

Many questions of the general theory of parabolic equations—that is, existence, uniqueness, and differentiability of solutions—are developed in detail in Friedman [9]. Bernstein [10], while an older source, is also concerned with the same problems of the general theory. A typical uniqueness theorem for the initial boundary value problem

$$Lu = g(x, t)u_{xx} - u_t = f(x, t, u, u_x) \text{ in } D + B_T \qquad (2\text{-}3)\dagger$$

$$u(x, 0) = \psi(x), \quad t = 0 \qquad (2\text{-}4)$$

is given below.

Uniqueness theorem: If the parabolic equation [Eqn (2-3)] has a continuous bounded coefficient $g(x, t)$ in $D + B_T$ (D: $a < x < b$, B_T: $0 < t < T$) and if $f(x, t, u, w)$ is monotonic decreasing in u, then there exists at most one solution of Eqns (2-3) and (2-4).

We now turn to the development of finite difference methods for parabolic equations. To explain the basic ideas and to illustrate the process of developing improved difference equations for a given problem, the simple dimensionless diffusion equation

$$u_t = u_{xx} \qquad (2\text{-}5)$$

will be used as the initial model. An orderly refinement of difference approximations, beginning with the simple *explicit* method for which the truncation error is $O(\Delta t)$ and $O[(\Delta x)^2]$, will be developed. During this discussion the terms *explicit* and *implicit* methods will be introduced and made precise. Basically, an explicit formula provides for a noniterative 'marching' process for obtaining the solution at *each* present point in terms of known preceding and boundary values. Since parabolic (and hyperbolic) equations characteristically have open integration domains, explicit methods are applicable to these problems. Stability questions are critical in these situations. On the other hand, implicit procedures generally involve iterative simultaneous calculations of many present values in terms of known preceding and boundary values. Stability difficulties are not as serious in implicit methods.

2-1   Simple explicit methods

The numerical treatment of parabolic equations is initiated with the dimensionless initial boundary value problem in one space variable

---

† This equation is clearly quasilinear.

$$u_t = u_{xx}, \quad 0 < x < 1, \quad 0 < t \le T$$

$$u(x, 0) = f(x), \quad 0 < x < 1$$

$$u(0, t) = g(t), \quad 0 < t \le T$$

$$u(1, t) = h(t), \quad 0 < t \le T$$

(2-6)

Development of a finite difference analog for this problem necessitates the introduction of a net whose mesh points are denoted by $x_i = ih$, $t_j = jk$ where $i = 0, 1, 2, \ldots, M$; $j = 0, 1, \ldots, N$ with $h = \Delta x = 1/M$, $k = \Delta t = T/N$.† The boundaries are specified by $i = 0$ and $i = M$ and any 'false' boundaries (see Problem 1-29) by $i = -1, -2, \ldots, i = M + 1, M + 2, \ldots$, and so forth. The initial line is denoted by $j = 0$ and the discrete approximation at $x_i = ih$, $t_j = jk$ is designated $U_{i,j}$. The exact solution to Eqns (2-6), denoted by $u_{i,j}$, is assumed to exist and to have four continuous derivatives with respect to x and two continuous derivatives with respect to t; that is, $u \in C^{4,2}$.‡

If an approximate solution $U_{i,j}$ is assumed to be known at all mesh points up to time t_j, a method must be specified to advance the solution to time t_{j+1}. The value of $U_{i,j+1}$ at $x = 0$ and $x = 1$ should be selected as those boundary conditions specified on u in Eqns (2-6), that is

$$U_{0,j+1} = g(t_{j+1}), \qquad U_{M,j+1} = h(t_{j+1}).$$

(2-7)

At other points $0 < i < M$ the partial differential equation will be replaced by some difference equation. The simplest replacement, developed in Section 1-5 [Eqns (1-45) and (1-51)], consists of approximating the space derivative by the centered second difference and the time derivative by a forward difference at (x_i, t_j). The resulting difference equation is

$$\frac{1}{k} [U_{i,j+1} - U_{i,j}] = \frac{1}{h^2} [U_{i+1,j} - 2U_{i,j} + U_{i-1,j}], \quad i = 1, \ldots, M - 1.$$ (2-8)

Upon solving Eqn (2-8) for $U_{i,j+1}$ we obtain the *explicit* equation for 'marching' ahead in time

$$U_{i,j+1} = rU_{i-1,j} + (1 - 2r)U_{i,j} + rU_{i+1,j}, \quad i = 1, \ldots, M - 1$$ (2-9)

where
$$r = \frac{k}{h^2} = \frac{\Delta t}{(\Delta x)^2}.$$ (2-10)

---

† The quantities i and j are not necessarily integers. As we shall see, midvalues such as $U_{i,j+1/2}$ will be useful. Further, the space and/or time increments can vary with the index but this will not be so unless indicated.

‡ We say $f(x_1, x_2, \ldots, x_n) \in C^{\alpha_1, \alpha_2, \ldots, \alpha_n}$ if f has α_i continuous derivatives with respect to x_i.

The computational molecule for Eqn (2-9) is illustrated in Fig. 2-1. Note that Eqn (2-9) takes an especially simple form when $r = \frac{1}{2}$.

The approximate solution is computed as follows.

At $t = 0$ the solution is prescribed by the initial condition in Eqns (2-6). Advance to time $t = \Delta t = k$ is carried out by employing Eqns (2-7) and (2-9), whereupon the steps are repeated to advance to time $t = 2k$, and so forth. The explicit relation [Eqn (2-9)] is often called the *forward difference equation*.

Now we must consider the question 'How accurate is this approximation?'

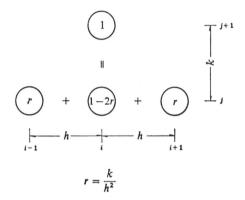

$$r = \frac{k}{h^2}$$

Fig. 2-1 Computational molecule for explicit difference approximation of $u_{xx} = u_t$

To accept the results without examining these considerations is folly. Early studies revolved around the selection of examples for which exact solutions of both the differential and difference problems were obtainable. Studies of this type were made by Hildebrand [11], Juncosa and Young [12, 13, 14], and others. While these provided early guidance, the method lacks the necessary generality required to treat more complex linear or nonlinear problems and will therefore be omitted.

A useful analysis based upon the concept of a maximum principle can be applied to analyzing the difference between u and U. This method has evolved from the works of several authors, including Laasonen [15], Douglas [16], and Collatz [17]. Maximum principle analysis has broad application and is therefore presented here.

Using the results of Section 1-5, Eqns (2-6) are written in the form

$$u_{i,j+1} = ru_{i-1,j} + (1 - 2r)u_{i,j} + ru_{i+1,j} + O[k^2 + kh^2],$$
$$i = 1, 2, \ldots, M - 1 \quad (2\text{-}11)$$

which emphasizes the local truncation error. Since U is only defined at a finite number of points, no differential equation for U can be found. However, the difference equation for

$$z_{i,j} = u_{i,j} - U_{i,j} \quad (2\text{-}12)$$

is obtainable by subtracting Eqn (2-9) from Eqn (2-11), thus

$$z_{i,j+1} = rz_{i-1,j} + (1 - 2r)z_{i,j} + rz_{i+1,j} + O[k^2 + kh^2],$$
$$i = 1, 2, \ldots, M - 1. \quad (2\text{-}13)$$

Since U agrees with u initially and on the boundary,

$$z_{i,0} = 0, \quad i = 0, \ldots, M$$
$$z_{0,j} = z_{M,j} = 0, \quad j = 0, \ldots, N. \quad (2\text{-}14)$$

The three coefficients on the right-hand side of Eqn (2-13) sum to one for all values of r and all are nonnegative if

$$0 < r \leq \tfrac{1}{2}. \quad (2\text{-}15)$$

Let Eqn (2-15) hold. Then

$$|z_{i,j+1}| \leq r\,|z_{i-1,j}| + (1 - 2r)|z_{i,j}| + r\,|z_{i+1,j}| + A[k^2 + kh^2]$$
$$\leq \|z_j\| + A[k^2 + kh^2], \quad i = 1, \ldots, M - 1 \quad (2\text{-}16)$$

where
$$\|z_j\| = \max_{i=0,\ldots,M} |z_{i,j}|. \quad (2\text{-}17)$$

Consequently,
$$\|z_{j+1}\| \leq \|z_j\| + A[k^2 + kh^2] \quad (2\text{-}18)$$

and since $\|z_0\| = 0$ we easily calculate that

$$\|z_j\| \leq Aj[k^2 + kh^2]$$
$$\leq AT[k + h^2] = AT[\Delta t + (\Delta x)^2] \quad (2\text{-}19)$$

as $jk = j\Delta t \leq T$.

Equation (2-19) means that the error $z_{i,j}$ tends to zero as Δx and Δt tend to zero. Thus the solution of the finite difference analog converges to the solution of the differential equation as Δx and Δt tend to zero. This is the most important property required of a finite difference approximation to a partial differential equation. We state the result as a theorem.

Theorem: Let $u \in C^{4,2}$† be the solution of Eqn (2-6) and let U be the solution of Eqns (2-7) and (2-9). If $0 < r \leq \tfrac{1}{2}$, then $\max |u_{i,j} - U_{i,j}| \leq AT[(\Delta t) + (\Delta x)^2]$, $0 \leq x_i \leq 1$, $0 \leq t_j \leq T$. (The value of A depends upon upper bounds for u_{tt} and u_{xxxx}.)

Another conclusion can be drawn from this theorem. From Section 1-8 we observe that the above boundedness condition implies stability in the sense of John. Thus, for the problem described by Eqns (2-6), the *stability condition* is

$$0 < r = \frac{\Delta t}{(\Delta x)^2} \leq \tfrac{1}{2}. \quad (2\text{-}20)$$

† This condition can be relaxed to $C^{2,1}$; see, for example, Douglas [16, 18].

In the next section we shall see that Eqn (2-20) is also the stability condition of von Neumann (see, for example, Richtmyer [19] and Section 1-8).

In Section 1-5 the truncation error for Eqn (2-11) was found to be

$$O[k^2 + kh^2] = O(k^2) + O(kh^2) \tag{2-21}$$

where the terms of order k^2 and kh^2, evaluated at i, j, are

$$\frac{k^2}{2} \frac{\partial^2 u}{\partial t^2} - \frac{kh^2}{12} \frac{\partial^4 u}{\partial x^4} = \frac{k}{2} \left\{ k \frac{\partial^2 u}{\partial t^2} - \frac{h^2}{6} \frac{\partial^4 u}{\partial x^4} \right\}. \tag{2-22}$$

But u satisfies the diffusion equation $u_t = u_{xx}$, hence

$$u_{tt} = u_{xxt} = u_{txx} = u_{xxxx} \tag{2-23}$$

that is, u also satisfies $u_{tt} = u_{xxxx}$ if it is $C^{4,2}$. Equation (2-22) now takes the value

$$\frac{k}{2} \left\{ k - \frac{h^2}{6} \right\} \frac{\partial^4 u}{\partial x^4}. \tag{2-24}$$

Therefore, if k and h go to zero in such a way that

$$\frac{k}{h^2} = \frac{1}{6} \tag{2-25}$$

the truncation error will be $O[k^3]$ or, which is the same thing, $O[h^6]$. In this special case, first discovered by Milne [20], the error goes to zero even faster than indicated in the Theorem. Indeed, if $r = \frac{1}{6}$, one can replace the bound in that theorem by

$$\max |u_{i,j} - U_{i,j}| \leq A_1 T[(\Delta t)^2 + (\Delta x)^4]. \tag{2-26}$$

The solution of the difference equation approaches the solution of the differential equation with special rapidity, a result that has been used by some to obtain highly accurate numerical solutions of the diffusion equation.

PROBLEMS

2-1   By application of Taylor series expansion for $u(x, t)$ establish the validity of Eqns (2-22) and (2-24). Find the coefficients of the terms of order k^3 and kh^4 if $r = \frac{1}{6}$.

2-2   Upon setting $U_{i,j} = X_i T_j$ verify that the method of separation of variables can be applied to Eqn (2-9). What are the separation equations for X_i and T_j?

2-3   Obtain solutions for the difference equations describing X_i and T_j in Problem 2-2 by setting $X_i = e^{\lambda i}$ and determining two values of λ (one value for T_j), say λ_1, λ_2. The solution for X_i can then be obtained by superposition as $X_i = C_1 e^{i\lambda_1} + C_2 e^{i\lambda_2}$, where C_1 and C_2 are constants.

2-4   With $f(x) = x(1 - x), g(t) = 0, h(t) = 0$ in Eqn (2-6) compute the first row of solutions, away from the boundary, if $r = \frac{1}{6}$ and $\Delta t = \frac{1}{54}$.

2-2 Fourier stability method

In Section 1-8 an alternative stability definition to that of John was introduced. For easy reference we restate that definition here and recall that it is useful only when the corresponding continuous solutions remain bounded. A finite difference process within the strip $0 < x < 1$, $t > 0$ is unstable if for a fixed network and fixed homogeneous boundary conditions there exist initial conditions for which the finite difference solutions $U_{i,j}$ become unbounded as $j \to \infty$.

There are at least two well-known methods for determining stability criteria. Unfortunately, neither of these methods can be used directly to study stability in the nonlinear case without a prior 'linearization'.

A useful and simple method of finding a stability criterion is to examine the propagating effect of a single row of errors, say along the line $t = 0$. These are represented by a finite Fourier series of the form $\sum_i A_i \exp [(-1)^{1/2}\lambda_i x]$ in which the number of terms is equal to the number of mesh points on the line. Usually the effect of a single term, $\exp [(-1)^{1/2}\lambda x]$ with λ any real number, is analyzed and the complete effect is then obtained by linear superposition. This *Fourier method* effectively ignores boundary conditions and, insofar as these may affect the stability criterion, a second procedure called the *matrix method* is preferable. First we shall illustrate the Fourier method for various difference approximations. In this section we consider the explicit scheme specified by Eqn (2-9).

Let the value on the initial line be designated by

$$\exp [(-1)^{1/2}\lambda x] = \exp [(-1)^{1/2}\lambda ih]. \qquad (2\text{-}27)$$

We then observe (see Problem 2-3) that

$$\exp [\alpha jk] \exp [(-1)^{1/2}\lambda ih] \qquad (2\text{-}28)$$

is a solution of the difference equation [Eqn (2-9)] obtained by separation of variables. We now seek the condition that the term in $t = jk$ does not increase as $j \to \infty$. Upon substituting Eqn (2-28) into Eqn (2-9) and cancelling common factors we obtain

$$e^{\alpha k} = r \{\exp [(-1)^{1/2}\lambda h] + \exp [-(-1)^{1/2}\lambda h]\} + (1 - 2r)$$

$$= 2r \cos \lambda h + (1 - 2r) \qquad (2\text{-}29)$$

$$= 1 - 4r \sin^2 \tfrac{1}{2}\lambda h.$$

Since r is nonnegative, $e^{\alpha k}$ is always less than 1. Thus, if $e^{\alpha k} \geq 0$ the solution [Eqn (2-28)] will decay steadily as $j \to \infty$. If $-1 < e^{\alpha k} < 0$ the solution will have a decaying amplitude of oscillating sign as $j \to \infty$. Finally, if $e^{\alpha k} < -1$ the solution oscillates with increasing amplitude as $j \to \infty$. In the last case Eqn (2-9) is unstable. For stability we must have $-1 \leq 1 - 4r \sin^2 \frac{1}{2}\lambda h$ or

$$r \leq \frac{1}{2 \sin^2 \frac{1}{2}\lambda h} = \frac{1}{1 - \cos \lambda h} \qquad (2\text{-}30)\dagger$$

thus demonstrating the dependency of the stability limit on h. If r is smaller than $1/2(1 - \cos \lambda h)$ the finite difference solution will decay, but in an oscillating manner. To ensure nonoscillatory decay, it is necessary to have $1 - 4r \sin^2 \frac{1}{2}\lambda h \geq 0$, a result that is ensured if $r \leq 1/[2(1 - \cos \lambda h)]$.

A second or *matrix method* automatically includes the effects of boundary conditions on the stability. We shall discuss this procedure following Section 2-4. While stability criteria are not drastically changed by boundary conditions they do have an effect. For example, Crandall [21] notes that for the problem

$$u_{xx} = u_t, \qquad u(x, 0) = 1, \qquad u(0, t) = u_x(1, t) = 0 \qquad (2\text{-}31)$$

the value $r = \frac{1}{2}$ is just on the stable side when Eqn (2-9) is applied. On the other hand, for the problem

$$u_{xx} = u_t, \qquad u(x, 0) = 1, \qquad u_x(0, t) = u(0, t), \qquad u_x(1, t) = 0 \qquad (2\text{-}32)$$

the value $r = \frac{1}{2}$ is just on the unstable side.

PROBLEMS

2-5 Carry out a Fourier stability analysis, using an explicit method similar to Eqn (2-9), for the parabolic equation $u_t = u_{xx} + u_x$.

2-6 Consider the problem specified by Eqn (2-31). Find the solution to the explicit finite difference approximation [Eqn (2-9)] with the discrete boundary conditions

$$U_{0,j} = 0, \qquad U_{M+1,j} = U_{M-1,j} \qquad \text{(see Problem 1-29)}.$$

$$\textit{Ans:} \quad U_{i,j} = \sum_{n=1}^{M} d_n \left(1 - 4r \sin^2 \frac{(2n-1)}{2M} \frac{\pi}{2} \right)^j \sin \frac{(2n-1)}{2} \frac{i\pi}{M}.$$

(The d_n are constants to fit initial conditions.)

2-7 Compare the result of Problem 2-6 with the exact solution to Eqn (2-31), that is

$$u(x, t) = \sum_{n=1}^{\infty} c_n \exp \left[-\pi^2 \left(\frac{2n-1}{2} \right)^2 t \right] \sin \frac{(2n-1)}{2} \pi x$$

where c_n are constants to fit initial conditions.

$\dagger$ If $r \leq \frac{1}{2}$ this inequality holds for $\frac{1}{2} \leq 1/(1 - \cos \lambda h)$.

2-8 Show that the stability limit for Problem 2-6 is

$$r \le \frac{1}{1 + \cos{(\pi/2M)}}.$$

Find the value for $M = 2, 4, 6$ and $M \to \infty$.

2-9 Carry out Problems 2-6 through 2-8 for the problem described by Eqns (2-32).

2-3 Implicit methods

With the imposition of a stability limit on $\Delta t/(\Delta x)^2$, explicit finite difference formulae for the diffusion equation can give useful approximations. However, the requirement that $\Delta t/(\Delta x)^2 < \frac{1}{2}$ places a severe restriction on the interval size in the t direction with a resultant increase in computation. Why does this 'trouble' occur? A qualitative argument, due to Crandall [21], can be based upon Fig. 2-2. If $U_{i,j}$ is known at each mesh point on EF then

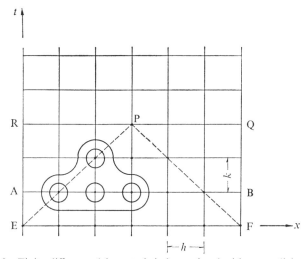

Fig. 2-2 Finite difference 'characteristics' associated with an explicit method

$U_{i,j}$ can be calculated at *every* point within the triangle EFP by successive use of the explicit formula, Eqn (2-9), shown in computational molecule form in Fig. 2-2. Moreover, in this computation, *knowledge of values of $U_{i,j}$ on ER and* FQ *is not required.* That is to say, the explicit finite difference approximation acts like a hyperbolic system with *two* real finite difference characteristics (the lines EP and FP). On the other hand, the continuous system is parabolic with the single characteristic ($t = $ constant) EF or RQ. The continuous solution at P is not determined until boundary information along ER and FQ is prescribed. Now, if r is held constant as $h \to 0$, the slope of EP (which is h/k) goes to zero and hence in the limit EP does tend to the true characteristic. Our argument must therefore be qualitative.

Consequently, an explicit finite difference formula provides a somewhat imperfect model for a parabolic equation. A *natural* 'parabolic' finite difference approximation would be incapable of producing any values at mesh points on AB if the boundary conditions at A and B were not specified. Values at each mesh point along AB would be given as functions of the values along EF and the boundary conditions at A and B. Such behaviour results from the use of *implicit* finite difference approximations which seem to have been first used by Crank and Nicolson [22] and O'Brien, Hyman, and Kaplan [23].

An *implicit* formula is one in which two or more unknown values in the $j + 1$ row (see Fig. 2-1) are specified in terms of known values in the j row (and $j - 1, j - 2, \ldots$, if necessary) by a single application of the expression. If there are M unknown values in the $j + 1$ row the formula must be applied M times across the length of the row. The resulting system of M simultaneous equations specifies the M net values implicitly.

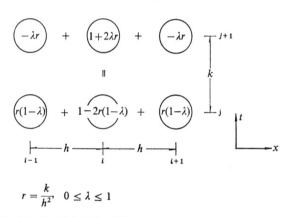

$$r = \frac{k}{h^2}, \quad 0 \leq \lambda \leq 1$$

Fig. 2-3   Implicit finite difference approximation for $u_{xx} = u_t$

The simplest implicit method is that suggested first by O'Brien *et al.* [23]. Upon approximating the derivative u_{xx} of Eqns (2-6) in the $j + 1$ row instead of in the j row we obtain

$$\frac{1}{k} [U_{i,j+1} - U_{i,j}] = \frac{1}{h^2} [U_{i+1,j+1} - 2 U_{i,j+1} + U_{i-1,j+1}] \qquad (2\text{-}33)$$

or in terms of the unknowns in the $j + 1$ row, $r = k/h^2$,

$$-rU_{i-1,j+1} + (1 + 2r)U_{i,j+1} - rU_{i+1,j+1} = U_{i,j}. \qquad (2\text{-}34)$$

Crank and Nicolson [22] used an average of approximations in the j and $j + 1$ row instead of Eqn (2-33). More generally one can introduce a *weighting factor* λ and replace Eqn (2-33) by

$$U_{i,j+1} - U_{i,j} = r \{\lambda [U_{i-1,j+1} - 2U_{i,j+1} + U_{i+1,j+1}]$$
$$+ (1 - \lambda)[U_{i-1,j} - 2U_{i,j} + U_{i+1,j}]\} \qquad (2\text{-}35)$$

with $0 \leq \lambda \leq 1$. A single application of Eqn (2-35) equates a linear combination of three unknowns in the $j + 1$ row to a linear combination of three known values in the j row. Thus we obtain

$$-r\lambda U_{i-1,j+1} + (1 + 2r\lambda)U_{i,j+1} - r\lambda U_{i+1,j+1}$$

$$= r(1 - \lambda)U_{i-1,j} + [1 - 2r(1 - \lambda)]U_{i,j} + r(1 - \lambda)U_{i+1,j}. \quad (2\text{-}36)$$

Equation (2-36) is shown in computational molecule form in Fig. 2-3. If $\lambda = 1$ Eqn (2-36) becomes the O'Brien *et al.* form, Eqn (2-34). If $\lambda = \frac{1}{2}$ we obtain the Crank–Nicolson formula. On the other hand, if $\lambda = 0$ the explicit relation [Eqn (2-9)] is recovered.

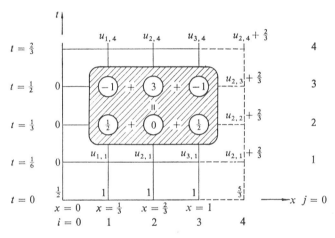

Fig. 2-4 Finite difference net for problem of Eqn (2-37) with implicit equation [Eqn (2-36)] and 'false' boundary

Before further discussion let us give the details of a simple example. Consider the problem

$$u_t = u_{xx}$$

$$u(x, 0) = 1 \quad (2\text{-}37)$$

$$u(0, t) = 0, \qquad \partial u/\partial x(1, t) = 1.$$

Let $h = \frac{1}{3}$, $k = \frac{1}{6}$ so $r = k/h^2 = \frac{3}{2}$ and $\lambda = \frac{2}{3}$. The finite difference approximation for $u_t = u_{xx}$ will be Eqn (2-36), and for the boundary condition $\partial u/\partial x$ we will use the central difference $U_{4,j} - U_{2,j} = 1 \cdot 2h = \frac{2}{3}$. Thus, the addition of a 'false' boundary to our integration domain (Fig. 2-4)† is

---

† At $(0, 0)$ the value $\frac{1}{2}$ is selected to decrease error. Singularity problems will be discussed in Section 5-1.

required. Upon successively setting $i = 1, 2$, and 3 we obtain the matrix form of the resulting three equations as

$$\begin{bmatrix} 3 & -1 & 0 \\ -1 & 3 & -1 \\ 0 & -2 & 3 \end{bmatrix} \begin{bmatrix} U_{1,j+1} \\ U_{2,j+1} \\ U_{3,j+1} \end{bmatrix} = \begin{bmatrix} 0 & \frac{1}{2} & 0 \\ \frac{1}{2} & 0 & \frac{1}{2} \\ 0 & 1 & 0 \end{bmatrix} \begin{bmatrix} U_{1,j} \\ U_{2,j} \\ U_{3,j} \end{bmatrix} + \begin{bmatrix} 0 \\ 0 \\ 1 \end{bmatrix} \quad (2\text{-}38)$$

Equations (2-38) take a *tridiagonal form* (elements occur only on the main diagonal and on one subdiagonal above and below). Equations (2-38) demonstrate that at each step of the marching process it is necessary to solve a set of simultaneous equations which are linear in this linear example. If the coefficients in the matrix of the unknowns are independent of j, as in this example, we can find the inverse of the matrix on the left-hand side of Eqns (2-38) and thereby obtain an explicit formula. In this example we have

$$\begin{bmatrix} U_{1,j+1} \\ U_{2,j+1} \\ U_{3,j+1} \end{bmatrix} = \frac{1}{12} \begin{bmatrix} 1 & 3 & 1 \\ 3 & 3 & 3 \\ 2 & 6 & 2 \end{bmatrix} \begin{bmatrix} U_{1,j} \\ U_{2,j} \\ U_{3,j} \end{bmatrix} + \frac{1}{18} \begin{bmatrix} 1 \\ 3 \\ 8 \end{bmatrix} \quad (2\text{-}39)$$

Application of this formula is not essentially different from the explicit expression [Eqn (2-9)] except that each point has a different *molecule.*

If the system of equations to be solved is very large, the matrix inversion method that we have just discussed may use excessive computer time. This is especially true when the equations are tridiagonal, and tridiagonal systems frequently occur. Note that Eqn (2-36) is tridiagonal for all $\lambda \neq 0, r \neq 0$. Using Eqn (2-36) as a model, let us write the general tridiagonal system of n equations in the form

$$b_1 u_1 + c_1 u_2 = d_1$$

$$a_i u_{i-1} + b_i u_i + c_i u_{i+1} = d_i, \quad i = 2, 3, \ldots, n - 1. \quad (2\text{-}40)$$

$$a_n u_{n-1} + b_n u_n = d_n.$$

As we shall see, this system can be solved explicitly for the unknowns, thereby eliminating any matrix operations. The method we describe was discovered independently by many and has been called the Thomas algorithm (see [24]) by Young. Its general description first appeared in widely distributed published form in an article by Bruce *et al.* [25].

The Gaussian elimination process—that is, successive subtraction of a suitable multiple of each equation from the following equations—transforms the system into a simpler one of *upper bidiagonal* form. We designate the coefficients of this new system by a_i', b_i', c_i', d_i' and in particular we note that

$$a_i' = 0, \quad i = 2, 3, \ldots, n$$

$$b_i' = 1, \quad i = 1, 2, \ldots, n. \quad (2\text{-}41)$$

The coefficients c_i', d_i' are calculated successively from the relations

$$c_1' = \frac{c_1}{b_1}, \qquad d_1' = \frac{d_1}{b_1} \tag{2-42}$$

$$\left.\begin{aligned} c_{i+1}' &= \frac{c_{i+1}}{b_{i+1} - a_{i+1}c_i'} \\ d_{i+1}' &= \frac{d_{i+1} - a_{i+1}d_i'}{b_{i+1} - a_{i+1}c_i'} \end{aligned}\right\} i = 1, 2, \ldots, n-1 \tag{2-43}$$

and, of course, $c_n = 0$.

Having completed the elimination we examine the new system and see that the nth equation is now

$$u_n = d_n'. \tag{2-44}$$

Substituting this value into the $(n-1)$st equation

$$u_{n-1} + c_{n-1}'u_n = d_{n-1}'$$

we have

$$u_{n-1} = d_{n-1}' - c_{n-1}'u_n.$$

Thus, starting with u_n (Eqn 2-44) we have successively the solution for u_i as

$$u_i = d_i' - c_i'u_{i+1}, \qquad i = n-1, n-2, \ldots, 1. \tag{2-45}$$

When performing this calculation one must be sure that $b_1 \neq 0$ and that $b_{i+1} - a_{i+1}c_i' \neq 0$. If $b_1 = 0$ we can solve for u_2 and thus reduce the size of our system. If $b_{i+1} - a_{i+1}c_i' = 0$ we can again reduce the system size by solving for u_{i+2}. A serious source of round-off error can arise if $|b_{i+1} - a_{i+1}c_i'|$ is small. Modifications of the Gauss elimination process, often called pivoting, can help to avoid this error.† It has been shown by Douglas [26] that this method does not introduce large errors in the solution due to round-off errors in the calculation. Cuthill and Varga [27] propose and use an alternative algorithm which is more efficient for self-adjoint equations with time-independent coefficients. If the coefficients are time-dependent more calculation is required. Blair [28] has analysed the round-off error for this method.

---

† The reality of some of these problems is easily seen in the following simple example. Let the nonsingular system of equations be:

$$u_1 + u_2 = 2; \qquad u_1 + u_2 + u_3 = 3; \qquad u_2 + u_3 = 2.$$

Here $a_1 = 0$, $a_2 = 1$, $a_3 = 1$, $b_1 = b_2 = b_3 = 1$, $c_1 = c_2 = 1$, $c_3 = 0$, and $d_1 = d_3 = 2$, $d_2 = 3$. Then $c_1' = 1$, $d_1' = 2$ while c_2' and d_2' are *not defined*.

When the tridiagonal algorithm is applied to the Crank–Nicolson form of Eqn (2-36) ($\lambda = \frac{1}{2}$), the specific formulas for $M - 1$ unknowns become

$$d_i = \frac{1 - r}{1 + r} U_{i,j} + \frac{r}{2(1 + r)} (U_{i-1,j} + U_{i+1,j}),$$

$$i = 1, 2, \ldots, M - 1$$

$$c_1' = -\frac{r}{2(1 + r)}, \quad d_1' = d_1$$

$$c_i' = -\frac{r/[2(1 + r)]}{1 + \{r/[2(1 + r)]\}c_{i-1}'}, \quad (2\text{-}46)$$

$$i = 2, 3, \ldots, M - 2, (c_{M-1} = 0)$$

$$d_i' = \frac{d_i + \{r/[2(1 + r)]\}d_{i-1}'}{1 + \{r/[2(1 + r)]\}c_{i-1}'}, \quad i = 2, 3, \ldots, M - 1$$

and

$$U_{M-1,j+1} = d_{M-1}',$$

$$U_{i,j+1} = d_i' - c_i' U_{i+1,j+1}, \quad i = M - 2, M - 3, \ldots, 1$$

PROBLEMS

2-10 For a general 4×4 tridiagonal system verify the Thomas algorithm.

2-11 Apply the Thomas algorithm to Eqn (2-36) for arbitrary λ and r. Verify Eqns (2-46) by setting $\lambda = \frac{1}{2}$.

2-12 Apply the Crank–Nicolson method to the problem $u_t = u_{xx}; u_x - u = e^{-t}$, $x = 0, t > 0; u = x(1 - x), 0 < x < 1, t = 0; u_x = 3, x = 1, t > 0$. Let $h = \frac{1}{4}$, $k = \frac{1}{4}$ and write the implicit formula at the $j + 1$ row. Invert the matrix and find the explicit form.

2-13 Nonlinear equations of the form $u_t = u_{xx} + \phi(x, t, u)$ arise in problems of diffusion with chemical reaction. Typical auxiliary condition are $u(0, t) = f(t)$, $u_x(1, t) = g(t), u(x, 0) = h(x)$. Use the general implicit formula [Eqn (2-36)], with arbitrary λ and r, to develop an implicit finite difference approximation for this problem. Does the Thomas algorithm apply?

2-14 Ignoring the boundary conditions, use the Fourier method to investigate the stability of Eqn (2-36). What can you say about stability if $\lambda = 1, \frac{1}{2}, \frac{1}{4}$?

Ans: With $U_{i,j} = \exp[\alpha j k] \exp[(-1)^{1/2} \beta i h]$ we find

$$e^{\alpha k} = \frac{1 - 4r(1 - \lambda) \sin^2 \frac{1}{2}\beta h}{1 + 4r\lambda \sin^2 \frac{1}{2}\beta h}. \quad (2\text{-}47)$$

2-15 Let a, b, c, and d be functions of x, t, u, u_x, and u_t. If $au_{xx} + bu_{xt} + cu_{tt}$ $= d$ is parabolic, under what conditions will the adoption of an implicit method analogous to Eqn (2-36) lead to a system of linear equations? When can the Thomas algorithm be applied?

2-16 Crandall [33] has examined the stability, oscillation, and truncation error of the two-level implicit formula of Eqn (2-36). The results are shown in Fig. 2-5. Each coordinate point (r, λ) represents a different finite scheme for integrating the heat conduction equation. Precise values of stability and oscillation limits depend on the number of spatial subdivisions M and on the problem boundary conditions. As M increases there is a rapid approach to limiting values. These limiting values are shown in Fig. 2-5. If the boundary conditions are approximated to $O(h^4)$, then superior accuracy can be expected from those formulas that are $O(h^4)$. Develop one or more of these results by Fourier stability analysis.

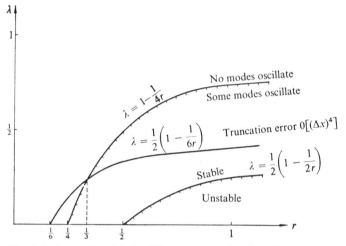

Fig. 2-5 Properties of finite difference approximations for $u_{xx} = u_t$

2-4 An unconditionally unstable difference equation

Another fairly obvious attempt to improve the local truncation error of the approximation for u_t was proposed by Richardson [29]. This 'overlapping steps' method

$$\frac{U_{i,j+1} - U_{i,j-1}}{2k} = \frac{1}{h^2} [U_{i+1,j} - 2U_{i,j} + U_{i-1,j}] \qquad (2-48)$$

is unstable for *all* values of $r = k/h^2$! To establish this result we use the Fourier method and set

$$U_{i,j} = \exp [\alpha jk] \exp [(-1)^{1/2}\beta ih]$$

into Eqn (2-48), with the result

$$e^{\alpha k} - e^{-\alpha k} = -8r \sin^2 \tfrac{1}{2}\beta h. \qquad (2-49)$$

Upon multiplying Eqn (2-49) by $e^{\alpha k}$ and solving the resulting quadratic in $e^{\alpha k}$ we find

$$e^{\alpha k} = -4r \sin^2 \tfrac{1}{2}\beta h \pm [1 + 16r^2 \sin^4 \tfrac{1}{2}\beta h]^{1/2}$$
$$= -4r \sin^2 \tfrac{1}{2}\beta h \pm [1 + 8r^2 \sin^4 \tfrac{1}{2}\beta h + O(r^4)]. \qquad (2-50)$$

Upon selection of the negative sign it follows that

$$e^{\alpha k} = -1 - 4r \sin^2 \tfrac{1}{2}\beta h(1 + 2r \sin^2 \tfrac{1}{2}\beta h) - O(r^4)$$

or

$$|e^{\alpha k}| > 1 + 4r \sin^2 \tfrac{1}{2}\beta h. \tag{2-51}$$

Consequently, for all $r > 0$, $|e^{\alpha k}| > 1$ and the finite difference approximation [Eqn (2-48)] is always unstable. The computed solution would bear little resemblance to the exact solution!

This analysis serves as an excellent example of what can happen if no attention is paid to the mathematical properties of the finite difference approximation.

2-5 Matrix stability analysis

As previously observed, boundary conditions are ignored when the Fourier stability method is applied. On the other hand, the matrix method for analyzing stability will automatically include the effects of the boundaries. To illustrate the procedure we will utilize the Crank–Nicolson formula

$$U_{i-1,j+1} - 2(1 + r^{-1})U_{i,j+1} + U_{i+1,j+1}$$
$$= -\{U_{i-1,j} - 2(1 - r^{-1})U_{i,j} + U_{i+1,j}\} \tag{2-52}$$

obtained from Eqn (2-36) with $\lambda = \tfrac{1}{2}$.

Let U take the boundary values at $x = 0$ and $x = 1$, prescribed for u at the boundary mesh points and denote by V_j the vector of other than boundary mesh point values along the jth line. Since the boundary values are fixed we need to use Eqn (2-52) only at $M - 1$ interior points.† In particular, if the boundary values are zero, then Eqn (2-52) generates the matrix form

$$AV_{j+1} = BV_j. \tag{2-53}$$

In any case this is the equation governing stability; with other than zero boundary conditions a vector will be added to Eqn (2-53) which can, at most, depend upon i [see, for example, Eqns (2-38) and (2-39)]. The matrices A and B are

$$A = C - 2r^{-1}I, \qquad B = -C - 2r^{-1}I \tag{2-54}$$

where I is the identity matrix and C has the tridiagonal form

$$C = \begin{bmatrix} -2 & 1 & 0 & 0 & 0 & \ldots & 0 \\ 1 & -2 & 1 & 0 & 0 & \ldots & 0 \\ 0 & 1 & -2 & 1 & 0 & \ldots & 0 \\ & & & \ldots & & & \end{bmatrix} \tag{2-55}$$

Stability of the finite difference approximation is ensured if all the eigenvalues of $A^{-1}B$ are, in absolute value, less than or equal to 1. This result is made plausible in the following paragraph. Other more rigorous arguments are given in Richtmyer [19] and Forsythe and Wasow [30].

† If the boundary conditions involve derivatives, the application of false boundaries will lead to more unknowns on the jth line.

The nonsingular nature of A allows us to rewrite Eqn (2-53) as

$$V_{j+1} = PV_j, \qquad P = A^{-1}B. \qquad (2\text{-}56)$$

Upon repeated application of Eqn (2-56)

$$V_{j+1} = PV_j = P^2 V_{j-1} = \cdots = P^j V_1 = P^{j+1} V_0 \qquad (2\text{-}57)$$

where V_0 is the vector of initial values. Let the eigenvalues of P be denoted by μ_n and the corresponding eigenvectors by $v^{(n)}$, $n = 1, 2, \ldots, M - 1$. From matrix theory we recall that

$$P v^{(n)} = \mu_n v^{(n)}. \qquad (2\text{-}58)$$

If the eigenvalues of P are distinct, we can write

$$V_1 = \sum_{n=1}^{M-1} \alpha_n v^{(n)}$$
$$\vdots \qquad\qquad\qquad\qquad (2\text{-}59)$$
$$V_{j+1} = P^j V_1 = \sum_{n=1}^{M-1} \alpha_n \mu_n^j v^{(n)}$$

a result that follows from successive application of Eqn (2-58). As $j \to \infty$ errors will not grow in magnitude if all eigenvalues μ_n, $n = 1, 2, \ldots, M - 1$ are, in absolute value, less than or equal to 1.

In our example problem determination of the eigenvalues of $P = A^{-1}B$ is accomplished from the general form of Eqn (2-58), $Pv = \mu v$, or

$$(B - A\mu)v = 0 \qquad (2\text{-}60)$$

which does not require inversion of A. Upon substituting Eqn (2-54) into Eqn (2-60) we obtain

$$\left\{ C - 2r^{-1} \frac{\mu - 1}{\mu + 1} I \right\} v = 0. \qquad (2\text{-}61)$$

Thus the eigenvalues μ of $A^{-1}B$ are related to the eigenvalues γ of C by the equation

$$\gamma = 2r^{-1} \frac{\mu - 1}{\mu + 1} \qquad \text{or} \qquad \mu = \frac{2 + r\gamma}{2 - r\gamma} \qquad (2\text{-}62)$$

and the eigenvectors, v, of C are those of $A^{-1}B$.

For the matrix C of Eqn (2-55) we find from $|C - \gamma I| = 0$ the eigenvalues

$$\gamma_n = -4 \sin^2 \frac{n\pi h}{2}, \quad n = 1, 2, \ldots, M - 1 \qquad (2\text{-}63)$$

and the corresponding eigenvectors

$$v^{(n)} = \begin{bmatrix} \sin n\pi h \\ \sin 2n\pi h \\ \vdots \\ \sin (M-1)\, n\pi h \end{bmatrix} = \{\sin in\pi h\}, \quad n = 1, 2, \ldots, M - 1. \qquad (2\text{-}64)$$

Thus from Eqn (2-62) the eigenvalues of $A^{-1}B$ are

$$\mu_n = \frac{1 - 2r \sin^2 (n\pi h/2)}{1 + 2r \sin^2 (n\pi h/2)}, \quad n = 1, 2, \ldots, M - 1. \tag{2-65}$$

This result is essentially identical with Eqn (2-47), with $\lambda = \frac{1}{2}$, obtained by application of the Fourier method.

Stability seems to be guaranteed since $|\mu_n| < 1$ for all $r > 0$, but a rather trivial restriction on r occurs as a consequence of the form of the solution for large t. To discover this restriction note that Eqn (2-53) has the solution given by Eqn (2-59), that is

$$U_{i,j} = \sum_{n=1}^{M-1} \alpha_n \mu_n^j \sin in\pi h. \tag{2-66}$$

This may be compared with the exact solution of the problem $u_t = u_{xx}$, $u(0, t) = u(1, t) = 0, u(x, 0) = f(x)$. Upon applying the separation of variables method one obtains

$$u(x, t) = \sum_{n=1}^{\infty} a_n e^{-n^2\pi^2 t} \sin n\pi x. \tag{2-67}$$

The terms $\exp(-n^2\pi^2 t)$ and μ_n^j occupy analogous positions. As h and $k \to 0$, for fixed r, we see that

$$\mu_n^j = \left\{ \frac{1 - 2(k/h^2) \sin^2 (n\pi h/2)}{1 + 2(k/h^2) \sin^2 (n\pi h/2)} \right\}^j \to (1 - kn^2\pi^2)^{t/k}$$

$$\to e^{-n^2\pi^2 t}. \tag{2-68}$$

Thus, if r is held fixed, the finite difference solution converges to that of the differential equation.

Now, for large t, the analytic solution [Eqn (2-67)] is dominated by the term $e^{-\pi^2 t}$. Similar domination in the finite difference solution should be expected. To ensure this we require

$$\mu_1 > 0 \quad \text{and} \quad \mu_1 > |\mu_n|, \quad \text{for all } n > 1. \tag{2-69}$$

An elementary computation shows that both of these requirements are satisfied if

$$r < (\sin \pi h)^{-1}. \tag{2-70}$$

For the interval sizes that occur this is a trivial restriction indeed.

When a boundary condition involving the partial derivative $\partial u/\partial x$ is specified, Eqn (2-52) must also be used on the boundaries. The matrix A (and B is of order $M + 1$, and the determinant $|A\mu - B|$ is proportional to

$$\begin{vmatrix} \frac{1}{2}q & 1 & 0 & 0 & \ldots & 0 \\ 1 & q & 1 & 0 & \ldots & 0 \\ 0 & 1 & q & 1 & \ldots & 0 \\ & & & \ldots & 1 & \frac{1}{2}q \end{vmatrix} \tag{2-71}$$

$$q = -\frac{2(1 + \mu) + 2r^{-1}(1 - \mu)}{1 + \mu}.$$

Utilizing the results of Rutherford [31] we find the zeros are

$$q_n = 2 \cos \frac{n\pi}{M}, \quad n = 0, 1, 2, \ldots, M$$

so that

$$\mu_n = \frac{1 - 2r \cos^2 (n\pi/2M)}{1 + 2r \cos^2 (n\pi/2M)}, \quad n = 0, 1, 2, \ldots, M. \tag{2-72}$$

Clearly, stability is unrestricted, except that $q_M = -2$ and $\mu_M = 1$ when $n = M$.

PROBLEMS

2-17 Generalize Eqn (2-53) if the equation $u_t = u_{xx}$ is subject to the boundary conditions $u(0, t) = e^{-t}$ and $\partial u/\partial x \, (1, t) = 0$.

2-18 Verify Eqn (2-68).

2-19 Apply the two-level implicit formula, Eqn (2-36), with arbitrary λ ($0 \le \lambda \le 1$) to $u_t = u_{xx}$ with zero boundary conditions. Develop the equations corresponding to Eqns (2-53) through (2-62).

2-20 Carry out a Fourier stability analysis when the simple explicit method is applied to $u_t = u_{xx} + u_x + u$. How do the lower-order space derivatives affect the analysis?

2-21 Apply the two-level implicit formula, Eqn (2-36), to the equation in Problem 2-19 and write the algorithm in matrix form.

2-6 Extension of matrix stability analysis

Richtmyer [19] gives a large selection of finite difference formulae for the diffusion equation $u_t = u_{xx}$. Several of these are three-level formulae. Stability analysis of such procedures by matrix methods necessitates modification of the previous technique. To have a specific reference let us herein use an adaptation of Richardson's formula, Eqn (2-48), which has unrestricted stability while preserving the numerical advantages of explicit methods. Du

Fort and Frankel [32] replace $2U_{i,j}$ in Eqn (2-48) by $U_{i,j-1} + U_{i,j+1}$ thereby generating the three time level formula

$$\frac{U_{i,j+1} - U_{i,j-1}}{2k} = \frac{U_{i+1,j} - U_{i,j-1} - U_{i,j+1} + U_{i-1,j}}{h^2}. \qquad (2\text{-}73)$$

Upon designating V_j as the vector of mesh values along the j line we can write the general three-level recurrence relation as

$$AV_{j+1} = BV_j + CV_{j-1}. \qquad (2\text{-}74)$$

For the specific case under consideration, the matrices are

$$A = (2r + 1)I, \qquad C = (1 - 2r)I$$

and

$$B = 2r \begin{bmatrix} 0 & 1 & 0 & \cdots \\ 1 & 0 & 1 & \cdots \\ & \ddots & & \\ & & 1 & 0 \end{bmatrix}. \qquad (2\text{-}75)$$

Upon writing Eqn (2-74) as

$$V_{j+1} = A^{-1}BV_j + A^{-1}CV_{j-1} \qquad (2\text{-}76)$$

and setting

$$W_j = \begin{bmatrix} V_j \\ V_{j-1} \end{bmatrix} \qquad (2\text{-}77)$$

Eqn (2-76) can be reduced to the two-level† formula

$$W_{j+1} = PW_j, \qquad P = \begin{bmatrix} A^{-1}B & A^{-1}C \\ I & 0 \end{bmatrix} \qquad (2\text{-}78)$$

Here P is expressed in partitioned matrix notation.

If the characteristic roots μ_n of P are distinct and the characteristic vectors are $w^{(n)}$, we can write

$$W_1 = \sum \alpha_n w^{(n)}, \qquad W_{j+1} = P^j W_1 = \sum \alpha_n \mu_n^j w^{(n)}. \qquad (2\text{-}79)$$

For stability we require $|\mu_n| \leq 1$ for all n.

The characteristic roots of P are, by definition, the zeros of the equation $|P - \mu I| = 0$. Upon using the definition of P, Eqn (2-78), we find the characteristic roots are the zeros of the equation

$$|\mu^2 A - \mu B - C| = 0. \qquad (2\text{-}80)$$

If the boundary values are specified, the simplicity of the matrices in the

† Generalization to a $(q + 1)$ level is given by Richtmyer [19].

example problem [Eqn (2-75)] allows an easy computation. In fact, the determinant vanishes when

$$(2r + 1)\mu^2 - 4r\mu \cos\{i\pi/(M + 1)\}$$
$$+ (2r - 1) = 0, \quad i = 1, 2, \ldots, M. \quad (2\text{-}81)$$

Clearly $|\mu| < 1$ for all r, thereby establishing stability.

Three-level formulae have the disadvantage of requiring a special starting procedure since one line of values, in addition to the initial line, must be known before the formulae can be applied.

PROBLEMS

2-22 Carry out the computation leading to Eqn (2-81).

2-23 For $\theta \geq 0$ examine the three-level formula

$$(1 + \theta)\frac{U_{i,j+1} - U_{i,j}}{k} - \theta\frac{U_{i,j} - U_{i,j-1}}{k} = \frac{U_{i+1,j+1} - 2U_{i,j+1} + U_{i-1,j+1}}{h^2} \quad (2\text{-}82)$$

for stability using the matrix method.

2-24 Compare the computational molecules of Eqns (2-73), (2-82), and (2-36) with regard to advantages and disadvantages.

2-7 Consistency, stability, and convergence

Finite difference formulae for partial differential equations arising in initial value problems may display a phenomenon which has no counterpart in ordinary differential equations. *Successive refinement of the interval* $\Delta t = k$ *may generate a finite difference solution which is stable, but which may converge to the solution of a different differential equation.*

To illustrate this highly undesirable situation let us write down the local truncation errors associated with some of our finite difference approximations for $u_t = u_{xx}$. These can be obtained by using Taylor's series in a manner analogous to that of Section 1-5. Denoting the finite difference operator by F we have, for $u_{i,j}$:

Explicit equation [Eqn (2-11)]:

$$Fu_{i,j} - (u_t - u_{xx}) = \left(\frac{1}{2}ku_{tt} - \frac{1}{12}h^2\frac{\partial^4 u}{\partial x^4}\right) + \cdots \quad (2\text{-}83)$$

Crank–Nicolson equation [Eqn (2-52)]:

$$Fu_{i,j} - (u_t - u_{xx}) = \frac{1}{2}k\frac{\partial}{\partial t}\left(\frac{\partial u}{\partial t} - \frac{\partial^2 u}{\partial x^2}\right) - \frac{1}{12}h^2\frac{\partial^4 u}{\partial x^4}$$
$$+ \frac{1}{6}k^2\left(\frac{\partial^3 u}{\partial t^3} - \frac{3}{2}\frac{\partial^4 u}{\partial x^2 \partial t^2}\right) + \cdots \quad (2\text{-}84)$$

Du Fort–Frankel equation [Eqn (2-73)]:

$$Fu_{i,j} - (u_t - u_{xx}) = \frac{1}{6}k^2\frac{\partial^3 u}{\partial t^3} - \frac{1}{12}h^2\frac{\partial^4 u}{\partial x^4} + \frac{k^2}{h^2}\frac{\partial^2 u}{\partial t^2} + \cdots \qquad (2\text{-}85)$$

The finite difference equation is said to be *consistent* (compatible) with the differential equation if the local truncation errors [the right-hand sides of Eqns (2-83)–(2-85)] tend to zero as $h, k \to 0$. The explicit and Crank–Nicolson formulae are clearly compatible while the always stable Du Fort–Frankel formula is consistent if *k goes to zero faster than h*. If they go to zero at the same rate—that is, $k/h = \beta$, β fixed—then this approximation is consistent not with the diffusion equation but with the hyperbolic equation

$$u_t - u_{xx} + \beta^2 u_{tt} = 0! \qquad (2\text{-}86)$$

Lax (see Richtmyer [19]) studied the relation between consistency, stability, and convergence of the approximations of linear initial value problems by finite difference equations. The major result of that study is termed the *Lax equivalence theorem.* The proof is omitted.

Theorem: Given a properly posed initial boundary value problem and a finite difference approximation to it that satisfies the consistency condition, then stability is the necessary and sufficient condition for convergence.

PROBLEMS

2-25 Find the truncation error for the three-level formula associated with Eqn (2-82) and examine the consistency of the finite difference formula.

Ans: O$[k^2 + kh^2]$; consistent.

2-8 Pure initial value problems

Before moving on to equations with variable coefficients let us examine the pure initial value problem

$$u_t = u_{xx}, \quad -\infty < x < \infty, \quad 0 < t \le T$$
$$u(x, 0) = f(x), \quad -\infty < x < \infty. \qquad (2\text{-}87)$$

If $u \in C^{4,2}$ and the forward difference formula [Eqn (2-9)] is applied, the analysis of Section 2-1 is essentially unaltered. Hence, convergence and stability are assured if $r \le \frac{1}{2}$.

In unbounded domains implicit methods are not readily applicable because they generate infinite matrices with all the attendant difficulties. Thus one usually seeks improved difference equations of explicit type. If $u \in C^{6,3}$, then by Taylor series expansion we have

$$u_{i,j+1} = u_{i,j} + k(u_t)_{i,j} + \frac{k^2}{2}(u_{tt})_{i,j} + O(k^3)$$

$$= u_{i,j} + k(u_{xx})_{i,j} + \frac{k^2}{2}(u_{xxxx})_{i,j} + O(k^3) \qquad (2\text{-}88)$$

since $u_t = u_{xx}$ and $u_{tt} = u_{xxxx}$. Upon approximating u_{xx} and u_{xxxx} by second and fourth differences respectively, that is,

$$(u_{xx})_{i,j} = \frac{1}{h^2}\delta_x^2 u_{i,j} + O(h^2)$$

$$= \frac{1}{h^2}[u_{i+1,j} - 2u_{i,j} + u_{i-1,j}] + O(h^2)$$

and $\quad (u_{xxxx})_{i,j} = \dfrac{1}{h^4}\delta_x^4 u_{i,j} + O(h^2)$

$$= \frac{1}{h^4}\{u_{i-2,j} - 4u_{i-1,j} + 6u_{i,j} - 4u_{i+1,j} + u_{i+2,j}\} + O(h^2)$$

we obtain

$$u_{i,j+1} = u_{i,j} + \frac{k}{h^2}\delta_x^2 u_{i,j} + \frac{1}{2}\frac{k^2}{h^4}\delta_x^4 u_{i,j} + O(k^3 + kh^2). \qquad (2\text{-}89)$$

The difference equation obtained by omitting the local truncation error $O(k^3 + kh^2)$ is analyzed in the same manner as that employed for Eqn (2-9). The restriction, $r = k/h^2 \leq \frac{1}{2}$, is again required and the error is

$$\|u_{i,j} - U_{i,j}\| = \|z_j\| = O[h^2 + k^2]. \qquad (2\text{-}90)$$

Now $k = rh^2$, so k^2 is dominated by h^2 and the local truncation error is $O(h^2)$, as before. No improvement was obtained; while the error in the time direction was improved, that in the space direction was not!

The local space truncation error can be improved by utilizing the fact that

$$(u_{xx})_{i,j} = \frac{1}{h^2}\delta_x^2 u_{i,j} - \frac{1}{12}h^2(u_{xxxx})_{i,j} + O(h^4)$$

$$= \frac{1}{h^2}\delta_x^2 u_{i,j} - \frac{1}{12}\frac{1}{h^2}\delta_x^4 u_{i,j} + O(h^4). \qquad (2\text{-}91)$$

When Eqn (2-91) is introduced into Eqn (2-88), and the truncation error neglected, we find the equation

$$U_{i,j+1} = U_{i,j} + \frac{k}{h^2}\delta_x^2 U_{i,j} + \frac{1}{h^4}\left[\frac{1}{2}k^2 - \frac{1}{12}h^2k\right]\delta_x^4 U_{i,j}. \qquad (2\text{-}92)$$

When the maximum analysis is employed it is easily shown that

$$\|z_j\| = O[h^4 + h^2k + k^2] \tag{2-93}$$

if $r \leq \frac{2}{3}$.† The truncation error has been significantly reduced and the limitation on r slightly eased.

Since Eqn (2-92) cannot be evaluated at the mesh points next to the boundaries it cannot be applied to the bounded region problem unless the solution can be extended across the boundaries by symmetry considerations involving the boundary conditions. Use of backward differences of order $O(h^4)$ can remove this limitation.

PROBLEMS

2-26   Verify Eqn (2-93) and the resulting stability limit $r \leq \frac{2}{3}$.

2-27   Can the Fourier method be applied to Eqn (2-92)? If so, carry out the necessary arguments.

2-9   Variable coefficients

Most of the previously discussed methods can be generalized to the case of linear equations with variable coefficients. In many respects the equations

$$u_t = au_{xx} + bu_x + cu + d \tag{2-94}$$

where a, b, c, and d are functions of x and t only, and with boundary conditions

$$pu_x + qu = v \tag{2-95}$$

where p, q, and v are functions of t only, involve very little extra difficulty. This is especially true if small intervals are to be used together with finite difference formulae of simple form.

A simple explicit formula for Eqn (2-94), developed using the approximations of Section 2-1, is given by

$$U_{i,j+1} = c_1 U_{i+1,j} + c_0 U_{i,j} + c_{-1} U_{i-1,j} + kd_{i,j} \tag{2-96}$$

where
$$c_{-1} = ra_{i,j}$$
$$c_0 = 1 - 2ra_{i,j} - rhb_{i,j} + rh^2c_{i,j} \tag{2-97}$$
$$c_1 = ra_{i,j} + rhb_{i,j}$$

and the notation $(\ )_{i,j}$ means evaluation at (ih, jk). If the coefficients do not involve t, a stability criterion can be obtained by the matrix method. This is

† This analysis requires a lower limit $r \geq \frac{1}{6}$, but a refined argument removes this limitation.

easily done since a line of errors z_{j+1} is again being considered which satisfies an equation of the form

$$z_{j+1} = Pz_j \qquad (2\text{-}98)$$

while the matrix P does not depend upon j. If the coefficients are constants we can also use the Fourier method.

Forsythe and Wasow [30] describe a general explicit finite difference approximation for Eqn (2-94), of which Eqn (2-96) is a special case. This formula is

$$U_{i,j+1} = \sum_s (c_s)_{i,j} U_{i+s,j} + kd_{i,j} \qquad (2\text{-}99)$$

with the summation extending over some finite set of grid points. The coefficients c_s must satisfy the conditions

$$\left. \begin{array}{l} \displaystyle\lim_{h\to 0} \frac{1}{k}\left[\sum_s (c_s)_{i,j} - 1\right] = c_{i,j}, \qquad \lim_{h\to 0}\frac{h}{k}\sum_s s(c_s)_{i,j} = b_{i,j} \\[2ex] \displaystyle\lim_{h\to 0}\frac{1}{2}\frac{h^2}{k}\sum_s s^2(c_s)_{i,j} = a_{i,j} \end{array} \right\} \qquad (2\text{-}100)$$

The authors assume the c_s are twice continuously differentiable with respect to h.

Of particular interest are those finite difference approximations for which all c_s are nonnegative for $h \le h_1$ $(h_1 > 0)$ in the computational domain R of the (x, t) plane. Such formulas will be called approximations of *positive type*. Equations (2-97) are of positive type if

$$a(x, t) > 0, \qquad r < \frac{1}{2a(x, t)} \qquad (2\text{-}101)$$

in R. When $a \ge 1$, the second inequality of Eqn (2-101) implies our previous stability condition. The importance of approximations of positive type is summarized in the following result.

Theorem: Difference approximations taking the form of Eqn (2-99), of positive type, are stable. If the formulae are also consistent then convergence follows from the equivalence theorem of John (to follow), modified to a bounded domain. Development of these results is found in the treatise of Forsythe and Wasow [30, pp. 108–112].

The previous discussion should not leave one with the inference that it is essential for a difference equation to be of positive type if it is to be computationally useful. Stability and convergence of the difference equation are direct consequences of certain boundedness properties which exist for wider classes of difference equations than those of positive type.

John [34], in an outstanding contribution to numerical methods for parabolic equations, develops a sufficient condition for the convergence of explicit finite difference analogs of the linear pure initial value problem described by Eqn (2-94), $-\infty < x < \infty$, $u(x, 0) = f(x)$ with $a(x, t) > 0$. His results are extensive and the arguments are beyond the scope of this work. Hence, we shall only summarize the results.

John considers the general explicit formula [Eqn (2-99)] with $-\infty < i < \infty$ and with $(c_s)_{i,j}$ dependent upon h and k as well as the indicated x and t. We shall assume $(c_s)_{i,j} = 0$ unless $|s| \le m$, where m is independent of h, i, and j, and that $r = k/h^2$ is fixed throughout the discussion.

Consistency of Eqn (2-99) with the partial differential equation [Eqn (2-94)] is easily seen to be equivalent to Eqn (2-100). The important result is that stability and consistency imply convergence. Thus, we have the following theorem:

Theorem: Let $u \in C^{2,1}$ be the solution of the pure initial value problem [Eqn (2-94)] subject to $u(x, 0) = f(x)$, and let U be the solution of Eqn (2-99) subject to $U_{i,0} = f_i$. If Eqn (2-99) is both consistent and stable, then U converges uniformly to u as $h \to 0$ (recall $k = rh^2$).

Difference equations should be logically derived to ensure consistency. If this is the case the *important content of this result, and that of Lax (Section 2-7), is the reduction of the convergence problem to that of determining stability.* Various steps of the development are helpful in examining individual problems and are therefore presented here.

Since $k = rh^2$ we hereafter assume that k is a function of h and expand the coefficients $(c_s)_{i,j}$ in terms of h. Thus,

$$(c_s)_{i,j} = (\alpha_s)_{i,j} + h(\beta_s)_{i,j} + \tfrac{1}{2}h^2(\gamma_s)_{i,j}(h) \qquad (2\text{-}102)$$

where $(\alpha_s)_{i,j}$, $(\beta_s)_{i,j}$, and $(\gamma_s)_{i,j}(h)$ are uniformly bounded in $0 \le t \le T$ and

$$\lim_{h \to 0} (\gamma_s)_{i,j}(h) = (\gamma_s)_{i,j}(0) \qquad (2\text{-}103)$$

uniformly in the strip $0 \le t \le T$. The consistency conditions [Eqn (2-100)] are equivalent to the six relations

$$\left.\begin{aligned}
&\sum_s (\alpha_s)_{i,j} = 1, && \sum_s s(\alpha_s)_{i,j} = 0, \\[4pt]
&\sum_s (\beta_s)_{i,j} = 0, && \sum_s s^2(\alpha_s)_{i,j} = 2ra_{i,j} \\[4pt]
&\sum_s s(\beta_s)_{i,j} = rb_{i,j}, && \sum_s (\gamma_s)_{i,j}(0) = 2rc_{i,j}
\end{aligned}\right\} \qquad (2\text{-}104)$$

A necessary condition for stability is that

$$\left| \sum_s (\alpha_s)_{i,j} \exp\left[(-1)^{1/2} s\theta\right] \right| \le 1 \qquad (2\text{-}105)$$

for all real θ and all (i, j) in the region. The slightly stronger condition

$$\left| \sum_s (\alpha_s)_{i,j} \exp\left[(-1)^{1/2} s\theta\right] \right| \le e^{-\delta\theta^2}, \quad |\theta| \le \pi, \text{ for some } \delta > 0 \qquad (2\text{-}106)$$

is sufficient for stability† provided that the quantities $(i, j$ subscripts omitted)

$$\alpha_s, \quad (\alpha_s)_x, \quad (\alpha_s)_{xx}, \quad (\beta_s), \quad (\beta_s)_x, \quad \gamma_s(h)$$

exist and are uniformly bounded in the integration region for sufficiently small h.

Equation (2-106) is satisfied if

$$\left.\begin{array}{ll}(\alpha_s)_{i,j} \geq 0, & (\alpha_0)_{i,j}, (\alpha_1)_{i,j} \geq \epsilon > 0 \\[2mm] \text{and} \quad \sum_s (\alpha_s)_{i,j} = 1, & \sum_s s(\alpha_s)_{i,j} = 0 \end{array}\right\} \tag{2-107}$$

Since the consistency conditions [Eqn (2-104)] are assumed to hold for Eqn (2-99), conditions (2-107) hold for that difference equation. Hence, a convergence theorem similar to that of Section 2-1, can be proved provided a, a_x, a_{xx}, b, b_x, and c are bounded.

Lastly, a direct appeal to the definition establishes that stability is ensured if

$$(c_s)_{i,j} \geq 0, \qquad \sum_s (c_s)_{i,j} \leq 1 + O(k). \tag{2-108}$$

No such general results appear to be available for implicit methods. A Crank–Nicolson formula for Eqn (2-94) in a finite x domain is

$$\frac{1}{k}[U_{i,j+1} - U_{i,j}] = \frac{1}{2h^2}[a_{i,j}\delta_x^2 + hb_{i,j}\mu\delta_x + h^2 c_{i,j}]$$
$$\times (U_{i,j+1} + U_{i,j}) + \tfrac{1}{2}(d_{i,j+1} + d_{i,j}) \tag{2-109}$$

where we have again used the difference operators

$$\delta_x U_{i,j} = U_{i+1/2,j} - U_{i-1/2,j}, \qquad \mu U_{i,j} = \tfrac{1}{2}[U_{i+1/2,j} + U_{i-1/2,j}]. \tag{2-110}$$

By expansion of Eqn (2-109) the elements on the $j + 1$ line are determinable by solving a tridiagonal system of linear equations for each time step. Generally the coefficients do depend upon time and hence upon j, but there are many problems in which this is not the case. If the coefficients are independent of t, Eqn (2-109) can be written in the matrix form

$$AV_{j+1} = BV_j + d_j. \tag{2-111}$$

Stability is governed by the characteristic roots of $A^{-1}B$.

PROBLEMS

2-28   Describe an explicit algorithm taking the form of Eqn (2-96) for the equation

$$u_t = e^{-x}u_{xx} + xu_x + e^x$$

$$0 \leq x \leq 1, \qquad u(x, 0) = 0, \qquad u(0, t) = 1/(t + 1), \qquad u_x(1, t) = u(1, t).$$

† This stability is with respect to the maximum norm used previously in this chapter.

2-29 Write a two-line method analogous to Eqn (2-35) for Eqn (2-94).

2-30 Let $a(x) = 1$, $b(x) = x(1 - x)$, $c(x) = e^{-x}$. Expand Eqn (2-109) for this case and put the result in matrix form. Carry out a matrix stability analysis.

2-31 Show that the diffusion equation with spherical symmetry, $u_t = u_{xx} + 2x^{-1}u_x$, transforms to $w_t = w_{xx}$ under the change of dependent variable $w = ux$.

2-32 Show that the diffusion equation with circular cylindrical symmetry, $u_t = u_{xx} + x^{-1}u_x$ transforms to $e^{2y}u_t = u_{yy}$ under the change of independent variable $y = \ln x$, $x > 0$. This transformation is then applicable to hollow cylinder problems.

2-10 Examples of equations with variable coefficients

In this section we examine some often-occurring examples possessing variable coefficients. Unless otherwise stated the integration domains are bounded.

(a) Diffusion in circular cylindrical coordinates

One of the simplest equations of this class is that of diffusion in circular cylindrical coordinates with cylindrical symmetry

$$u_t = u_{xx} + x^{-1}u_x \qquad (2\text{-}112)$$

where x is the radial space variable. When the Crank–Nicolson concept is applied we obtain the implicit form

$$\frac{U_{i,j+1} - U_{i,j}}{k} = \frac{1}{2h^2}\{\delta_x^2 + i^{-1}\mu\delta_x\}(U_{i,j+1} + U_{i,j}). \qquad (2\text{-}113)$$

The matrix form [Eqn (2-111)] of this system possesses tridiagonal matrices A and B, so the Thomas algorithm is applicable.

There is an apparent difficulty in $(1/x)(\partial u/\partial x)$ at $x = 0$, but this is eliminated by noting that there is symmetry about the line $x = 0$ and hence $\lim_{x\to 0}(\partial u/\partial x) = 0$. Thus the indeterminate form $x^{-1}u_x$ takes the value

$$\lim_{x\to 0}\frac{1}{x}\frac{\partial u}{\partial x} = \frac{\partial^2 u}{\partial x^2}. \qquad (2\text{-}114)$$

Albasiny [35] examined the stability of Eqn (2-113) with boundary conditions of the form

$$u_x = 0 \quad \text{at} \quad x = 0, \qquad u_x + uf(t) = 0 \quad \text{at} \quad x = 1 \qquad (2\text{-}115)$$

by applying the matrix method. He confirmed that this finite difference approximation was stable for all values of x.

The local truncation error of the approximation [Eqn (2-113)] can be found by expanding $u_{i,j+1}$ about $u_{i,j}$ to obtain

$$k^{-1}(u_{i,j+1} - u_{i,j}) - \left[\frac{\partial}{\partial t} + \frac{1}{2}k\frac{\partial^2}{\partial t^2} + \frac{1}{6}k^2\frac{\partial^3}{\partial t^3} + \cdots\right]u_{i,j}$$

$$= \left\{\left(\frac{\partial^2}{\partial x^2} + \frac{1}{x}\frac{\partial}{\partial x}\right) + \frac{1}{2}k\left(\frac{\partial^2}{\partial x^2} + \frac{1}{x}\frac{\partial}{\partial x}\right)^2\right.$$

$$\left. + \frac{1}{6}k^3\left(\frac{\partial^2}{\partial x^2} + \frac{1}{x}\frac{\partial}{\partial x}\right)^3 + \cdots\right\}u_{i,j}. \quad (2\text{-}116)\dagger$$

Upon truncating after the first term on the right-hand side we obtain the local error [see Eqn (2-84)] $O[k^2 + h^2]$. This cannot be easily reduced by special techniques such as that employed in Section 2-8 because of the presence of the term $(1/x)(\partial u/\partial x)$. This term makes the differentiation formidable with attendant complications in the resulting finite difference formula.

(b) Equations with reducible error

Certain special equations can be treated to produce a smaller error. One general class is

$$a(x)u_t = u_{xx} + bu \quad (2\text{-}117)$$

where b is constant. This can be approximated by

$$a_i k^{-1}(u_{i,j+1} - u_{i,j}) = \frac{1}{2h^2}\left(bh^2 + \delta_x^2 - \frac{1}{12}\delta_x^4 + \cdots\right)(u_{i,j+1} + u_{i,j}). \quad (2\text{-}118)$$

Operation throughout with $1 + \frac{1}{12}\delta_x^2$ preserves the tridiagonal form and reduces the local truncation error to $O[k^2 + h^4]$. Development of the resulting finite difference equation is left for the exercises.

If the equation has the form

$$u_t = a(x)[b(x)u_x]_x \quad (2\text{-}119)$$

it becomes
$$(a^{-1}b)u_t = u_{XX} \quad (2\text{-}120)$$

under the independent variable transformation specified by $dX = b^{-1}\,dx$.

In Problem 2-32 we noted that $u_t = u_{xx} + x^{-1}u_x$ transforms to an equation of the form of Eqn (2-117) when one sets $x = e^X$. The resulting equation

$$e^{2X}u_t = u_{XX} \quad (2\text{-}121)$$

is only applicable when the integration domain does not include $x = 0$—that is, for a ring cross-section.

(c) Diffusion with spherical symmetry

Diffusion in three dimensions with spherical symmetry is modeled by the equation

$$u_t = u_{xx} + 2x^{-1}u_x. \quad (2\text{-}122)$$

$\dagger$ The notation $\left(\dfrac{\partial^2}{\partial x^2} + \dfrac{1}{x}\dfrac{\partial}{\partial x}\right)^r$ means r applications of the operator.

From Problem 2-31 we observe that Eqn (2-111) becomes

$$w_t = w_{xx} \tag{2-123}$$

under the transformation $w = ux$.

PROBLEMS

2-33 Operate throughout Eqn (2-118) with $1 + \frac{1}{12}\delta_x^2$, obtain the resulting finite difference equation and place it in matrix form. Find the truncation error.

2-34 Develop a simple explicit algorithm for Eqn (2-121) subject to $u(1, t) = 0$, $u_x(2, t) = u(2, t), u(X, 0) = X$.

2-11 General concepts of error reduction

In several sections we have employed the ideas of Crank–Nicolson [22] and their generalizations. These have been important not only for their implicit nature but because they provide a decrease in the local truncation error in the time direction. The global truncation error in most of our examples is of the same order as the local error (see Sections 2-1, 2-7, 2-8, 2-10) provided the stability requirements are satisfied. It is therefore reasonable to suppose that an increase in the local accuracy would lead to a similar increase in the global accuracy. This is usually the case. We initiate these discussions by von Neumann's [see 22] derivation of the Crank–Nicolson equation [Eqn (2-35)] (with $\lambda = \frac{1}{2}$).

The difference $(1/k)(u_{i,j+1} - u_{i,j})$ is $O(k)$ at any point $(x_i, t), t_j \le t \le t_{j+1}$. At the particular choice $t = t_{j+1/2}$ it becomes centered and is $O(k^2)$ provided u_{ttt} is bounded. To take advantage of this increase in accuracy one must approximate u_{xx} at $(x_i, t_{j+1/2})$. If $u \in C^{4,3}$, then $u_{xxxx} = u_{tt}$ and $u_{xxtt} = u_{ttt}$ are bounded. Thus

$$u_{xx}|_{i,j+1/2} = \frac{1}{2}\{u_{xx}|_{i,j+1} + u_{xx}|_{i,j}\}$$

$$= \frac{1}{2h^2}\delta_x^2(u_{i,j+1} + u_{i,j}) + O(h^2)$$

$$= \frac{\partial u}{\partial t}\Big|_{i,j+1/2}$$

$$= \frac{1}{k}\{u_{i,j+1} - u_{i,j}\} + O(k^2) \tag{2-124}$$

or, upon rearrangement,

$$k^{-1}\{u_{i,j+1} - u_{i,j}\} = (2h^2)^{-1}\delta_x^2[u_{i,j+1} + u_{i,j}] + O(h^2 + k^2) \tag{2-125}$$

which we recognize as the Crank–Nicolson equation, Eqn (2-35).

Flatt [36] has shown that stability, in a uniform sense, does *not* hold for $r = k/h^2 > R$ where R depends upon the length L of the bar in a heat conduction problem. For a bar of length $L = 1$, $R = 4 - 2^{3/2}$. For separate treatment of convergence of the Crank–Nicolson equation one must use a procedure based upon a combination of Duhamel's principle and harmonic analysis (Douglas [18]).

Two general observations guide the development of improvements to the basic Crank–Nicolson idea. First, it is frequently convenient to restrict the number of time levels employed in a finite difference approximation to two, although multilevel equations are possible (Du Fort–Frankel is an example) and useful. Second, it is advantageous to restrict the difference equations to those leading to a tridiagonal matrix. Difference equations involving the solution value at mesh points that are more than one spatial interval from the center term cannot, in general, be applied at the grid points next to the boundary. Thus special 'boundary methods' are required leading to increases in computational complexity.

Certainly the Crank–Nicolson equation is a considerable improvement over the forward and backward equations, but it does not possess the highest order local accuracy that can be obtained employing six mesh points. Crandall [33] (see Fig. 2-5 and Problem 2-16) and Douglas [37] have investigated this problem in detail. A sketch of the results is presented in Fig. 2-5. Here we discuss a portion of that work.

We suppose $u \in C^{6,3}$ and attempt to derive a difference equation employing the six mesh points of Eqn (2-125), which is $O(h^4)$ in space and $O(k^2)$ in time. Since $(u_{i,j+1} - u_{i,j})/k$ is $O(k^2)$ at $(x_i, t_{j+1/2})$ it will be retained. In a manner analogous to Eqn (2-91) we have, for $u_{xx}|_{i,j+1/2}$,

$$u_{xx}|_{i,j+1/2} = \frac{1}{2h^2} \delta_x^2(u_{i,j+1} + u_{i,j}) - \frac{h^2}{12} u_{xxxx}|_{i,j+1/2} + O[h^4 + k^2]$$

$$= \frac{1}{2h^2} \delta_x^2(u_{i,j+1} + u_{i,j}) - \frac{h^2}{12} u_{xxt}|_{i,j+1/2} + O[h^4 + k^2] \qquad (2\text{-}126)$$

since $u_{xxxx} = u_{xxt}$. By Taylor series arguments we find

$$u_{xxt}|_{i,j+1/2} = \frac{1}{kh^2} \delta_x^2(u_{i,j+1} - u_{i,j}) + O[h^2 + k^2 + k^{-1}h^4]. \qquad (2\text{-}127)$$

If r is fixed and Eqn (2-127) substituted into Eqn (2-126) we find

$$\frac{u_{i,j+1} - u_{i,j}}{k} = \frac{1}{2h^2}\left(1 - \frac{1}{6r}\right)\delta_x^2 u_{i,j+1}$$

$$+ \frac{1}{2h^2}\left(1 + \frac{1}{6r}\right)\delta_x^2 u_{i,j} + O[k^2 + h^4]. \qquad (2\text{-}128)$$

Thus we find the difference equation

$$\frac{U_{i,j+1} - U_{i,j}}{k} = \frac{1}{2h^2}\left(1 - \frac{1}{6r}\right)\delta_x^2 U_{i,j+1} + \frac{1}{2h^2}\left(1 + \frac{1}{6r}\right)\delta_x^2 U_{i,j} \quad (2\text{-}129)$$

to be $O[h^4]$ and $O[k^2]$.

More general parabolic equations can be treated by employing an extension of Eqn (2-128). If we wish to solve

$$a(x, t)u_t = u_{xx} \quad (2\text{-}130)$$

the term $u_{xxxx}|_{i,j+1/2}$ in Eqn (2-126) must be replaced using the differential equation. Thus

$$u_{xxxx}|_{i,j+1/2} = \frac{\partial^2}{\partial x^2}[a(x, t)u_t]_{i,j+1/2}$$

$$= \frac{1}{kh^2}\delta_x^2[a_{i,j+1/2}(u_{i,j+1} - u_{i,j})] + O(h^2) \quad (2\text{-}131)$$

if $u \in C^{6,3}$, $a \in C^4$ and r is fixed. The resulting difference equation

$$a_{i,j+1/2}\frac{U_{i,j+1} - U_{i,j}}{k} = \frac{1}{2h^2}\delta_x^2\left[\left(1 - \frac{a_{i,j+1/2}}{6r}\right)U_{i,j+1}\right]$$

$$+ \frac{1}{2h^2}\delta_x^2\left[\left(1 + \frac{a_{i,j+1/2}}{6r}\right)U_{i,j}\right] \quad (2\text{-}132)$$

is locally accurate to $O(h^4)$. Equation (2-132), introduced by Douglas [38], was shown by Lees [39], using energy methods, to have a global truncation error of $O(h^4)$ for sufficiently smooth a and u. This equation is easily generalized to include lower-order derivatives.

The simplest three-level formula for the diffusion equation results from replacing u_{xx} by the average value of the second centered differences at the $j - 1, j$, and $j + 1$ levels and u_t by a centered first difference. The resulting difference formula

$$\frac{U_{i,j+1} - U_{i,j-1}}{2k} = \frac{1}{3h^2}\delta_x^2(U_{i,j+1} + U_{i,j} + U_{i,j-1}) \quad (2\text{-}133)$$

leading to a tridiagonal system, is not applicable for $j = 1$. The local error is $O(h^2 + k^2)$ as is expected. The first line, $j = 1$, must be computed by another method. Douglas [18] shows that the simple explicit method produces the required accuracy for one time step.

Finite difference formulae of higher-order correctness than Eqn (2-133) are obtained by methods similar to those already employed in this section. Thus, for example, if $u \in C^{6,3}$, we have

$$u_{xx}|_{i,j} = \frac{1}{3h^2}\delta_x^2(u_{i,j+1} + u_{i,j} + u_{i,j-1}) - \frac{h^2}{12}u_{xxxx}|_{i,j} + O(h^4). \quad (2\text{-}134)$$

Replacing u_{xxxx} by u_{tt} we find

$$u_{xx}|_{i,j} = \frac{1}{3h^2} \delta_x^2(u_{i,j+1} + u_{i,j} + u_{i,j-1}) - \frac{h^2}{12} u_{tt}|_{i,j} + O(h^4)$$

so the difference equation becomes

$$\frac{U_{i,j+1} - U_{i,j-1}}{2k} = \frac{1}{3h^2} \delta_x^2(U_{i,j+1} + U_{i,j} + U_{i,j-1}) - \frac{h^2}{12k^2} \delta_t^2 U_{i,j} \quad (2\text{-}135)$$

which is $O[h^4 + k^2]$. $U_{i,1}$ must be determined by some other method.

More than three time levels can be used. For example, the difference equation

$$\frac{U_{i,j+1} - U_{i,j-1}}{2k} = \frac{1}{6h^2} \delta_x^2[3U_{i,j+1} + 2(U_{i,j} + U_{i,j-1} + U_{i,j-2})] \quad (2\text{-}136)$$

is a stable, second-order analog of the diffusion equation (see Douglas [18]).

PROBLEMS

2-35 Carry out the steps leading from Eqn (2-131) to Eqn (2-132).

2-36 An alternative substitute for u_{xxxx} in Eqn (2-134) is u_{xxt}. Make this substitution and find the difference equation and the local error.

Ans:

$$\frac{U_{i,j+1} - U_{i,j-1}}{2k} = \frac{1}{3h^2} \delta_x^2 (U_{i,j+1} + U_{i,j} + U_{i,j-1}) - \frac{1}{24k} \delta_x^2 (U_{i,j+1} - U_{i,j-1}).$$

$$(2\text{-}137)$$

2-37 Write the generalization of Eqn (2-137) for $a(x)u_t = u_{xx}$.

2-12 Explicit methods for nonlinear problems

A large collection of physical problems having nonlinear parabolic equations as models is given in Ames [8]. That survey also lists extensive exact, approximate and numerical methods for those examples. We shall cover fewer examples, but give more details.

Many of the numerical methods and techniques of proof for linear equations with constant coefficients carry over to nonlinear equations. Questions of stability and convergence are more complicated. Richtmyer [19] considered the nonlinear problem $u_t = (u^5)_{xx}$ which is typical of the equation

$$u_t = (u^n)_{xx}. \quad (2\text{-}138)$$

If the simplest explicit scheme is applied one obtains the finite difference equation

$$U_{i,j+1} = U_{i,j} + r\{U_{i+1,j}^n - 2U_{i,j}^n + U_{i-1,j}^n\}. \quad (2\text{-}139)$$

The heuristic approach to the stability question is as follows: If $u_t = u_{xx}$ is replaced by $u_t = (\sigma u_x)_x$, σ constant, the stability argument of Section 2-1 generates the stability criterion $\sigma r \leq \frac{1}{2}$, instead of $r \leq \frac{1}{2}$. The effective diffusion coefficient for Eqn (2-139) is nu^{n-1}, as can be seen by writing that equation as

$$u_t = (nu^{n-1}u_x)_x. \tag{2-140}$$

This suggests that Eqn (2-139) might be stable if

$$nu^{n-1}r \leq \frac{1}{2} \tag{2-141}$$

that is, *for nonlinear problems, stability depends not only on the form of the finite difference system but also generally upon the solution being obtained.* The system may be stable for some values of t and not for others. In practice it is highly desirable to monitor stability (if the difference equations are not unconditionally stable) by checking the inequality in Eqn (2-141) and to alter $k = \Delta t$ to restore stability when necessary.

Several authors have reported that the expectations of the heuristic stability argument seemed to be verified. Among these one finds the case $n = 5$ by Richtmyer [19] and by Laganelli, Ames, and Hartnett [40] for the case $n = 2$. The latter example will be discussed in the next section.

The maximum principle analysis can be applied to examine convergence and stability for the general initial boundary value problem

$$u_t = \phi(x, t, u, u_x, u_{xx}), \quad 0 < x < 1, \quad 0 < t \leq T \tag{2-142}$$

subject to smooth initial and boundary conditions. This problem is well posed in the region (see, for example, Friedman [9]) if

$$\frac{\partial \phi}{\partial u_{xx}} \geq a > 0. \tag{2-143}$$

Then we have

$$u_{i,j+1} = u_{i,j} + ku_t|_{i,j} + \frac{k^2}{2}u_{tt}|_{i,j} + O(k^3)$$

$$= u_{i,j} + k\phi\left[ih, jk, u_{i,j}, \frac{1}{2h}\mu\delta_x u_{i,j}, \frac{1}{h^2}\delta_x^2 u_{i,j}\right] + O[k^2 + kh^2] \tag{2-144}$$

where it has been assumed that

$$|\phi_u| + |\phi_{u_x}| + \phi_{u_{xx}} \leq b \tag{2-145}$$

and $u \in C^{4,2}$. Thus an explicit finite difference formula for Eqn (2-142) is

$$U_{i,j+1} = U_{i,j} + k\phi\left[ih, jk, U_{i,j}, \frac{1}{2h}\mu\delta_x U_{i,j}, \frac{1}{h^2}\delta_x^2 U_{i,j}\right]. \tag{2-146}$$

Setting $z_{i,j} = u_{i,j} - U_{i,j}$ we have, after subtracting Eqn (2-146) from Eqn (2-144) and applying the mean value theorem, the relation

$$z_{i,j+1} = z_{i,j} + k\left\{\frac{\partial\phi}{\partial u}z_{i,j} + \frac{\partial\phi}{\partial u_x}\left(\frac{1}{2h}\mu\delta_x z_{i,j}\right) + \frac{\partial\phi}{\partial u_{xx}}\frac{1}{h^2}\delta_x^0 z_{i,j}\right\}$$

$$+ O(h^2 k + k^2] \quad (2\text{-}147)$$

where the partial derivatives in Eqn (2-147) are evaluated at a point between the arguments of Eqns (2-144) and (2-146), as specified in the mean value theorem. Upon carrying out the specified finite differences, Eqn (2-147) can be rearranged as†

$$z_{i,j+1} = r\left[\frac{\partial\phi}{\partial u_{xx}} - \frac{1}{2}\frac{\partial\phi}{\partial u_x}h\right]z_{i-1,j} + \left[1 + \frac{\partial\phi}{\partial u}k - 2r\frac{\partial\phi}{\partial u_{xx}}\right]z_{i,j}$$

$$+ r\left[\frac{\partial\phi}{\partial u_{xx}} + \frac{1}{2}\frac{\partial\phi}{\partial u_x}h\right]z_{i+1,j} + O[k^2 + h^2 k]. \quad (2\text{-}148)$$

The argument used for the diffusion equation in Section 2-1 is applicable here, provided h and k can be chosen so that the coefficients of Eqn (2-148) are nonnegative. For the first and third coefficients we have

$$\frac{\partial\phi}{\partial u_{xx}} \pm \frac{1}{2}\frac{\partial\phi}{\partial u_x}h \geq a - \frac{1}{2}bh \geq 0 \quad (2\text{-}149)$$

if
$$h \leq 2a/b. \quad (2\text{-}150)$$

Also,
$$1 + \frac{\partial\phi}{\partial u}k - 2r\frac{\partial\phi}{\partial u_{xx}} \geq 1 - bk - 2br \geq 0 \quad (2\text{-}151)$$

if
$$0 < r \leq \frac{1 - bk}{2b}. \quad (2\text{-}152)$$

If Eqns (2-150) and (2-152) hold, and $u \in C^{4,2}$, we have

$$|z_{i,j+1}| \leq r\left(\frac{\partial\phi}{\partial u_{xx}} - \frac{1}{2}\frac{\partial\phi}{\partial u_x}h\right)|z_{i-1,j}| + \left(1 + \frac{\partial\phi}{\partial u}k - 2r\frac{\partial\phi}{\partial u_{xx}}\right)|z_{i,j}|$$

$$+ r\left(\frac{\partial\phi}{\partial u_{xx}} + \frac{1}{2}\frac{\partial\phi}{\partial u_x}h\right)|z_{i,j}| + O[k^2 + h^2 k]$$

$$\leq (1 + bk)\|z_j\| + A[k^2 + h^2 k] \quad (2\text{-}153)$$

where, again, $\|z_j\| = \max_{i=0,\dots,M}|z_{i,j}|$. Consequently

$$\|z_{j+1}\| \leq (1 + bk)\|z_j\| + A[k^2 + h^2 k]. \quad (2\text{-}154)$$

Now $\|z_0\| = 0$, so we have

$$\|z_j\| \leq \{1 + (1 + bk) + (1 + bk)^2 + \cdots + (1 + bk)^{j-1}\}A\{k^2 + h^2 k\}$$

$$\leq e^{bT}Aj[k^2 + h^2 k] \leq Te^{bT}A[k + h^2] \quad (2\text{-}155)$$

since
$$(1 + bk)^n \leq \left(1 + \frac{bT}{n}\right)^n < e^{bT}.$$

† If $\phi = u_{xx}$ we recover Eqn (2-13).

Consequently, the error $z_{i,j}$ goes to zero as Δx and Δt tend to zero if $h \le 2a/b$ and $0 < r \le (1 - bk)/2b$. Convergence and stability are assured if these relations hold. As an example of their application we consider the Burgers' equation (see Ames [8])

$$u_t = u_{xx} - uu_x \qquad (2\text{-}156)$$

with $0 \le u \le 1$, $0 \le u_x \le 1$. Here

$$\phi = u_{xx} - uu_x, \qquad \frac{\partial \phi}{\partial u_{xx}} = 1, \qquad \frac{\partial \phi}{\partial u_x} = -u, \qquad \frac{\partial \phi}{\partial u} = -u_x$$

so that $a = 1$, and $|\phi_u| + |\phi_{u_x}| + \phi_{u_{xx}} \le 3 = b$. Consequently, Eqn (2-150) becomes $h \le \frac{2}{3}$ and Eqn (2-152) becomes

$$0 < \frac{9k}{4} \le \frac{1 - 3k}{6} \qquad \text{or} \qquad k \le \frac{2}{33} = 0.067.$$

PROBLEMS

2-38 Describe a simple explicit method for the pure initial value problem governed by Burgers' equation [Eqn (2-156)] subject to $u(x, 0) = f(x)$.

2-39 The almost linear equation $u_t = u_{xx} + f(u)$ occurs in diffusion-reaction problems. If $f(u) = u^2$, apply the stability-convergence argument of this section to find the stability criterion.

2-40 The burning of a gas in a rocket is described by Forsythe and Wasow [30] in terms of the equation

$$u_t = -\tfrac{1}{2}u_x^2 + \lambda u_{xx} + d(x, t) \qquad (2\text{-}157)$$

for $-\infty < x < \infty$, $t > 0$, and d is periodic of period 2π in x. The auxiliary conditions are $u(x, 0) = f(x)$, $u(x + 2\pi, t) = u(x, t)$ with $f(x + 2\pi) = f(x)$. Apply the stability-convergence analysis of this section to obtain stability criterion.

2-41 Let $u = -2\lambda \ln \psi$ in Eqn (2-157) and find the equation in the variable ψ. Note that it is *linear*.

$$Ans: \ \psi_t = \lambda \psi_{xx} - (2\lambda)^{-1} d(x, t)\psi$$

$$\psi(x, 0) = \exp[-f(x)/2\lambda], \qquad \psi(x + 2\pi, t) = \psi(x, t).$$

2-42 Beginning with Burgers' equation, Eqn (2-156), perform the following operations:

(a) Set $u = v_x$ and integrate the resulting equation with respect to x, discarding an arbitrary function of t.
(b) Find the equation for w if $\psi = F(w)$ is substituted into $\psi_t = \lambda \psi_{xx}$.
(c) Compare the equations of parts (a) and (b), thus determining F. The result, called the Hopf transformation by Ames [8], generalizes to the three-dimensional momentum equations of fluid mechanics.

2-13 An application of the explicit method

Experimental studies of transpiration cooling are sometimes conducted with the porous test section preceded by a solid section. As a result there is a run of boundary-layer flow prior to the transpiration section, as shown in Fig. 2-6.

This problem, including heat transfer and skin friction, has been studied by Laganelli, Ames, and Hartnett [40]. A simpler version is included in Ames [8] and Crandall [21]. We shall not include the heat transfer calculations here.

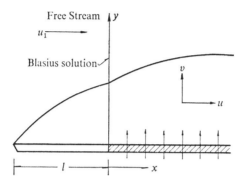

Fig. 2-6   Transpiration cooled flat plate

We suppose that a flat plate of nonporous length l (see Fig. 2-6) is fixed in a steady incompressible laminar stream followed downstream by a porous transpiration region. The boundary layer equations with zero pressure gradient in cartesian coordinates (Schlichting [41]) are

$$u_x + v_y = 0 \tag{2-158a}$$

$$uu_x + vu_y = \nu u_{yy}. \tag{2-158b}$$

The classical Blasius solution describes the boundary layer growth over the nonporous section. This solution evaluated at l provides the 'initial' condition, at $x = 0$, for continuation of the solution into the transpiration region. The boundary and initial conditions for that region are

$$
\left.
\begin{array}{lll}
y = 0: & u = 0, \quad v = v_c, \text{ for } x > 0 \\
y \to \infty: & u \to u_1, \text{ for } x > 0 \\
x = 0: & u = \text{Blasius solution}\dagger
\end{array}
\right\} \tag{2-159}
$$

A stream function ψ, defined through the relations

$$u = \frac{\partial \psi}{\partial y}, \qquad v = -\frac{\partial \psi}{\partial x} \tag{2-160}$$

† These values were obtained from Rosenhead and Simpson [42] or Howarth [43].

guarantees that the continuity equation, (2-158a), is satisfied. Upon introduction of these relations into Eqn (2-158b) we obtain the *third-order* equation

$$\psi_y\psi_{xy} - \psi_x\psi_{yy} = \nu\psi_{yyy} .$$ (2-161)

Such a form is undesirable for several reasons. Among these is our relative lack of information regarding their numerical analysis. Thus, a way is sought to maintain the *second-order* nature of the equations. This goal can be accomplished by an application of the *von Mises transformation* (see Ames [8]). The goal in this transformation is to change from (x, y) coordinates to (x, ψ) coordinates. To accomplish this we find

$$\frac{\partial u}{\partial x}\bigg|_y = \frac{\partial u}{\partial x}\bigg|_\psi + \frac{\partial u}{\partial \psi}\frac{\partial \psi}{\partial x} = \frac{\partial u}{\partial x}\bigg|_\psi - v\frac{\partial u}{\partial \psi}$$

$$\frac{\partial u}{\partial y}\bigg|_x = \frac{\partial u}{\partial \psi}\frac{\partial \psi}{\partial y} = u\frac{\partial u}{\partial \psi}$$

$$\frac{\partial^2 u}{\partial y^2}\bigg|_x = u\frac{\partial}{\partial \psi}\left(u\frac{\partial u}{\partial \psi}\right)$$

whereupon Eqn (2-158b) becomes

$$u\frac{\partial u}{\partial x}\bigg|_\psi = \nu u\frac{\partial}{\partial \psi}\left(u\frac{\partial u}{\partial \psi}\right)\bigg|_x .$$ (2-162)

With the understanding that $u = u(x, \psi)$, we drop the subscripts in Eqn (2-162) and treat

$$u_x = \nu(uu_\psi)_\psi$$ (2-163)

which is second order!

The boundary and initial conditions must now be transformed to (x, ψ) coordinates. Upon integrating Eqn (2-160) we have

$$\psi = -\int v(x, y)\, dx + f(y)$$

$$\psi = \int u(x, y)\, dy + g(x)$$ (2-164)

where f and g are arbitrary functions. Upon employing the boundary conditions at $y = 0$ there results

$$\psi(x, 0) = -v_c x + f(0), \qquad \psi(x, 0) = g(x).$$

The equality of these two relations implies that $g(x) = -v_c x + f(0)$, thereby satisfying the stream function at $x = 0$.† The value $f(0)$ is chosen to be zero, thereby establishing the reference $\psi = 0$.

---

† If $v_c = F(x)$, the method is easily generalized. Throughout this section v_c is held constant.

As a consequence of the foregoing arguments, Eqn (2-163) is subject to the following boundary and initial conditions:

$$\psi = -v_c x: \qquad u = 0, \qquad v = v_c, \quad \text{for } x > 0$$

$$\psi \to \infty: \qquad u \to u_1, \quad \text{for } x > 0 \tag{2-165}$$

$$x = 0: \qquad u = \text{Blasius solution}$$

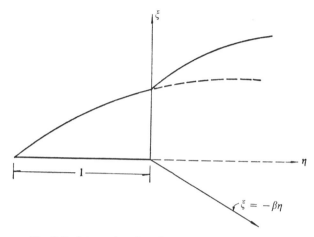

Fig. 2-7 Integration domain for transpiration problem

Finally, the dimensionless variables

$$\eta = x/l, \qquad \xi = \frac{\psi}{u_1 l}\left(\frac{u_1 l}{\nu}\right)^{1/2}, \qquad \Gamma = 1 - (u/u_1)^2 \tag{2-166}$$

are introduced, whereupon Eqn (2-163) becomes

$$\frac{\partial \Gamma}{\partial \eta} = (1 - \Gamma)^{1/2}\frac{\partial^2 \Gamma}{\partial \xi^2} \tag{2-167}$$

subject to the auxiliary data

$$\xi \to \infty: \ \Gamma \to 0,$$

$$\xi = -\frac{v_c}{u_1}\left(\frac{u_1 l}{\nu}\right)^{1/2}\eta = -\beta\eta: \qquad \Gamma = 1 \tag{2-168}$$

$\Gamma(0, \xi) = $ Blasius solution. (See Lagonelli, Ames, and Hartnett [40] for data.) Note that the integration domain, shown in Fig. 2-7, does not have orthogonal boundaries (for $v_c \neq 0$) since the second boundary condition is applied along the sloping boundary $\xi = -\beta\eta$. Irregular mesh point techniques must be employed near that boundary.

Since the integration domain is infinite, with propagation in the η direction, an explicit method analogous to Eqn (2-139) is employed. Thus, setting $G_{i,j} = G(i\,\Delta\eta, j\,\Delta\xi)$† we have the finite difference equation

$$G_{i+1,j} = G_{i,j} + r(1 - G_{i,j})^{1/2}[G_{i,j+1} - 2G_{i,j} + G_{i,j-1}]. \quad (2\text{-}169)$$

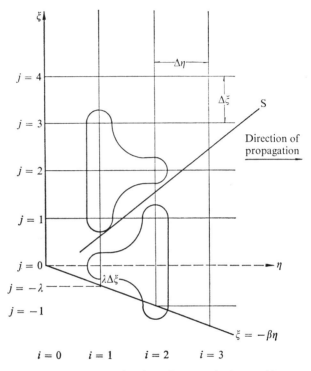

Fig. 2-8   Computational net for transpiration problem

The heuristic stability argument of Section 2-12 for Eqn (2-140) may be employed here. Thus we find that the quantity

$$(1 - G_{i,j})^{1/2}\frac{\Delta\eta}{(\Delta\xi)^2} \le \frac{1}{2} \quad (2\text{-}170)$$

should suffice for the computational stability of the finite difference approximation. Now initially, and in fact everywhere except at $(0, 0)$, $0 \le G < 1$ so that a choice of $\Delta\eta/(\Delta\xi)^2 = \frac{1}{2}$ should be safe. This was found, in actual practice, to be quite satisfactory.

Beginning with the initial values along the ξ-axis there is no difficulty in obtaining that part of the solution above the line S in Fig. 2-8 using the explicit algorithm, Eqn (2-169). The calculation below the line S must proceed by employing the boundary values on $\xi = -\beta\eta$, but the explicit molecule

† $G_{i,j}$ is the approximate solution obtained from Eqn (2-169) for $\Gamma_{i,j}$.

cannot be employed since there is insufficient information on the previous line. The irregular boundary involves only modifications in the ξ direction, giving rise to irregular boundary points such as the one at $i = 1, j = -\lambda$ $(0 < \lambda < 1)$, which we label $P_{1,-\lambda}$, and regular points such as that at $i = 2$. $j = -1$.

Calculation of the solution *adjacent to an irregular boundary point* is accomplished by quadratic† interpolation; thus, for $i = 1$ we have

$$G_{1,0} = -\frac{\lambda}{\lambda + 2} G_{1,2} + \frac{2\lambda}{1 + \lambda} G_{1,1} + \frac{2}{(\lambda + 1)(\lambda + 2)} G_{1,-\lambda}$$

which becomes

$$G_{1,0} = -\frac{\lambda}{\lambda + 2} G_{1,2} + \frac{2\lambda}{\lambda + 1} G_{1,1} + \frac{2}{(\lambda + 1)(\lambda + 2)} \qquad (2\text{-}171)$$

because the boundary value of $G_{1,-\lambda}$ is unity. More generally, if the irregular point is at $i, j = j_0 - \lambda$ we have

$$G_{i,j_0} = -\frac{\lambda}{\lambda + 2} G_{i,j_0+2} + \frac{2\lambda}{\lambda + 1} G_{i,j_0+1} + \frac{2}{(\lambda + 1)(\lambda + 2)}. \qquad (2\text{-}172)$$

As the solution propagates the value of λ increases being determined, say, by a similar triangle calculation. If the boundary point $i, j_0 - 1$ is regular, calculation of the solution at the point adjacent to it is accomplished by use of an implicit algorithm

$$G_{i,j_0} = G_{i-1,j_0} + r(1 - G_{i-1,j_0})^{1/2}(G_{i,j_0-1} - 2G_{i,j_0} + G_{i,j_0-1})$$

which, however, contains only the unknown G_{i,j_0}. Upon setting $G_{i,j_0-1} = 1$ and solving for G_{i,j_0} we have

$$G_{i,j_0} = \frac{G_{i-1,j_0} + r(1 - G_{i-1,j_0})^{1/2}(G_{i,j_0+1} + 1)}{1 + 2r(1 - G_{i-1,j_0})^{1/2}} \qquad (2\text{-}173)$$

for the value of the solution at a mesh point adjacent to a regular boundary point. Placement of the implicit molecule is shown in Fig. 2-8.

PROBLEM

2-43   Let

$$v_c = \begin{cases} v_0 x(1 - x), & 0 \le x \le 1 \\ 0, & x > 1 \end{cases}$$

in Eqn (2-159). Reformulate the problem, determine the integration domain, and set up a computational algorithm near irregular boundary points.

---

† Quadratic interpolation is used for increased accuracy, instead of linear interpolation

$$G_{i+1,0} = \frac{\lambda}{1 + \lambda} G_{i+1,1} + \frac{1}{1 + \lambda} G_{i+1,-\lambda}.$$

2-14   Implicit methods for nonlinear problems

The analysis of Section 2-12 led to a limitation, Eqn (2-152), on $0 < r = k/(h)^2$ $\leq (1 - bk)/2b$ that can be extremely inconvenient, particularly if b is large. In such cases the number of time steps and the resulting computation can be prohibitively large. Having large partial derivatives of ϕ in a small portion of the region enforces the use of small time steps over all the region, unless a rather complicated machine program is constructed. Such effects, called *boundary layers* in engineering parlance, place severe restrictions on the numerical methods. As was previously observed in Section 2-5, implicit methods for linear problems had certain stability advantages. It is therefore natural to turn to implicit methods in seeking to avoid the aforementioned restrictions. Several systems will be investigated.

(a) Application of the backward difference

If Eqn (2-143) holds, then the implicit relation [Eqn (2-142)] may be solved for u_{xx}. Thus, we assume the partial differential equation to have the form

$$u_{xx} = \psi(x, t, u, u_x, u_t) \tag{2-174}$$

where the properly posed requirement is

$$\frac{\partial \psi}{\partial u_t} \geq a > 0. \tag{2-175}$$

Upon application of backward differences we find the implicit algorithm, for calculation on the $j + 1$ row, to be

$$\frac{1}{h^2}\{U_{i+1,j+1} - 2U_{i,j+1} + U_{i-1,j+1}\} = \psi\bigg[ih, (j + 1)k, U_{i,j+1},$$
$$\frac{1}{2h}(U_{i+1,j+1} - U_{i-1,j+1}), \frac{1}{k}(U_{i,j+1} - U_{i,j})\bigg] \tag{2-176a}$$

or, in the operator notation,

$$\frac{1}{h^2}\delta_x^2 U_{i,j+1} = \psi\bigg[ih, (j + 1)k, U_{i,j+1},$$
$$\frac{1}{2h}\mu\delta_x U_{i,j+1}, (U_{i,j+1} - U_{i,j})/k\bigg]. \tag{2-176b}$$

Convergence of the solution of Eqn (2-176a) to that of Eqn (2-174), for boundary value problems, has been proved by Douglas [16] and Lees [44] by arguments similar to those already employed. The error is O$[h^2 + k]$ and *no restrictions on r occur*. Unfortunately, the algebraic problem, depending upon the complexity of ψ, may have become complicated since the algebraic equations are nonlinear, but an *inner* iteration is easily devised by a *method of successive approximations*.

To motivate the successive approximation procedure let us consider the simple system

$$Lu = f(u) \tag{2-177}$$

where L is a linear partial differential operator. One may linearize Eqn (2-177) by a Picard method which introduces a sequence of functions $\{u^{(n)}\}$ which satisfy the boundary conditions specified for u and the linear partial differential equations

$$Lu^{(n+1)} = f[u^{(n)}]. \tag{2-178}$$

When the sequence $\{u^{(n)}\}$ converges, the convergence is linear—that is, as $n \to \infty$,

$$u^{(n+1)} - u = O[u^{(n)} - u]. \tag{2-179}$$

Alternatively, if f is differentiable, Bellman, Juncosa, and Kalaba [45] suggest a different (Newton) linearization, used also by others (e.g. Douglas [18]). If one replaces the right-hand side of Eqn (2-178) by the expansion about $u^{(n)}$

$$f[u^{(n)}] + [u^{(n+1)} - u^{(n)}]f'[u^{(n)}] \tag{2-180}$$

then a new linear partial differential equation results, namely

$$Lu^{(n+1)} - f'[u^{(n)}]u^{(n+1)} = f[u^{(n)}] - u^{(n)}f'[u^{(n)}] \tag{2-181}$$

and this sequence, when convergent, is usually quadratically convergent—that is,

$$u^{(n+1)} - u = O[(u^{(n)} - u)^2] \quad \text{as } n \to \infty. \tag{2-182}$$

The generalization of this Newton second-order method to Eqn (2-174) is

$$u_{xx}^{(n+1)} = \psi[x, t, u^{(n)}, u_x^{(n)}, u_t^{(n)}] + [u^{(n+1)} - u^{(n)}]\frac{\partial\psi}{\partial u}$$

$$+ [u_x^{(n+1)} - u_x^{(n)}]\frac{\partial\psi}{\partial u_x} + [u_t^{(n+1)} - u_t^{(n)}]\frac{\partial\psi}{\partial u_t} \tag{2-183}$$

where the partial derivatives of ψ have the same arguments as ψ—that is, they are evaluated at the nth step.

Successive approximations to $U_{i,j+1}$ can now be computed by solving the linear, tridiagonal system

$$\frac{1}{h^2}\delta_x^2 U_{i,j+1}^{(n+1)} - \left\{\frac{\partial\psi}{\partial u} + \frac{1}{2h}\frac{\partial\psi}{\partial u_x}\mu\delta_x + \frac{1}{k}\frac{\partial\psi}{\partial u_t}\right\}U_{i,j+1}^{(n+1)}$$

$$= -\left\{\frac{\partial\psi}{\partial u} + \frac{1}{2h}\frac{\partial\psi}{\partial u_x}\mu\delta_x + \frac{1}{k}\frac{\partial\psi}{\partial u_t}\right\}U_{i,j+1}^{(n)}$$

$$+ \psi\left[ih, (j+1)k, U_{i,j+1}^{(n)}, \frac{1}{2h}\mu\delta_x U_{i,j+1}^{(n)}, (U_{i,j+1}^{(n)} - U_{i,j})/k\right]. \tag{2-184}$$

As they stand the expressions in braces ($\{\ \}$) need to be evaluated at each step in the sequence, which may be excessive of computer time. The authors show that one can replace those expressions by a *constant*, A, whose optimum value is

$$A = \frac{1}{2}k^{-1}\left\{\max\frac{\partial\psi}{\partial u_t} + \min\frac{\partial\psi}{\partial u_t}\right\}.$$

If $k = o(h)$ and k is sufficiently small, the sequence $U_{i,j+1}^{(n)}$ converges to the solution of Eqn (2-176a). The condition $k = o(h)$ is required only for rapid convergence of the inner iteration and not for convergence of the solution of Eqn (2-176a) to the solution of the differential equation.

In some instances Eqn (2-174) can be written in quasilinear form†

$$u_{xx} + f(x, t, u)u_x + g(x, t, u) = p(x, t, u)u_t. \qquad (2\text{-}185)$$

The finite difference equation

$$\frac{1}{h^2} \delta_x^2 U_{i,j+1} + \frac{1}{2h} f[ih, (j+1)k, U_{i,j}]\mu\delta_x U_{i,j+1} + g[ih, (j+1)k, U_{i,j}]$$
$$= p[ih, (j+1)k, U_{i,j}] \frac{U_{i,j+1} - U_{i,j}}{k} \qquad (2\text{-}186)$$

contains $U_{i,j+1}$ only linearly. Thus, the algebraic problem is linear and tri-diagonal at each time step. The convergence proof remains valid (see Douglas [16]).

A special case of Eqn (2-185) bears mention. It is frequently convenient to leave the equation in self-adjoint form rather than write it in the expanded form of Eqn (2-185). Thus, for

$$\frac{\partial}{\partial x}\left[p(x, t)\frac{\partial u}{\partial x}\right] + q(x, t, u) = s(x, t, u)\frac{\partial u}{\partial t} \qquad (2\text{-}187)$$

a useful implicit finite difference equation, generating a linear algebraic problem, is

$$\frac{1}{h^2}\{p[(i+\tfrac{1}{2})h, (j+1)k][U_{i+1,j+1} - U_{i,j+1}]$$
$$- p[(i-\tfrac{1}{2})h, (j+1)k][U_{i,j+1} - U_{i-1,j+1}]\}$$
$$+ q[ih, (j+1)k, U_{i,j}]$$
$$= s[ih, (j+1)k, U_{i,j}] \frac{U_{i,j+1} - U_{i,j}}{k}. \qquad (2\text{-}188)$$

The same convergence proof applies to this special case of Eqn (2-174).

(b) Crank–Nicolson form‡

Application of the Crank–Nicolson method is easily formulated for nonlinear equations. For Eqn (2-174) we have

$$\frac{1}{2h^2}\delta_x^2(U_{i,j+1} + U_{i,j}) = \psi\left[ih, (j+\tfrac{1}{2})k, \tfrac{1}{2}(U_{i,j+1} + U_{i,j}),\right.$$
$$\left. \frac{1}{4h}\mu\delta_x(U_{i,j+1} + U_{i,j}), \frac{1}{k}(U_{i,j+1} - U_{i,j})\right]. \qquad (2\text{-}189)$$

---

† Burgers' equation, $u_t + uu_x = u_{xx}$, is of this form.
‡ Here we use $\lambda = \tfrac{1}{2}$ on the right-hand side of Eqn (2-35). Similar extensions are possible with arbitrary λ, where $\lambda \geq \tfrac{1}{2}$ for inferred stability.

Lees [39, 44] has shown by an application of energy estimates that the solution of the difference equation, (2-189), converges to that of the differential equation, (2-174), with *no* restriction on *r*. The algebraic equations of Eqn (2-189) are nonlinear and require an iterative solution similar to that of Eqn (2-178).

When the Crank–Nicolson concept is applied to Eqn (2-185), the finite difference algorithm assumes the simpler form

$$\frac{1}{2h^2}\delta_x^2[U_{i,j+1} + U_{i,j}] + \frac{1}{4h}f[ih, (j + \tfrac{1}{2})k, \tfrac{1}{2}(U_{i,j+1} + U_{i,j})]\delta_x(U_{i,j+1} + U_{i,j})$$

$$+ g[ih, (j + \tfrac{1}{2})k, \tfrac{1}{2}(U_{i,j+1} + U_{i,j})]$$

$$= p[ih, (j + \tfrac{1}{2})k, \tfrac{1}{2}(U_{i,j+1} + U_{i,j})]\frac{U_{i,j+1} - U_{i,j}}{k}. \qquad (2\text{-}190)$$

These nonlinear equations must be solved by iteration. One obvious way to accomplish this is to evaluate the coefficients *f*, *g*, and *p* using the old value $U_{i,j+1}^{(k)}$ thus generating a linear tridiagonal system.

(c) Predictor–corrector methods

The aforementioned nonlinear algebraic equations which arise when finite differences are applied to Eqn (2-174) can be solved as discussed, or they can be avoided in several large classes of nonlinear equations by using predictor–corrector methods.

Predictor–corrector methods have been successfully used by manv in the numerical solution of *ordinary* differential equations. A discussion of some of these is to be found in Hamming [46, 47] and Fox [7] variously labeled as the Adams–Bashforth method, methods of Milne, and so forth. The general approach begins from known or previously computed results at previous pivotal values, up to and including the point x_n, by 'predicting' results at x_{n+1} with formulae which need no knowledge at x_{n+1}. These predicted results are relatively inaccurate. They are then improved by the use of more accurate 'corrector' formulae which require information at x_{n+1}. Generally this amounts to computing results at x_{n+1} from a nonlinear algebraic equation, to the solution of which the 'predictor' gives a first approximation and the 'corrector' is used repeatedly, if necessary, to obtain the final result.

Douglas and Jones [48] have considered Eqn (2-174) on $0 < x < 1$, $0 < t \le T$ with $u(x, 0)$, $u(0, t)$ and $u(1, t)$ as specified boundary conditions. If either

$$\psi = f_1(x, t, u)\frac{\partial u}{\partial t} + f_2(x, t, u)\frac{\partial u}{\partial x} + f_3(x, t, u) \qquad (2\text{-}191)$$

or

$$\psi = g_1\left(x, t, u, \frac{\partial u}{\partial x}\right)\frac{\partial u}{\partial t} + g_2\left(x, t, u, \frac{\partial u}{\partial x}\right) \qquad (2\text{-}192)$$

a predictor–corrector modification of the Crank–Nicolson procedure is possible *so that the resulting algebraic problem is linear.* This is significant since the class equation [Eqn (2-191)] includes Burgers' equation

$$u_{xx} = uu_x + u_t$$

of turbulence and suggests extension to higher-order systems in fluid mechanics. The class equation, (2-192), includes the equation of nonlinear diffusion:

$$\frac{\partial}{\partial x}\left(K(u)\frac{\partial u}{\partial x}\right) = \alpha(u)\cdot\frac{\partial u}{\partial t}.$$

If ψ is of the form of Eqn (2-191), the following predictor–corrector analog (combined with the boundary data $u_{i,0}$, $u_{0,j}$, and $u_{M,j}$) leads to *linear* algebraic equations. The predictor is

$$\frac{1}{h^2}\,\delta_x^2 U_{i,j+1/2} = \psi\left[ih, (j + \tfrac{1}{2})k, U_{i,j}, \frac{1}{2h}\mu\delta_x U_{i,j}, \frac{2}{k}(U_{i,j+1/2} - U_{i,j})\right] \quad (2\text{-}193)$$

for $i = 1, 2, \ldots, M - 1$. This is followed by the corrector

$$\tfrac{1}{2}\delta_x^2[U_{i,j+1} + U_{i,j}] = \psi\left[ih; (j + \tfrac{1}{2})k, U_{i,j+1/2},\right.$$
$$\left.\frac{1}{4h}\mu\delta_x(U_{i,j+1} + U_{i,j}), \frac{1}{k}(U_{i,j+1} - U_{i,j})\right]. \quad (2\text{-}194)$$

Equation (2-193) is a backward difference equation utilizing the intermediate time points $(j + \tfrac{1}{2})\Delta t$. Since Eqn (2-191) only involves $\partial u/\partial t$ linearly, the calculation into the $(j + \tfrac{1}{2})$ time row is a linear algebraic problem. To move up to the $(j + 1)$ time row we use Eqn (2-194), and by virtue of the linearity of Eqn (2-191) in $\partial u/\partial x$, this problem is also a linear algebraic problem. As an alternative to Eqn (2-193) one may use the predictor

$$\frac{1}{2h^2}\delta_x^2[U_{i,j+1/2} + U_{i,j}]$$
$$= \psi\left[ih, (j + \tfrac{1}{2})k, U_{i,j}, \frac{1}{2h}\mu\delta_x U_{i,j}, \frac{2}{k}(U_{i,j+1/2} - U_{i,j})\right]. \quad (2\text{-}195)$$

When ψ is given by Eqn (2-192), and when we replace the corrector by

$$\frac{1}{2h^2}\delta_x^2[U_{i,j+1} + U_{i,j}]$$
$$= \psi\left[ih, (j + \tfrac{1}{2})k, U_{i,j+1/2}, \frac{1}{2h}\mu\delta_x U_{i,j+1/2}, \frac{1}{k}(U_{i,j+1} - U_{i,j})\right] \quad (2\text{-}196)$$

then the predictor–corrector system [Eqns (2-193) and (2-196)] generates linear algebraic equations for the calculation of the finite difference approximation.

Douglas and Jones [48] have demonstrated that the predictor–corrector scheme defined by Eqns (2-193) and (2-194) converges to the solution of Eqn (2-174) when ψ is specified by Eqn (2-191). The truncation error is $O[h^2 + k^2]$.

When ψ is given by Eqn (2-192), convergence is also established when the corrector adopted is Eqn (2-196). In this case the error is $O[h^2 + k^{3/2}]$.

Miller [49] has investigated and compared this predictor–corrector method with the explicit method and the exact solution for a problem given by Burgers' equation

$$u_t + uu_x = \nu u_{xx}, \quad 0 < x < 1, \quad 0 < t \le T$$
$$u(0, t) = u(1, t) = 0 \qquad\qquad\qquad (2\text{-}197)$$
$$u(x, 0) = \sin \pi x.$$

The exact solution is obtained by transforming this into a linear diffusion problem (see Problem 2-42). Solutions are obtained for $0 < \nu \le 1.0$, with emphasis on small values of ν.† For values of ν, $0.01 \le \nu \le 1.0$, all three

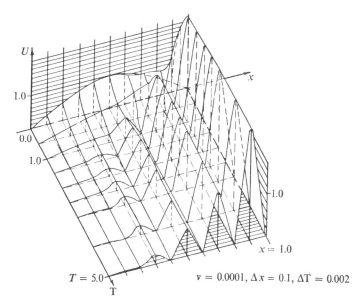

Fig. 2-9 Oscillation of the numerical solution for Burgers' equation using an explicit method

solutions were in excellent agreement. For $\nu < 0.01$, computation by means of the exact solution is not practical because of the slow convergence of the Fourier series.

As ν decreases from 10^{-2} to 10^{-4} a definite consistent pattern emerged on all grids when the explicit method was employed. A disturbance (shock?) appears at $x = 0.5$ for small t and passes to the right with steepened front as t increases (see Fig. 2-9). After the disturbance reaches a maximum near

† As will be observed in Chapter 4, numerical methods for $u_t + uu_x = 0$ and their generalizations often employ an artificial viscosity or dissipation mechanism to control instability. νu_{xx} plays a suitable role for small ν.

$x = 1.0$ for some time t, all values of u tend to decrease in a uniform manner. As h is reduced, these disturbances increase always maximizing as close to $x = 1.0$ as the grid allows. The ripples, shown in Fig. 2-9, become more exaggerated as $\nu \to 10^{-4}$. More ripples appear as ν decreases, propagating back toward $x = 0$. As $\nu \to 10^{-4}$ the disturbance is larger and decays more slowly as is physically expected. Computer time limitations imposed by stability requirements prevented further use of the explicit method.

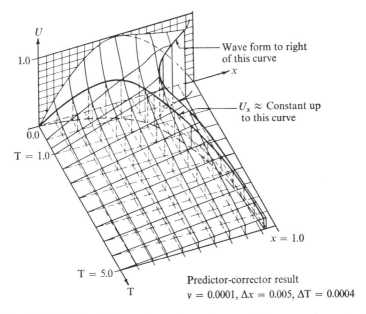

Fig. 2-10   Solution of Burgers' equation using a predictor–corrector method

The known property of unconditional stability of the predictor–corrector method was employed to refine the grid in the x direction while maintaining the same step $k = 4 \times 10^{-4}$. The computation at $h = 5 \times 10^{-3}$ ($r = 16.00$) is shown in Fig. 2-10. The ripples are gone! The computed solutions, with the predictor–corrector, tend to the asymptotic approximate profile of Cole [50].

PROBLEMS

2-44   Prove that Eqn (2-186) converges to the solution of Eqn (2-185).

2-45   Write out the Crank–Nicolson form of $u_t + u u_x = \nu u_{xx}$ and describe an inner iteration for the solution of the nonlinear algebraic problem.

2-46   Develop a predictor-corrector algorithm for the equation $u_t = (u^n u_x)_x$ of nonlinear diffusion.

2-47 Employ the three-level formula [Eqn (2-133)] in developing a finite difference approximation for the nonlinear equation

$$\frac{\partial}{\partial x}\left[a(x)\frac{\partial u}{\partial x}\right] = b(x, t, u)\frac{\partial u}{\partial t} + c(x, t, u).$$

Are the algebraic equations linear? Douglas [18] reports numerous successful applications of such trilevel formulae, although convergence proofs are not yet available.

2-15 Concluding remarks

A large number of nonlinear examples involving parabolic equations are presented in Ames [8]. Examples include boundary layer flow, heat conduction and chemical reaction, thermal ignition, nonlinear heat conduction, moving boundary problems (melting and freezing), diffusion and reaction, higher order equations, and percolation problems.

Simultaneous equations of parabolic type, mixed systems, and problems in higher dimensions will be considered in Chapter 5.

This chapter closes with a discussion of some numerical experiments by Bellman, Juncosa, and Kalaba [45] using Eqn (2-178) (Picard approximation) and Eqn (2-181) (Newton approximation) on a parabolic equation.

Consider the parabolic equation

$$u_t - u_{xx} = (1 + u^2)(1 - 2u) \tag{2-198}$$

over the two triangles: $0 \le t \le 1 - x$, $0 \le x \le 1$, and $0 \le t \le 1.5 - x$, $0 \le x \le 1.5$. In each case boundary conditions were chosen to give the unique exact solution $u = \tan(x + t)$. To avoid stability questions a finite difference analog of Eqn (2-198) is constructed by the Crank–Nicolson method with $h = k = 0.01$. Consequently, $r = 100$. The solution of the Picard and Newton forms of Eqn (2-198) will be obtained by iteration. We let n_j be the number of iterations required to obtain an acceptable approximation to $U_{i,j}$ at the grid points on the line jk. The criterion for acceptance of an approximation was

$$\max_j \left| \frac{U_{i,j}^{(n)} - U_{i,j}^{(n-1)}}{U_{i,j}^{(n)}} \right| \le 10^{-6} \tag{2-199}$$

so that $n_j = \min n$ such that Eqn (2-199) holds.

The function $f[u^{(n)}]$ on the right-hand side of Eqns (2-178) and (2-181) was replaced by the average value $\frac{1}{2}\{f[u_{i,j+1}^{(n)}] + f[u_{i,j}^{(n)}]\}$ which sometimes provides improved accuracy. The iteration procedure for the Picard method was

$$\frac{1}{k}[U_{i,j+1}^{(n)} - U_{i,j}^{(n_j)}] - \frac{1}{2h^2}[U_{i+1,j+1}^{(n+1)} - 2U_{i,j+1}^{(n+1)}$$
$$+ U_{i-1,j+1}^{(n+1)} + U_{i+1,j}^{(n_j)} - 2U_{i,j}^{(n_j)} + U_{i-1,j}^{(n_j)}]$$
$$= \{1 + \frac{1}{2}[(U_{i,j+1}^{(n)})^2 + (U_{i,j}^{(n_j)})^2]\}\{1 - U_{i,j+1}^{(n)} - U_{i,j}^{(n_j)}\} \tag{2-200}$$

with a corresponding expression for the Newton form.

The significant results were that while both methods required about the same amount of work in the triangle $0 \leq t \leq 1 - x$, $0 \leq x \leq 1$, where there are *no steep gradients*, the situation is radically different in the larger triangle. In $0 \leq t \leq 1.5 - x$, $0 \leq x \leq 1.5$, the boundary condition at $t = 1.5 - x$ is $u = \tan 1.5$, which is a large number. In this triangle the Picard method required nine times as many iterations as the Newton method for the same accuracy. A similar result occurred for the nonlinear elliptic equation $u_{xx} + u_{yy} = e^u$.

Based on the above 'experimental' evidence we conclude that if a nonlinear problem has no steep gradients there appears to be no advantage of the Newton method over the Picard approach. However, when steep gradients occur there is a decided advantage on the side of the Newton method.

REFERENCES

1. Hadamard, J. *Lectures on Cauchy's Problem in Linear Partial Differential Equations.* Yale University Press, New Haven, Connecticut, 1923.
2. Courant, R. and Hilbert, D. *Methods of Mathematical Physics*, vol. 2. Wiley (Interscience), New York, 1962.
3. Lavrentiev, M. M. *Some Improperly Posed Problems of Mathematical Physics.* Springer, Berlin and New York, 1967.
4. Moon, P. and Spencer, D. E. *Field Theory Handbook.* Springer, Berlin, 1961.
5. Crank, J. *Mathematics of Diffusion.* Oxford University Press, London and New York, 1956.
6. Carslaw, H. S. and Jaeger, J. C. *Conduction of Heat in Solids*, 2nd edit. Oxford University Press, London and New York, 1960.
7. Fox, L. (ed.). *Numerical Solution of Ordinary and Partial Differential Equations.* Macmillan (Pergamon), New York, 1962.
8. Ames, W. F. *Nonlinear Partial Differential Equations in Engineering.* Academic Press, New York, 1965.
9. Friedman, A. *Partial Differential Equations of Parabolic Type.* Prentice-Hall Inc., Englewood Cliffs, N.J., 1964.
10. Bernstein, D. L. *Existence Theorems in Partial Differential Equations.* Princeton University Press, Princeton, N.J., 1950.
11. Hildebrand, F. B. *J. Math. Phys.*, **31**, 35, 1952.
12. Juncosa, M. L. and Young, D. M. *Proc. Am. math. Soc.*, **5**, 168, 1954.
13. Juncosa, M. L. and Young, D. M. *J. Soc. ind. appl. Math.*, **1**, 111, 1953.
14. Juncosa, M. L. and Young, D. M. *Proc. Camb. phil. Soc.*, **53**, 448, 1957.
15. Laasonen, P. *Acta Math.*, **81**, 309, 1949.
16. Douglas, J. *Pacif. J. Math.*, **6**, 35, 1956.
17. Collatz, L. *The Numerical Treatment of Differential Equations.* Springer, Berlin and New York, 1960.
18. Douglas, J. Survey of numerical methods for parabolic differential equations, in *Advances in Computers*, vol. 2 (F. L. Alt, ed.). Academic Press, New York, 1961.
19. Richtmyer, R. D. *Difference Methods for Initial Value Problems*, 2nd edit. Wiley (Interscience), New York, 1967.

20. Milne, W. E. *Numerical Solution of Differential Equations.* Wiley, New York, 1953.
21. Crandall, S. H. *Engineering Analysis*, pp. 382, 395. McGraw-Hill, New York, 1956.
22. Crank, J. and Nicolson, P. *Proc. Camb. phil. Soc.*, **32**, 50, 1947.
23. O'Brien, G. G., Hyman, M. A., and Kaplan, S. *J. Math. Phys.*, **29**, 223, 1951.
24. Thomas, L. H. Elliptic problems in linear difference equations over a network, *Watson Sci. Comput. Lab. Rept.* Columbia University, New York, 1949.
25. Bruce, G. H., Peaceman, D. W., Rachford, H. H., and Rice, J. D. *Trans. Am. Inst. Min. Engrs (Petrol Div.)*, **198**, 79, 1953.
26. Douglas, J. *J. Ass. comput. Mach.*, **6**, 48, 1959.
27. Cuthill, E. H. and Varga, R. S., *J. Ass. comput. Mach.*, **6**, 236, 1959.
28. Blair, P. M. M.A. thesis, Rice University, Houston, Texas, 1960.
29. Richardson, L. F. *Phil. Trans. R. Soc.*, **A210**, 307, 1910.
30. Forsythe, G. E. and Wasow, W. R. *Finite Difference Methods for Partial Differential Equations.* Wiley, New York, 1960.
31. Rutherford, D. E. *Proc. R. Soc. Edinb.*, **A42**, 229, 1952.
32. DuFort, E. C. and Frankel, S. P. *Mathl. Tabl. natn. Res. Coun., Wash.*, **7**, 135, 1953.
33. Crandall, S. H. *Q. appl. Math.*, **13**, 318, 1955.
34. John, F. *Communs pure appl. Math.*, **5**, 155, 1952.
35. Albasiny, E. L. *Q. J. Mech.*, **13**, 374, 1960.
36. Flatt, H. P. Chain matrices and the Crank–Nicolson equation, in *Advances in Computers*, vol. 2 (F. L. Alt, ed.). Academic Press, New York, 1961.
37. Douglas, J., Jr. *J. Math. Phys.*, **35**, 145, 1956.
38. Douglas, J., Jr. *Trans. Am. math. Soc.*, **89**, 484, 1958.
39. Lees, M. *Duke Math. J.*, **27**, 297, 1960.
40. Laganelli, A. L., Ames, W. F., and Hartnett, J. P. *Am. Inst. Aeron. Astron. Jl*, **6**, 193, 1968.
41. Schlichting, H. *Boundary Layer Theory.* McGraw-Hill, New York, 1955.
42. Rosenhead, L. and Simpson, J. H. *Proc. Camb. phil. Soc.*, **32**, 385, 1936.
43. Howarth, L. *Proc. R. Soc.*, **A164**, 547, 1938.
44. Lees, M. *J. Soc. ind. appl. Math.*, **7**, 167, 1959.
45. Bellman, R., Juncosa, M., and Kalaba, R. Some numerical experiments using Newton's method for nonlinear parabolic and elliptic boundary value problems, *Rept. No. P-2200.* Rand Corp., Santa Monica, California, 1961.
46. Hamming, R. W. *Numerical Methods for Scientists and Engineers.* McGraw-Hill, New York, 1962.
47. Hamming, R. W. *J. Ass. comput. Mach.*, **6**, 37, 1959.
48. Douglas, J., Jr. and Jones, B. F. *J. Soc. ind. appl. Math.*, **11**, 195, 1963.
49. Miller, E. L. Predictor–corrector studies of Burgers' model of turbulent flow. Master's thesis, University of Delaware, Newark, Delaware, 1966.
50. Cole, J. D. *Q. appl. Math.*, **9**, 225, 1951.

3

Elliptic equations

3-0 Introduction

Equilibrium problems in two-dimensional, and higher, continua give rise to elliptic partial differential equations. A prototype is the famous equation of Laplace:

$$u_{xx} + u_{yy} + u_{zz} = 0. \tag{3-1}$$

This equation holds for the steady temperature in an isotropic medium, characterizes gravitational or electrostatic potentials at points of empty space, and describes the velocity potential of an irrotational, incompressible fluid flow. The two-dimensional counterpart of Eqn (3-1) lies at the foundation of the theory of analytic functions of a complex variable.

For elliptic equations in two dimensions the characteristics are complex—for example, those for

$$u_{xx} + u_{yy} = 0 \tag{3-2}$$

are $x + iy = \alpha = $ constant and $x - iy = \beta = $ constant, where α and β are complex. Equation (3-2) transforms into

$$\frac{\partial^2 u}{\partial \alpha \, \partial \beta} = 0 \tag{3-3}$$

which, upon integration, generates the *general*† solution

$$u = f(\alpha) + g(\beta) = f(x + iy) + g(x - iy) \tag{3-4}$$

where f and g are arbitrary functions.

Since α and β are complex, they can hardly be used for any computational process. Moreover, if we employ their real and imaginary parts as new coordinates, we merely reproduce Eqn (3-2) and accomplish nothing.

Nevertheless, these observations assist in determining the types of auxiliary conditions required for a *well-posed* problem (see Section 2-0). If, in the domain $y > 0$, we ask for the solution of Eqn (3-2) subject to

$$u = \phi(x), \qquad \frac{\partial u}{\partial y} = \psi(x) \quad \text{on } y = 0 \tag{3-5}$$

we obtain from Eqn (3-4) the result

$$u(x, y) = \text{Re}\,[\phi(z) - i \int_0^z \psi(s)\,ds], \qquad z = x + iy. \tag{3-6}$$

† Every solution has the form of the *general* solution or can be put in that form by a suitable transformation.

From function theory we know that the only analytic function that remains bounded on $z \to \infty$ is a constant, therefore Eqns (3-6) will either have singularities or will increase indefinitely. A well-posed practical problem is not expected to have these properties.

An alternative argument employs the so-called maximum (minimum) modulus theorem—that is, the solution of Eqn (3-2) has no maxima or minima at *interior* points of the domain of integration. If, then, boundary conditions (3-5) force the solution [Eqns (3-6)] to increase in the y direction, it will continue to increase *unless* another boundary intervenes. Thus, a closed boundary is desirable for a well-posed elliptic problem. Two conditions on the boundary are clearly too many. Before examining the well-posed question in yet a third light we digress for some further general remarks.

Much progress in partial differential equations has been achieved by series developments. For the differential equation of second order

$$u_{xy} = G(x, y, u, u_x, u_y, u_{xx}, u_{yy}) \tag{3-7}$$

subject to the auxiliary data

$$u(x, 0) = \phi(x), \qquad u_y(x, 0) = \psi(x) \tag{3-8}$$

we may attempt a solution in the form

$$u(x, y) = \sum_{i=0}^{\infty} a_i(x) y^i \tag{3-9}$$

and determine the coefficients $a_i(x)$ from the auxiliary conditions (3-8) and Eqn (3-7). It is not a difficult task to show that this procedure works if G, ϕ, and ψ are analytic functions of their arguments. The series (3-9) will converge for sufficiently small y.

For rather general systems the following existence–uniqueness theorem is a classic. A proof is available, for example, in Garabedian [1].

Cauchy–Kowalewski theorem

About any point at which the matrix A of coefficients $a_{jk}(\zeta, \eta, u)$ and the column vector h of functions $h_j(\eta)$ are analytic, a neighborhood can be found where there exists a unique vector u with analytic components u_k solving the initial value problem

$$u_\zeta = A(u)u_\eta, \qquad u(0, \eta) = h(\eta). \tag{3-10}$$

The requirement of analyticity on the elements of A and h does not, on the face of it, seem to be too serious a restriction, for we know by the Weierstrass approximation theorem that any continuous function can be approximated arbitrarily closely and uniformly in any given interval by polynomials, which are clearly analytic. Such reasoning would be valid *if* we could be sure that close approximation of the initial data always implies close approximation of the solution. That this is *not* the case follows from a simple example by Hadamard [2].

Consider Eqn (3-2) and ask for a solution such that

$$u(x, 0) = n^{-\alpha} \sin nx, \qquad u_y(x, 0) = 0 \qquad (3\text{-}11)$$

where $n > 0$ and $\alpha > 0$ is fixed. By the Cauchy–Kowalewski theorem a solution exists and is, by separation of variables,

$$u(x, y) = n^{-\alpha} \sin nx \cosh ny. \qquad (3\text{-}12)$$

With $\alpha > 0$ fixed, let $n \to \infty$, whereupon the initial data converge uniformly to

$$u(x, 0) = 0, \qquad u_y(x, 0) = 0 \qquad (3\text{-}13)$$

thereby determining the solution $u(x, y) = 0$. But for $y \neq 0$, the functions (3-12) do not converge to zero but become very large as $n \to \infty$ in an arbitrary neighborhood of the x-axis. Consequently, approximating the initial data arbitrarily closely does not guarantee a corresponding approximation for the solution! Stated alternatively, the solution does not always depend in a stable way upon the data provided in the Cauchy–Kowalewski theory.†

When a partial differential equation has accompanying auxiliary conditions which select among all possible solutions, one uniquely determined function, we call the data *properly posed*, provided that the solution depends continuously on these data. In Hadamard's example the data of Eqns (3-11) are *not properly posed for the Laplace equation*.

Thus we expect, and proof is possible, that Eqn (3-2) requires *one* condition at every point of the closed boundary. This condition may specify the function (Dirichlet problem) or the normal derivative (Neumann problem), or a combination of both (mixed problems). Further, all or part of the boundary may be at infinity if the function remains finite there.

3-1 Simple finite difference schemes

Any finite difference scheme for Eqn (3-2) must preserve the truth of the foregoing statements. Before proceeding it would be prudent to examine some schemes in the light of those remarks.

Consider a rectangular region with an internal net whose mesh points are denoted by $x_i = ih$, $y_j = jh$, where $i = 0, 1, 2, \ldots, M$; $j = 0, 1, 2, \ldots, N$. The exact solution is denoted by $u_{i,j}$ and the discrete approximation by $U_{i,j}$. The simplest replacement, developed in Section 1-5, consists of approximating both second derivatives of $u_{xx} + u_{yy} = 0$ by centered second differences so that

$$\frac{U_{i+1,j} - 2U_{i,j} + U_{i-1,j}}{h^2} + \frac{U_{i,j+1} - 2U_{i,j} + U_{i,j-1}}{h^2} = 0. \qquad (3\text{-}14)$$

---

† No known problem in science or engineering leads to an initial-value problem of this type for Laplace's equation. This is fortunate since, in that case, the differential system would be useless. A slight change of data would lead to an enormous change in the solution.

Equation (3-14) is expressible in the form

$$U_{i,j} = \tfrac{1}{4}[U_{i+1,j} + U_{i-1,j} + U_{i,j+1} + U_{i,j-1}] \qquad (3\text{-}15)$$

clearly demonstrating that the value at any point is the average of those at the surrounding points. Thus, the solution obtained from this representation, like the exact solution, has no maxima or minima at interior points of the domain of integration. That is to say, if extreme values exist they must lie on the boundary.

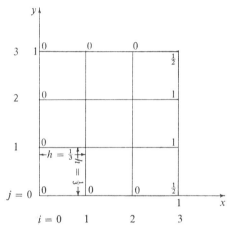

Fig. 3-1   Computational network for Laplace's equation

Equation (3-15), when applied repeatedly, ultimately expresses each $U_{i,j}$ as a sum of multiples of the boundary values. To illustrate this point more clearly we consider the network shown in Fig. 3-1 for the domain $0 \le x \le 1$, $0 \le y \le 1$. Here $h = \tfrac{1}{3}$ and there are four internal points $P_{1,1}$, $P_{1,2}$, $P_{2,1}$, and $P_{2,2}$. The boundary conditions for this Dirichlet problem are $u(0, y) = 0$, $u(1, y) = 1$, $u(x, 0) = 0$, $u(x, 1) = 0$. At the upper and lower right-hand corners we have anticipated a boundary singularity and minimized its influence by employing the average value $\tfrac{1}{2}$ at both points. We shall not need these values for this example, but a discussion of singularities is necessary and will be given later in this volume.

Equations for $U_{1,1}$, $U_{1,2}$, $U_{2,1}$, and $U_{2,2}$ are obtained by employing Eqn (3-15) at each interior point. Thus, we obtain

$$U_{2,1} + U_{0,1} + U_{1,2} + U_{1,0} - 4U_{1,1} = 0$$

$$U_{2,2} + U_{0,2} + U_{1,3} + U_{1,1} - 4U_{1,2} = 0$$

$$U_{3,1} + U_{1,1} + U_{2,2} + U_{2,0} - 4U_{2,1} = 0 \qquad (3\text{-}16)$$

$$U_{3,2} + U_{1,2} + U_{2,3} + U_{2,1} - 4U_{2,2} = 0$$

All values with i index 0 or 3, or j index 0 or 3 are known. Upon using those values, Eqns (3-16) can be recast in matrix form as follows:

$$
\begin{bmatrix}
-4 & 1 & 1 & 0 \\
1 & -4 & 0 & 1 \\
1 & 0 & -4 & 1 \\
0 & 1 & 1 & -4
\end{bmatrix}
\begin{bmatrix}
U_{1,1} \\
U_{1,2} \\
U_{2,1} \\
U_{2,2}
\end{bmatrix}
=
\begin{bmatrix}
-U_{0,1} - U_{1,0} \\
-U_{0,2} - U_{1,3} \\
-U_{3,1} - U_{2,0} \\
-U_{3,2} - U_{2,3}
\end{bmatrix}
=
\begin{bmatrix}
0 \\
0 \\
-1 \\
-1
\end{bmatrix}
\tag{3-17}
$$

Our problem is now reduced to solving a system of four linear equations in four unknowns. The vector of boundary conditions, on the right-hand side of Eqns (3-17), has a strong influence on the solution.

As the interval h is decreased, more internal mesh points appear thereby requiring more equations. Thus, more boundary values appear, and in the

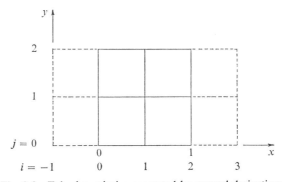

Fig. 3-2 False boundaries generated by normal derivatives

limit they are all present so that their complete knowledge is necessary for a unique solution.

On the other hand, suppose that the normal derivatives are specified on the boundaries (Neumann problem) and we examine the extreme situation of Fig. 3-2 wherein $0 \le x \le 1$, $0 \le y \le 1$, $h = \frac{1}{2}$. We wish to solve $u_{xx} + u_{yy} = 0$ subject to the conditions $u_x(0, y) = g_0(y)$, $u_x(1, y) = g_1(y)$, $u_y(x, 0) = h_0(x)$, $u_y(x, 1) = h_1(x)$. When Eqn (3-15) is applied we must solve nine algebraic equations typified by those along the line $j = 1$. To treat the boundary condition at $P_{0,1}$ a false boundary is introduced (see Problem 1-29) and $u_x(0, y) = g_0(y)$ is approximated to second order by

$$
\frac{1}{2h} [U_{1,1} - U_{-1,1}] = g_{0,1}
\tag{3-18}
$$

while $u_x(1, y) = g_1(y)$ is approximated by

$$
\frac{1}{2h} [U_{3,1} - U_{1,1}] = g_{1,1}.
\tag{3-19}
$$

Thus, *along the line $j = 1$* we find that Eqn (3-15) generates the set

$$U_{1,1} + U_{-1,1} + U_{0,2} + U_{0,0} - 4U_{0,1} = 0$$

$$U_{2,1} + U_{0,1} + U_{1,2} + U_{1,0} - 4U_{1,1} = 0 \qquad (3\text{-}20)$$

$$U_{3,1} + U_{1,1} + U_{2,2} + U_{2,0} - 4U_{2,1} = 0$$

But by virtue of the relations (3-18) and (3-19), Eqns (3-20) become

$$2U_{1,1} + U_{0,2} + U_{0,0} - 4U_{0,1} = 2hg_{0,1}$$

$$U_{2,1} + U_{0,1} + U_{1,2} + U_{1,0} - 4U_{1,1} = 0 \qquad (3\text{-}21)$$

$$2U_{1,1} + U_{2,2} + U_{2,0} - 4U_{2,1} = -2hg_{1,1}$$

Upon forming the matrix of coefficients for the nine equations it is found to be singular (Problem 3-1), so that we cannot have a unique solution. No solution is in fact possible unless the analytic requirement (Sneddon [3])

$$\int_c \frac{\partial u}{\partial n}\, ds = 0 \qquad (3\text{-}22)$$

is satisfied. This integral around the boundary curve is approximated to $O(h^2)$ by the trapezoidal rule to be†

$$h[\tfrac{1}{2}g_{0,0} + g_{0,1} + \tfrac{1}{2}g_{0,2} + \tfrac{1}{2}h_{0,1} + h_{1,1} + \tfrac{1}{2}h_{2,1}$$
$$+ \tfrac{1}{2}g_{1,2} + g_{1,1} + \tfrac{1}{2}g_{1,0} + \tfrac{1}{2}h_{2,0} + h_{1,0} + \tfrac{1}{2}h_{0,0}] = 0. \qquad (3\text{-}23)$$

PROBLEMS

3-1 Formulate the nine equations for the Neumann problem of Fig. 3-2. Show that the matrix of coefficients is singular. Ignore any possible discrepancy between the two limiting values of $\partial u/\partial n$ at corner points like $P_{0,0}$.

3-2 In the complementary energy formulation of Prandtl [4] (see also Timoshenko and Goodier [5]) the state of torsion of a twisted uniform elastic prism is characterized by a stress function $\psi(x, y)$ which satisfies a Poisson equation

$$\psi_{xx} + \psi_{yy} = -2G\theta \qquad (3\text{-}24)$$

throughout $|x| \le a$, $|y| \le a$, and the stress-free condition on the boundary becomes simply $\psi = 0$. In practice the *torsional rigidity*, $c = (2/\theta)\iint \psi\, dx\, dy$, is usually of more interest than the detailed description of stress and strain. The double integral extends over the cross-section of the prism. Render these equations dimensionless by setting $x' = x/a$, $y' = y/a$, $c' = c/Ga^4$, $\psi' = \psi/G\theta a^2$.

$$\textit{Ans: } \psi'_{x'x'} + \psi'_{y'y'} = -2, \quad c' = 2\int_{-1}^{1}\int_{-1}^{1} \psi'\, dx'\, dy'.$$

† In Eqn (3-23) we have emphasized that there may be a discrepancy at the corner values. If there is *no* discrepancy then $g_{0,0} = h_{0,0}$, $g_{0,2} = h_{0,1}$, $h_{2,1} = g_{1,2}$, and $g_{1,0} = h_{2,0}$. Hence Eqn (3-23) becomes: $h[g_{0,0} + g_{0,1} + g_{0,2} + h_{1,1} + g_{1,2} + g_{1,1} + g_{1,0} + h_{1,0}] = 0$.

3-3 Use a crude network, $h = 1$, for the dimensionless formulation of Problem 3-2. Write out the computational molecule. Evaluate c' by employing a two-dimensional Simpson rule (Fig. 1-5) and compare the results with the actual value $c' = 2.2495$.

3-4 Repeat Problem 3-3 with $h = \frac{1}{2}$. Compare the calculated value of c' with the actual value. (*Hint:* Is there any symmetry condition?)

3-2 Iterative methods

In Chapter 2 and the previous section we have observed that the application of finite difference methods very often generates an associated algebraic problem.† In many cases, even for some nonlinear problems, one must solve a large set of simultaneous linear equations

$$Au = v \qquad (3\text{-}25)$$

where A is a square (often sparse) matrix, v is a known column vector, and u is the unknown column vector. Methods of solution for general computational problems fall into two categories—the *direct* and *iterative* procedures. Direct methods, of which the solution of a tridiagonal system is typical (Section 2-3), are those which give the exact answer in a finite number of steps, if there were no round-off error. The algorithm for such a procedure is often complicated and nonrepetitive. Many direct methods for linear systems are available in the literature—see, for example, Householder [6] and Bodewig [7] (the latter is an exceptional compendium). These methods have usually been omitted from consideration in past efforts because of excessive computer storage requirements, both in program and in the necessity to store many intermediate results for later use. Recent studies by Gustavson *et al.* [8] indicate that direct methods are very useful indeed for very sparse matrices.

Iterative methods consist of repeated application of an often simple algorithm. They yield the exact answer only as a limit of a sequence, even without consideration of round-off errors. They have been much employed to solve Eqn (3-25) because, additionally, they can be programmed to take advantage of zeros in A, are self-correcting, and their very structure easily permits modifications such as under- and overrelaxation. Bodewig [7] provides an excellent source for methods to 1956. However, our present treatment will pass rapidly over these classical procedures and then introduce the more highly refined schemes of 1968.

In any iteration one begins with an initial approximation and then successively modifies the approximation according to some rule.‡ To be useful the iteration must converge but it is not considered to be effective unless the convergence is rapid.

---

† The engineer would say the continuous problem has been replaced by a lumped parameter problem.

‡ Many iterative processes are independent of the initial guess. A typical example (due to Boyer) is given in Ames [9, p. 169].

To solve the nonsingular equation $Au = v$ by iteration we require a sequence $u^{(k)}$ so defined that $u^{(k)} \to A^{-1}v$ (the solution) as $k \to \infty$. If $u^{(k)}$ is a function of $A, v, u^{(k-1)}, \ldots, u^{(k-r)}$ we say r is the *degree of the iteration*. To minimize computer storage requirements r is usually chosen as 1, 2, or 3; thus we could write, with $r = 1$,

$$u^{(k)} = F_k(A, v, u^{(k-1)}). \tag{3-26}$$

If F_k is independent of k the iteration is said to be *stationary*, and if F_k is linear in $u^{(k-1)}$ the iteration is termed *linear*.

The most general linear iteration is

$$u^{(k)} = G_k u^{(k-1)} + r_k \tag{3-27}$$

where G_k is a matrix depending upon A and v, and r_k is a column vector. For this to be useful the exact solution should be reproduced—that is,

$$A^{-1}v = G_k A^{-1}v + r_k. \tag{3-28}$$

Thus we obtain, as a *consistency* condition,

$$r_k = (I - G_k)A^{-1}v. \tag{3-29}$$

If we introduce the notation $M_k = (I - G_k)A^{-1}$, the general linear iteration is expressible as

$$u^{(k)} = G_k u^{(k-1)} + M_k v \tag{3-30}$$

where $M_k A + G_k = I$.

Convergence of the iteration, Eqn (3-30), is studied by examining an error vector e_k defined by

$$
\begin{aligned}
e_k &= u^{(k)} - A^{-1}v \\
&= G_k u^{(k-1)} + M_k v - A^{-1}v \\
&= G_k u^{(k-1)} + M_k v - G_k A^{-1}v - M_k v \\
&= G_k e_{k-1} \tag{3-31}
\end{aligned}
$$

Thus e_k satisfies the basic iteration equation [Eqn (3-30)] with $v = 0$. Applying Eqn (3-31) repetitively, we have

$$e_1 = G_1 e_0, \quad e_2 = G_2 G_1 e_0, \quad \ldots, \quad e_k = G_k G_{k-1} \ldots G_1 e_0. \tag{3-32}$$

Consequently, the convergence of the iteration, for a specified initial error e_0, depends on whether

$$H_k e_0 = G_k G_{k-1} \ldots G_1 e_0 \to 0, \quad \text{as } k \to \infty. \tag{3-33}$$

When the iteration is both stationary and linear $G_k = G$ and $H_k = G^k$ so the question of convergence revolves about consideration of $G^k e_0$.

Iterative methods fall quite naturally into two categories, *point iterative* and *block iterative.* Before giving a detailed discussion of convergence questions we describe some elementary point iterative methods as they apply to linear systems obtained from linear elliptic equations. Point iterative procedures are characterized by the explicit nature of the calculation of each component of successive approximation as opposed to the solution of several linear systems at each stage of the computation of the block iterative techniques.

3-3 Linear elliptic equations

Consider the linear elliptic equation†

$$au_{xx} + cu_{yy} + du_x + eu_y + fu = g(x, y) \qquad (3\text{-}34)$$

in the rectangular region $R: 0 \leq x \leq \alpha, 0 \leq y \leq \beta$, having Dirichlet boundary conditions. We suppose, for definiteness, that $a > 0, c > 0$ (see footnote), $f \leq 0$ and all are bounded in the region R and on its boundary B. Upon employing the second-order central differences of Eqns (1-49) and (1-51), with $h = k$, the finite difference approximation for Eqn (3-34) becomes

$$\beta_1 U_{i+1,j} + \beta_2 U_{i-1,j} + \beta_3 U_{i,j+1} + \beta_4 U_{i,j-1} - \beta_0 U_{i,j} = h^2 g_{ij} \qquad (3\text{-}35)$$

where the β_i are functions of $x_i = ih, y_j = jh$, given by

$$\beta_1 = a_{ij} + \tfrac{1}{2}hd_{ij}$$
$$\beta_2 = a_{ij} - \tfrac{1}{2}hd_{ij}$$
$$\beta_3 = c_{ij} + \tfrac{1}{2}he_{ij} \qquad (3\text{-}36)$$
$$\beta_4 = c_{ij} - \tfrac{1}{2}he_{ij}$$
$$\beta_0 = 2(a_{ij} + c_{ij} - \tfrac{1}{2}h^2 f_{ij})$$

The notation a_{ij} refers to $a(ih, jh)$, evaluated at the point where the computational molecule [Eqn (3-35)] is centered.

All the β_i will be positive if h is chosen so small that

$$0 < h < \min\left\{\frac{2a_{ij}}{|d_{ij}|}, \frac{2c_{ij}}{|e_{ij}|}\right\} \qquad (3\text{-}37)$$

where the minimum is taken over all points of the region and its boundary. Since $a > 0, c > 0, f \leq 0$, and all are bounded it follows that a positive minimum exists and for that h

$$\beta_0 \geq \sum_{m=1}^{4} \beta_m. \qquad (3\text{-}38)$$

† From Section 1-2 this equation is elliptic if $b^2 - 4ac < 0$; that is, $-4ac < 0$ since $b = 0$ here.

Let the number of interior mesh points be N, whereupon the approximation equation, (3-35), generates a system of N linear equations whose matrix form we write as

$$AU = V. \tag{3-39}$$

The vectors U and V consist of the N unknowns, $U_{i,j}$, and the quantities $-h^2 g_{ij}$ together with the boundary values, respectively. The matrix A has real coefficients whose main diagonal elements are the β_0 of Eqn (3-36) and whose off-diagonal elements are the *negatives* of the β_i which do not originate from boundary points. It is clear that the $N \times N$ matrix, $A = (\alpha_{ij})$, has the following properties†

$$\left.\begin{array}{l} \text{(i)} \quad \alpha_{ii} > 0, \quad \alpha_{ij} \leq 0, \quad i \neq j \\ \text{(ii)} \quad \alpha_{ii} \geq \sum_{j=1, i \neq j}^{N} |\alpha_{ij}| \quad \text{with strict inequality for some } i \\ \text{(iii)} \quad A \text{ is irreducible (see Section 3-4)} \end{array}\right\} \tag{3-40}$$

The first part of condition (3-40ii) follows directly from Eqn (3-38) while the second part is seen to be true by examining an interior mesh point *adjacent* to the boundary. For each such mesh point, at least one of the β_i of Eqn (3-36) will be the coefficient of a known boundary value and therefore will be included in the vector V instead of as an element of the matrix A. Condition (3-40iii) was discovered by Frobenius (see Geiringer [11]) and will be discussed in the next section.

The conditions of Eqn (3-40) are sufficient to prove, by contradiction, the existence of a unique solution to Eqn (3-39). To do this we will show that the determinant of $A \neq 0$. This will be the case provided the system $AU = 0$, obtained by setting $V = 0$, has only the solution which is identically zero everywhere in R. To force V to be zero we assume that the functions $g(x, y)$ and boundary values are zero. Let us now assume the homogeneous system has a nontrivial solution. Then for some point P of R we must have $U \neq 0$. Suppose, without loss of generality, that $U(P) > 0$, for otherwise consider $-U$ which would also be a solution of the homogeneous system $AU = 0$. Let M be the maximum positive value of U in R and let $U(Q) = M$. Then from Eqn (3-35) and condition (3-40ii) it follows that $U = M$ at each of the four adjacent points. By continuation of this process and employing the irreducibility of A, we conclude that $U = M$ for all points of R and the boundary B. This contradicts the original assumption that $U \equiv 0$ on B and therefore $U \equiv 0$ in $R + B$. Hence Eqn (3-39) has a unique solution. Additional remarks are given in Young and Frank [12].

As an example of the formulation, consider the equation

$$(x + 1)u_{xx} + (y^2 + 1)u_{yy} - u = 1$$

† A somewhat stronger result is sufficient to prove the existence of a unique solution to Eqn (3-39) if Eqns (3-40) are replaced by: (i) $|\alpha_{ii}| \neq 0$, $i = 1, 2, \ldots, N$; (ii) $|\alpha_{ii}| \geq \sum_{j=1, i \neq j}^{N} |\alpha_{ij}|$ and for some i strict inequality holds; (iii) A is irreducible. (See Taussky [10].)

in the region $0 \leq x \leq 1$, $0 \leq y \leq 1$ with the boundary values $u(0, y) = y$, $u(1, y) = y^2$, $u(x, 0) = 0$, $u(x, 1) = 1$. With $h = \frac{1}{3}$ we have four interior points. The β_k of this special case are

$$\beta_1 = \frac{i + 3}{3}, \qquad \beta_2 = \frac{i + 3}{3}, \qquad \beta_3 = \frac{j^2 + 9}{9}$$

$$\beta_4 = \frac{j^2 + 9}{9}, \qquad \beta_0 = 2 \left\{ \frac{i + 3}{3} + \frac{j^2 + 9}{9} + \frac{1}{18} \right\}$$

The matrix form of the four linear equations is

$$\begin{bmatrix} 5 & -\frac{10}{9} & -\frac{4}{3} & 0 \\ -\frac{13}{9} & \frac{17}{3} & 0 & -\frac{4}{3} \\ -\frac{5}{3} & 0 & \frac{17}{3} & -\frac{10}{9} \\ 0 & -\frac{5}{3} & -\frac{13}{9} & \frac{19}{3} \end{bmatrix} \begin{bmatrix} U_{1,1} \\ U_{1,2} \\ U_{2,1} \\ U_{2,2} \end{bmatrix} = \begin{bmatrix} \frac{1}{3} \\ \frac{20}{9} \\ \frac{2}{27} \\ \frac{56}{27} \end{bmatrix}$$

Several remarks about the previous equations are pertinent. First, *the main diagonal terms are dominant*—that is, in each case condition (3-4ii) holds with *strict inequality*. Second, the matrix is not symmetric—that is, $\alpha_{ij} \neq \alpha_{ji}$. Symmetric matrices have practical computational advantages. If $h = k$, the matrix A is symmetric for Laplace's equation $u_{xx} + u_{yy} = 0$ and for Poisson's equation $u_{xx} + u_{yy} = g(x, y)$ (Problem 3-5).

Let us assume that the linear elliptic equation is self-adjoint or can be made *self-adjoint* (Problem 3-6)—that is, can be written as

$$(au_x)_x + (cu_y)_y + fu = g(x, y). \tag{3-41}$$

If the region of integration has only regular mesh points, the symmetry of A can be assured if approximations of the form

$$(au_x)_x|_{i,j} = a_{i+1/2,j} h^{-2} [U_{i+1,j} - U_{i,j}] - a_{i-1/2,j} h^{-2} [U_{i,j} - U_{i-1,j}] + O(h^2) \tag{3-42}$$

$$(cu_y)_y|_{i,j} = c_{i,j+1/2} h^{-2} [U_{i,j+1} - U_{i,j}] - c_{i,j-1/2} h^{-2} [U_{i,j} - U_{i,j-1}] + O(h^2)$$

are employed. If irregular interior mesh points are present these approximations will not generally lead to symmetric matrices.

PROBLEMS

3-5 Consider the solution of the Dirichlet problem for Poisson's equation in the unit square, $0 \leq x \leq 1$, $0 \leq y \leq 1$. Let $h = \frac{1}{3} = k$ and the boundary values be $f(x, y)$. Find the matrix A and show that it is symmetric.

3-6 An equation of the form $au_{xx} + cu_{yy} + du_x + eu_y + fu = g(x, y)$ is said to be *essentially self-adjoint* when

$$\frac{\partial}{\partial y} [(d - a_x)/a] = \frac{\partial}{\partial x} [(e - c_y)/c].$$

Suppose this relation holds. Find the function ϕ so that the original equation can be put in the self-adjoint form of Eqn (3-41). (*Hint:* Multiply the original equation by ϕ.)

$$Ans: (\ln \phi)_x = (d - a_x)/a, \quad (\ln \phi)_y = (e - c_y)/c.$$

3-7 Convert the following equations to self-adjoint form by employing the results of Problem 3-6:

(a) $u_{xx} + u_{yy} + x^2 u_x + y^2 u_y + u = 0$

(b) $u_{xx} + u_{yy} + y^2 u_y = 0$

3-8 Apply the approximations of Eqns (3-42) to part (b) of Problem 3-7 in the region $0 \leq x \leq 1, 0 \leq y \leq 1$ and show that the resulting matrix is symmetric.

3-4 Some point iterative methods

Several of the best-known iterative methods are built around a partition of A into the form

$$A = R + D + S \tag{3-43}$$

where R, D, and S are matrices having the same elements as A respectively below the main diagonal, on the main diagonal, and above the main diagonal, and zeros elsewhere. We shall now suppose that the matrix A, of $Au = v$, has the properties of conditions (3-40).

(a) Jacobi method

This method, first employed by Jacobi [13] in 1844, is also called *iteration by total steps* and the *method of simultaneous displacements* (Geiringer [11]). Owing to its comparatively slow convergence it is little used today. Nevertheless its simplicity will serve to describe the general concepts.

The diagonal elements of D, the α_{ii}, are greater than zero, so D is nonsingular. Therefore D^{-1} exists and premultiplying $(R + D + S) u = v$ by D^{-1} we have

$$u = -D^{-1}(R + S)u + D^{-1}v. \tag{3-44}$$

Comparing this with the general linear iterative scheme [Eqn (3-30)] we see that the choice

$$G_k = -D^{-1}(R + S), \qquad M_k = D^{-1} \tag{3-45}$$

characterizes the Jacobi method

$$u^{(k)} = Gu^{(k-1)} + Mv. \tag{3-46}$$

In Eqn (3-46) the subscripts on G and M are discarded to emphasize the stationary nature of this iteration.

The actual execution of the Jacobi method is simple. In Eqn (3-46), with $G = [g_{ij}]$, note that since $G = -D^{-1}(R + S)$ we have $g_{ii} = 0$ for all i. A trial solution $u^{(0)}$ is chosen. The basic single step of the iteration consists in replacing the current value, $u^{(k-1)}$, by the improved value, $u^{(k)}$, obtained from the matrix operations

$$u^{(k)} = -D^{-1}(R + S)u^{(k-1)} + D^{-1}v. \tag{3-47}$$

As an example consider Eqn (3-35). The value of the improved solution $U_{ij}^{(k)}$ is obtained by an augmented weighted average of the old solution at the four neighbours of i, j:

$$U_{ij}^{(k)} = \frac{1}{\beta_0} [\beta_1 U_{i+1,j}^{(k-1)} + \beta_2 U_{i-1,j}^{(k-1)} + \beta_3 U_{i,j+1}^{(k-1)} + \beta_4 U_{i,j-1}^{(k-1)} - h^2 g_{ij}]. \tag{3-48}$$

The order in which one solves for the components $U_{i,j}^{(k)}$ is of no consequence in the computation, although for bookkeeping purposes one may wish to do this in some orderly fashion. On a computer the process requires the retention of $U^{(k-1)}$ until the kth iteration is completed. This is one of the disadvantages of the method.

Equation (3-47) can be expressed in *algebraic* form as

$$u_i^{(k)} = \sum_{j=1}^{N} g_{ij} u_j^{(k-1)} + c_i \tag{3-49}$$

where the u_i denote the elements of u and the c_i the elements of $c = D^{-1}v$.

The concept of matrix reducibility is important here. It was introduced in the previous section. With $A = (\alpha_{ij})$ we say that A, of $Au = v$, is *reducible* if some of the N_1 ($N_1 < N$) components of u are uniquely determined by some N_1 components of v. That is, a permutation matrix P exists, having elements which are either 0 or 1, with *exactly one* unit element in each row and column, such that

$$PAP^T = PAP^{-1} = \begin{bmatrix} A_1 & 0 \\ A_2 & A_3 \end{bmatrix} \tag{3-50}$$

that is, a partitioned representation. In a boundary value problem this means that the same N_1 values of u are independent of a portion of the boundary conditions. Reasonably posed elliptic difference equations should lead to irreducible matrices. Consequently, unless otherwise stated, our discussion will always assume the matrix to be irreducible. Further discussion is found in Forsythe and Wasow [14, p. 208].

(b) Gauss–Seidel method

This iteration was used by Gauss in his calculations (see Gerling [15] for an early description) and was independently discovered by Seidel [16] in 1874.†

---

† In neither case is a fixed cyclic computational order recommended. Today a cyclic order is recommended, as will be seen presently.

This scheme, also known as the method of *successive displacements or iteration by single steps*, is based upon *immediate* use of the improved values. To systematize such a computation the order in which one solves for the components of the kth approximation $u^{(k)}$ must be established beforehand. Such a sequential arrangement is called an *ordering* of the mesh points. For an arbitrary but fixed ordering, which we designate by u_i ($i = 1, 2, \ldots, N$ = number of mesh points), the method of successive displacements is derivable from Eqn (3-49) by employing the improved values when available—that is,

$$u_i^{(k)} = \sum_{j=1}^{i-1} g_{ij} u_j^{(k)} + \sum_{j=i+1}^{N} g_{ij} u_j^{(k-1)} + c_i. \qquad (3\text{-}51)$$

In matrix form Eqn (3-51) becomes

$$Iu^{(k)} = -D^{-1}Ru^{(k)} - D^{-1}Su^{(k-1)} + D^{-1}v$$

or

$$(R + D)u^{(k)} = -Su^{(k-1)} + v. \qquad (3\text{-}52)$$

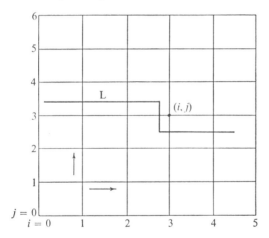

Fig. 3-3 Typical mesh point ordering for Gauss–Seidel method

Equation (3-52) may be rewritten in the form

$$u^{(k)} = -(R + D)^{-1}Su^{(k-1)} + (R + D)^{-1}v. \qquad (3\text{-}53)$$

Thus the *Gauss–Seidel method is a linear stationary iterative scheme.* The matrix $(R + D)^{-1}$ exists since the determinant of $R + D$ is nonzero.

To illustrate this concept we return to Eqn (3-35) and suppose the selected ordering to be $U_{11}, U_{12}, U_{13}, \ldots, U_{2,1}, \ldots$, that is, from bottom to top of the net moving from left to right as in Fig. 3-3. Thus, to calculate $U_{i,j}^{(k)}$ the values of the kth step are used at all points on lines below the jth and on the jth line for x coordinates less than i—that is, below the solid line L in Fig. 3-3. Consequently, the Gauss–Seidel form of Eqn (3-35) becomes

$$U_{ij}^{(k)} = \frac{1}{\beta_0} [\beta_1 U_{i+1,j}^{(k-1)} + \beta_2 U_{i-1,j}^{(k)} + \beta_3 U_{i,j+1}^{(k-1)} + \beta_4 U_{i,j-1}^{(k)} - h^2 g_{ij}]. \qquad (3\text{-}54)$$

Equation (3-54) clearly demonstrates that the latest estimates are employed immediately upon becoming available.

(c) Successive over-relaxation (SOR)

The relaxation expert of the pre-automatic computer days (Southwell [17, 18]) found that it was often desirable to *over-relax*—that is, make a larger change in an unknown than was required to reduce the corresponding residual to zero. Attempts to apply this idea automatically lead to the concept of successive over-relaxation (SOR) by Frankel [19] and Young [20]. This work has been expounded and enlarged by Friedman [21], Keller [22], and Arms, Gates, and Zondek [23]. The number of iterations needed to reduce the error of an initial estimate of the solution by a predetermined factor can often be substantially reduced by an extrapolation process from previous iterants of the Gauss–Seidel method.

The SOR method, as given by Young, carries this out in a straightforward manner. Let the ordering be as discussed in the Gauss–Seidel process; that is, u_i. Let $\bar{u}_i^{(k)}$ be the components of the kth Gauss–Seidel iteration. The SOR technique is defined by means of the relation

$$u_i^{(k)} = u_i^{(k-1)} + \omega[\bar{u}_i^{(k)} - u_i^{(k-1)}]$$
$$= (1 - \omega)u_i^{(k-1)} + \omega\bar{u}_i^{(k)} \tag{3-55}$$

that is to say, the *accepted value* at step k is extrapolated from the Gauss–Seidel value and the previous accepted value. If $\omega = 1$ the method reduces to that of Gauss–Seidel. The quantity ω is called a *relaxation parameter*, the choice of which determines the rapidity of convergence. We will discuss the choice of these parameters later.

Upon substituting Eqn (3-51) into Eqn (3-55) we have the algebraic form of SOR

$$u_i^{(k)} = (1 - \omega)u_i^{(k-1)} + \omega\left\{\sum_{j=1}^{i-1} g_{ij}u_j^{(k)} + \sum_{j=i+1}^{N} g_{ij}u_j^{(k-1)} + c_i\right\}. \tag{3-56}$$

In matrix notation we can write

$$Iu^{(k)} = (1 - \omega)Iu^{(k-1)} - \omega\{D^{-1}Ru^{(k)} + D^{-1}Su^{(k-1)} - D^{-1}v\}. \tag{3-57}$$

Solving for $u^{(k)}$ we have

$$(D + \omega R)u^{(k)} = [(1 - \omega)D - \omega S]u^{(k-1)} + \omega v$$

where I is the $N \times N$ unit matrix. On multiplying by $(D + \omega R)^{-1}$ we find

$$u^{(k)} = (D + \omega R)^{-1}\{[(1 - \omega)D - \omega S]u^{(k-1)} + \omega v\}. \tag{3-58}$$

Other point iterative methods have been developed, but before considering some of these we must examine the convergence question.

PROBLEMS

3-9 Solve the system

$$4x_1 + x_2 + x_3 = -1$$
$$x_1 + 6x_2 + 2x_3 = 0$$
$$x_1 + 2x_2 + 4x_3 = 1$$

by Gauss-Seidel and by SOR (use $\omega = 1.5$). Write both iterations in matrix form —that is, as Eqns (3-53) and (3-58).

3-10 Give the form analogous to Eqn (3-54) when SOR is applied.

3-11 Verify that the Jacobi iteration procedure converges (the new set is exactly the same as its predecessor) when applied to the system

$$x_1 + 2x_2 - 2x_3 = 1$$
$$x_1 + x_2 + x_3 = 3$$
$$2x_1 + 2x_2 + x_3 = 5$$

Show that the Gauss–Seidel method diverges. Thus it is possible to construct systems for which the Jacobi method converges but the Gauss–Seidel process diverges. This result is due to Collatz [24].

Ans: $x_1 = x_2 = x_3 = 1$.

3-12 Show that the Gauss–Seidel method converges for the system

$$5x_1 + 3x_2 + 4x_3 = 12$$
$$3x_1 + 6x_2 + 4x_3 = 13$$
$$4x_1 + 4x_2 + 5x_3 = 13$$

while the Jacobi method diverges.

Ans: $x_1 = x_2 = x_3 = 1$.

3-5 Convergence of point iterative methods

From Eqn (3-33) we see that the convergence of a linear iteration depends on whether

$$H_k e_0 = G_k G_{k-1} \ldots G_1 e_0 \to 0 \quad \text{as } k \to \infty.$$

An iterative method is said to *converge* if, for any given k, each component of the successive iterants $u^{(k)}$ tends to the corresponding component of the solution vector u for *all* initial vectors $u^{(0)}$. In machine computation one almost always has to round numbers thus introducing into $G_1 e_0$ a small component of some of the dominant eigenvectors. Thus $H_k e_0$ will fail to converge to zero for practical, but not for analytical, reasons. Hence, Forsythe and Wasow [14] suggest the following practical rule:

The iteration $u^{(k)} = G_k u^{(k-1)} + M_k v$ will ordinarily be convergent for a given initial error e_0 if and only if $H_k X \to 0$ as $k \to \infty$ for all vectors X. Henceforth, unless otherwise specified, we shall adopt this practical rule as our posture.

Now we recall in the matrix stability analysis of Section 2-5 that errors did not grow if the eigenvalues of P, in $V_{j+1} = PV_j$, were in absolute value less than or equal to 1. That result, made plausible in Section 2-5, is very similar to the convergence of iterative methods which concerns us here. We shall therefore be more precise employing results by various authors summarized, for example, in Forsythe and Wasow [14] and Varga [25].

In Section 3-2 we observed that for stationary linear iterations $G_k = G$, so that $H_k = G^k$ and the convergence question then depends on whether $G^k X \to 0$ for arbitrary X. The rapidity with which $e_k \to 0$ depends upon how large $G^k e_0$ is for large k. We must therefore review the behavior of $G^k X$.

Denote the eigenvalues of G by $\lambda_1, \lambda_2, \ldots, \lambda_N$ where some (or all) of these complex numbers may be equal. From matrix theory (Varga [25]) there exists a nonsingular matrix T such that $J = T^{-1}GT$ is a blockwise diagonal matrix called the *Jordan Canonical form*. Each of the diagonal blocks has the form

$$J_i = \begin{bmatrix} \lambda_i & 1 & & & & \\ & \lambda_i & 1 & & 0 & \\ & & \lambda_i & 1 & & \\ & & & \ddots & & \\ & 0 & & & \lambda_i & 1 \\ & & & & & \lambda_i \end{bmatrix} \tag{3-59}$$

with λ_i as each term on the main diagonal, 1's in the diagonal above the main diagonal, and 0's elsewhere. If the order of each Jordan block J_i is N_i we have $\sum N_i = N$. The importance of the Jordan form is that the oblique coordinate system defined by the columns of T reduces the matrix G to the simple form of a direct sum of, say, m components J_i. Of course, we may decompose the vector X, in the same coordinate system, into $X = X_1 + X_2 + \cdots + X_m$ where $X_i \neq 0$ corresponds to the Jordan block J_i.

We can write $J_i = \lambda_i I + W$, where W is of order N_i and consists of 1's on the diagonal above the main diagonal and 0's elsewhere. The matrix W is *nilpotent*—that is to say, a certain power of W is the zero matrix. In this case we easily see that $W^{N_i} = 0$ and $W^r (r = 1, 2, \ldots, N_i - 1)$ has 1's on the rth diagonal above the main diagonal and 0's elsewhere. Thus we find that for†
$k \geq N_i - 1$

$$J_i^k = (\lambda_i I + W)^k = \lambda_i^k I + \binom{k}{1}\lambda_i^{k-1}W + \cdots + \binom{k}{N_i - 1}\lambda_i^{k-N_i+1}W^{N_i-1}$$

and   (3-60)

$$J_i^k X_i = \lambda_i^k X_i + \binom{k}{1}\lambda_i^{k-1}WX_i + \cdots + \binom{k}{N_i - 1}\lambda_i^{k-N_i+1}W^{N_i-1}X_i.$$

† Here we use the standard notation $\binom{k}{j} = \dfrac{k!}{j!(k-j)!}$.

As $k \to \infty$, the last term, because of the binomial coefficient, is more dominant and we write

$$J_i^k X_i \sim \binom{k}{N_i - 1} \lambda_i^{k - N_i + 1} W^{N_i - 1} X_i \quad \text{as } k \to \infty \qquad (3\text{-}61)$$

with the following asymptotic meaning for vectors: As $k \to \infty$, the notation $U_k \sim V_k$; read as U_k is asymptotic to V_k as $k \to \infty$, means that for any norm (i) $\|U_k\| > 0$ and $\|V_k\| > 0$ for any sufficiently large k; and (ii) $\|U_k - V_k\|/\|V_k\| \to 0$ as $k \to \infty$.†

Any eigenvector Y of the block J_i must satisfy the defining relation $(J_i - \lambda_i I) Y = 0$, that is, $W Y = 0$. Consequently, it is easy to see that

$$Y^T = [k, 0, \ldots, 0] = k[1, 0, \ldots, 0] = k e_1 \qquad (3\text{-}62)$$

so that e_1, up to scalar multiples, is the only eigenvector of the block J_i.

We now examine Eqn (3-61). First, the matrix $W^{N_i - 1}$ is a matrix whose only nonzero element is a 1 in the upper right-hand corner. Thus, $W^{N_i - 1} X_i$ is a multiple of e_1 [see Eqn (3-62)] for all X_i with nonvanishing last element. Consequently, $Y_i = W^{N_i - 1} X_i$ is an eigenvector of J_i. From Eqn (3-61) it therefore follows that $J_i^k X_i \to 0$ as $k \to \infty$ for arbitrary X_i if and only if $|\lambda_i| < 1$. Each block of J behaves in the same manner. Thus we have established the following basic convergence theorem: *Convergence Theorem for Stationary Linear Iterations.*

The stationary linear iteration $u^{(k)} = G u^{(k-1)} + M v$ converges, that is to say $G^k X \to 0$ for an arbitrary X, if and only if each eigenvalue λ_i of G is less than 1 in absolute value.

From Eqn (3-61) we see that what is important is the behavior of those among the Jordan blocks J_i with a maximal value of $|\lambda_i|$, here called λ, which have the largest value of N_i, here called $\bar{N}$. In fact from Eqn (3-61) it follows that

$$e_k \sim \binom{k}{\bar{N} - 1} \sum_i \lambda_i^{k - \bar{N} + 1} Y_i \quad \text{as } k \to \infty \qquad (3\text{-}63)$$

where the sum is taken over all blocks with both $|\lambda_i| = \lambda$ and $N_i = \bar{N}$. Following Householder [26] we call $\lambda(G) = \max |\lambda_i|$ the *spectral radius* of G. In this notation we can restate the convergence theorem as follows: *the stationary linear iteration $u^{(k)} = G u^{(k-1)} + M v$ converges if and only if the spectral radius of G is less than 1.* If the matrix G is *symmetric*, then we know it has N linearly independent eigenvectors Y_i, even though not all the eigenvalues are necessarily distinct. In such a case we can write

$$e_0 = \sum_{i=1}^{N} \gamma_i Y_i \qquad (3\text{-}64)$$

that is, as a linear combination of these eigenvectors. Then also

$$e_k = G^k e_0 = \sum_{i=1}^{N} \gamma_i \lambda_i^k Y_i \qquad (3\text{-}65)$$

† If (i) and (ii) are true in one norm they are true in any norm.

since $GY_i = \lambda_i Y_i$. Thus, it is clearly necessary that the spectral radius be less than 1 for convergence. Moreover, the rate of convergence is *best* when the absolute value of the largest root is near zero and *poorest* when it is near 1.

We shall now examine the convergence of several of the simple iterative methods of Section 3-4. In all of these considerations, unless otherwise stated, it will be assumed that the matrix corresponding to G, of the iteration process, has the properties of the footnote to Eqns (3-40).

Convergence of the Jacobi method will be examined by two processes. First, we employ the maximum operation which proved so effective in Chapter 2, and this is followed by a proof by contradiction.

With the matrix $A = (\alpha_{ij})$ the Jacobi process for $Au = v$, $A = R + D + S$, was $u^{(k)} = Gu^{(k-1)} + Mv$, where $G = -D^{-1}(R + S)$ and $M = D^{-1}$. The error e_k satisfies $e_k = u^{(k)} - A^{-1}v = Ge_{k-1}$. In component form we write this expression as

$$e_{ki} = u_i^{(k)} - u_i = \frac{1}{\alpha_{ii}}\left[-\alpha_{i1}e_{k-1,1} - \alpha_{i2}e_{k-1,2} - \cdots - \alpha_{iN}e_{k-1,N}\right] \quad (3\text{-}66)$$

where u_i is the ith component of the exact solution. We set

$$\theta_i = \sum_{j=1}^{N}{}' |g_{ij}| = \sum_{j=1}^{N}{}' \left|\frac{\alpha_{ij}}{\alpha_{ii}}\right|, \quad i = 1, 2, \ldots, N \quad (3\text{-}67)$$

where the primed summation sign indicates that the term for $j = i$ is omitted. The set θ_i act as measures of the dominance of the coefficients on the main diagonal.

Let $\|e_k\| = \max_i |e_{ki}|$. From Eqn (3-66) we have, taking absolute values and employing the triangle inequality,

$$|e_{ki}| \leq \frac{1}{|\alpha_{ii}|} \sum_{j=1}^{N}{}' |\alpha_{ij}| |e_{k-1,j}| \leq \|e_{k-1}\|\theta_i, \quad i = 1, 2, \ldots, N. \quad (3\text{-}68)$$

Since Eqn (3-68) holds for all i it holds for the largest, from which it follows that

$$\|e_k\| \leq \theta_{\max} \|e_{k-1}\|. \quad (3\text{-}69)$$

Thus each cycle of the iteration accomplishes at least a fixed percentage decrease of the maximum error. The smaller $\theta_{\max}$ is, the faster the convergence will be. We can state this result as follows: *If*

$$\theta_{\max} = \max_i \sum_{j=1}^{N}{}' \frac{|\alpha_{ij}|}{|\alpha_{ii}|} < 1 \quad (3\text{-}70)$$

the Jacobi iteration converges. Strict inequality is required, so this is a more severe restriction on A than that of conditions (3-40). Eqn (3-70) states that a condition of diagonal dominance is *sufficient* for convergence—it is also sufficient for convergence of the Gauss–Seidel iteration (Problem 3-13).

The inequality requirement of Eqn (3-70) suggests that we re-examine the convergence of the Jacobi method. It will converge if the *spectral* radius is less than 1. Consequently we wish to show, with the conditions of the footnote to Eqns (3-40) on A, that any eigenvalue of $G = -D^{-1}(R + S)$ is less than 1 in absolute value. Let us assume that an eigenvalue μ of G exists such that $|\mu| \geq 1$. Corresponding to this eigenvalue there exists an eigenvector w such that $Gw = \mu w$ or $(G - \mu I)w = 0$. From this it follows that

$$\left(I - \frac{1}{\mu}G\right)w = 0. \tag{3-71}$$

If Eqn (3-71) holds, then the determinant of $I - (1/\mu)G$ must be zero since w is an eigenvector of G. From Eqns (3-45) we see that the main diagonal elements of G are all zero and the off-diagonal elements are $g_{ij} = -\alpha_{ij}/\alpha_{ii}$ $(i \neq j)$. Thus, from the conditions on A, the matrix G is irreducible and satisfies

$$1 \geq \sum_{j=1}^{N} |g_{ij}|, \quad i = 1, 2, \ldots, N \tag{3-72}$$

and for some i strict inequality holds.

The matrix $F = I - (1/\mu)G$ has unit diagonal elements and since G is irreducible, the matrix is also irreducible. Finally

$$\sum_{j=1}^{N}{}' |f_{ij}| = \sum_{j=1}^{N}{}' \left|\frac{g_{ij}}{\mu}\right| = \left|\frac{1}{\mu}\right| \sum_{j=1}^{N}{}' |g_{ij}| \leq 1 = |f_{ii}| \tag{3-73}$$

since $|\mu| \geq 1$. Thus the matrix F satisfies the same conditions as the matrix A, whereupon $\det F \neq 0$. Hence w must vanish. We now have a contradiction and the assumption that $|\mu| \geq 1$ is false. As a consequence we have the following result: *If the matrix A, of the linear system $Au = v$, has the properties of the footnote to Eqns (3-40), then the spectral radius of the Jacobi matrix G is less than 1 and the Jacobi method converges.* A similar result holds for the Gauss–Seidel iteration (Problem 3-14).

As previously noted the computational effectiveness of a convergent iterative method is directly related to the magnitude of the spectral radius of the matrix G of the iterative method. The relative magnitudes of the spectral radii of the matrices associated with the Jacobi and Gauss–Seidel methods have been examined by Stein and Rosenberg [27]. They obtained the following result: *If A satisfies conditions (3-40) and G_J, G_S are the matrices associated with the Jacobi and Gauss–Seidel iterations, respectively, then one and only one of the following mutually exclusive relations holds:*

$$
\begin{array}{ll}
\text{(i)} & \lambda(G_J) = \lambda(G_S) = 0\dagger \\
\text{(ii)} & 0 < \lambda(G_S) < \lambda(G_J) < 1 \\
\text{(iii)} & 1 = \lambda(G_S) = \lambda(G_J) \\
\text{(iv)} & 1 < \lambda(G_J) < \lambda(G_S)
\end{array} \tag{3-74}
$$

† Here we employ the notation $\lambda(G)$ = spectral radius of G.

The content of this theorem is that the Jacobi method and the Gauss–Seidel method are either both convergent or both divergent, and *if both converge then the Gauss–Seidel method converges faster.*

A happy circumstance occurs if the system $Au = v$ has a symmetric positive definite matrix A. If such is the case, Seidel [16] observed that the Gauss–Seidel method always converges, without further restrictions on A—of course det $A \neq 0$ and computational difficulties may arise due to such things as size, sparseness, and ill-conditioning.† We shall indicate the proof of this result in essentially the same manner as that employed by Seidel. This type of proof is the forerunner of the popular 'energy' proof currently (1968) in wide usage.

The class of problems known as *extremum* problems is related to linear and nonlinear algebra. An extremum problem consists in determining the set or sets of values $(u_1, u_2, \ldots, u_N)$ for which a given function $\phi(u_1, u_2, \ldots, u_N)$ is a maximum, minimum, or has a saddle point. If the solution domain is not restricted one can, at least in theory, set the N first partial derivatives of ϕ equal to zero and solve the resulting equation simultaneously. Thus we can associate with every extremum problem a set of simultaneous algebraic equations. The converse may not be true since, for a given set of simultaneous equations, it may not be possible to find a function ϕ whose partial derivatives have the same structure as the given set. One can easily verify that the linear system $Au = v$, $A = (\alpha_{ij})$, $u^T = (u_1, \ldots, u_N)$, $v^T = (v_1, \ldots, v_N)$, $i, j = 1, \ldots, N$ is equivalent to the conditions for an extremum of the function

$$\phi = \frac{1}{2} \sum_{i=1}^{N} \sum_{j=1}^{N} \alpha_{ij} u_i u_j - \sum_{i=1}^{N} v_i u_i$$

$$= Q - \sum_{i=1}^{N} v_i u_i. \tag{3-75}$$

The system $Au = v$ is said to be *positive* if all coefficients are real and if the quadratic form Q is nonnegative for all possible combinations of real u_j. The system is *positive-definite* if Q is nonnegative and takes the value zero only when every $u_j = 0$ (see Problem 3-17). Negative and negative-definite systems can be defined, but it is customary to make a preliminary change of sign throughout $Au = v$, whenever necessary, so that only the positive cases need be discussed.

The quadratic function [Eqn (3-75)] has only one extremum if the corresponding linear system $Au = v$ has a unique solution. If the linear system is positive-definite, then the extremum is a minimum.

In studying approximate methods for solving algebraic equations we often find that the function ϕ can be used to measure the accuracy of an approximation. Typical of these situations is the case where it is known that the exact solution corresponds to a minimum ϕ_0 of ϕ. If several approximate solutions

---

† This difficulty is not frequent in partial differential equations and will not be discussed here. For further information, see Todd [28].

are available, the corresponding values of ϕ may be used as a basis for comparison. The closer the value of ϕ to ϕ_0 the better the approximation. Crandall [29] enlarges upon this discussion and gives a number of physical examples.

We shall now prove that the *Gauss–Seidel method always converges when applied to a symmetric* ($\alpha_{ij} = \alpha_{ji}$) *positive-definite system* by employing an associated extremum problem. When the system $Au = v$ is symmetric and positive-definite, the function equation [Eqn (3-75)] has a single extremum which is a true minimum for the set of values that constitute the exact solution. The essence of the proof is to demonstrate that each single step of the iteration reduces ϕ.

From Eqn (3-51) we observe that the fundamental step consists in replacing the value $u_i^{(k-1)}$ by the improved value $u_i^{(k)}$. Thus, by subtracting $u_i^{(k-1)}$ from both sides of Eqn (3-51), we have

$$u_i^{(k)} - u_i^{(k-1)} = \frac{1}{\alpha_{ii}} \left[v_i - \sum_{j=1}^{i-1} \alpha_{ij} u_j^{(k)} - \sum_{j=i}^{N} \alpha_{ij} u_j^{(k-1)} \right]. \qquad (3\text{-}76)$$

With $\phi^{(k)}$ and $\phi^{(k-1)}$ denoting the values of the associated function [Eqn (3-75)] we have

$$\phi^{(k)} - \phi^{(k-1)} = \frac{1}{2} \sum_{i=1}^{N} \sum_{j=1}^{N}{}' \alpha_{ij}[u_i^{(k-1)}(u_j^{(k)} - u_j^{(k-1)}) + u_j^{(k)}(u_i^{(k)} - u_i^{(k-1)})]$$

$$+ \frac{1}{2} \sum_{i=1}^{N} \alpha_{ii}[(u_i^{(k)})^2 - (u_i^{(k-1)})^2] - \sum_{i=1}^{N} v_i(u_i^{(k)} - u_i^{(k-1)})$$

$$= - \frac{\alpha_{ii}}{2} \sum_{i=1}^{N} [u_i^{(k)} - u_i^{(k-1)}]^2 \qquad (3\text{-}77)$$

where the last step follows from the symmetry and Eqn (3-76). Now the main diagonal elements of a positive-definite system are positive (why?), so that Eqn (3-77) indicates a *decrease* in ϕ going from step $(k-1)$ to k. As the iteration continues the process must converge since ϕ, having a minimum, cannot decrease indefinitely. Clearly ϕ cannot stop short of its true minimum for it can cease decreasing only when every $u_i^{(k)} = u_i^{(k-1)}$ for all i—that is, when the iteration reproduces variables without change. Hence the process converges. Another proof is given by Reich [30].

Ostrowski [31] and Varga [25] give proofs of the following theorem:

If A is symmetric, $\alpha_{ii} > 0$, $i = 1, 2, \ldots, N$, then $\lambda(G_\omega) < 1$ if and only if A is positive-definite and $0 < \omega < 2$. Here G_ω is the matrix coefficient of $u^{(k-1)}$ in Eqn (3-58) and ω is the relaxation factor.

PROBLEMS

3-13 Prove that the diagonal dominance condition of Eqn (3-70) is *sufficient* for the convergence of the Gauss–Seidel iteration by employing the maximum operation. Is it a necessary condition?

3-14   If the matrix A has the properties of the footnote to Eqns (3-40) show that the spectral radius of the Gauss–Seidel matrix G is less than 1.

3-15   Does Problem 3-11 contradict the results of Eqn (3-74)? If so, why?

3-16   The general nonlinear system of equations $a_i(u_1, \ldots, u_N) = c_i, i = 1, 2, \ldots,$ N, is equivalent to the conditions for an extremum of $\phi = V - \sum_{i=1}^{N} c_i u_i$, where V is an integral of the a_i only if the a_i satisfy certain integrability relations. What are these integrability relations?

$$Ans: \frac{\partial a_i}{\partial u_j} = \frac{\partial a_j}{\partial u_i}, \quad i, j = 1, 2, \ldots, N.$$

3-17   The set $Au = v$ is positive-definite if it is symmetric and the determinant of the matrix A, and all of it principal minors, are positive; that is,

$$\alpha_{11} > 0, \quad \begin{vmatrix} \alpha_{11} & \alpha_{12} \\ \alpha_{21} & \alpha_{22} \end{vmatrix} > 0, \quad \ldots, \quad \det A > 0$$

(see Guillemin [32]). Use this result to show that the system of Problem 3-12 is positive-definite. Show that the main diagonal coefficients of a positive-definite system must be positive.

3-18   Show that any set of simultaneous linear equations may be transformed into an equivalent symmetric set. (*Hint:* Multiply the first equation by α_{11}, the second by $\alpha_{21}, \ldots,$ and the last by α_{N1} and add. This is the first equation.)

3-19   Compare the convergence of the Jacobi, Gauss–Seidel, and Gauss–Seidel with SOR on the system

$$4u_1 + u_2 = -1$$
$$u_1 + 6u_2 + 2u_3 = 0$$
$$2u_2 + 4u_3 = 0$$

Use $\omega = 1.8$ in the SOR computation.

3-20   Can SOR be used with the Jacobi method? Develop the algebraic and matrix forms of that iteration.

3-21   Use the five-point formula [Eqn (3-15)] for Laplace's equation in any rectangular region with specified boundary values [see, for example, Eqns (3-17)]. Does the Jacobi method converge? Does the Gauss–Seidel method converge?

3-6   Rates of convergence

The proofs of convergence in Section 3-5 gave little, if any, information regarding the rate of convergence. Even if a method converges, it may converge too slowly to be of practical value. Therefore, it is essential to determine the effectiveness of each method. To accomplish this we must consider both the *work required per iteration* and the *number of iterations necessary for convergence*. Within a factor of 2 or 3 the methods considered require essentially the same amount of work per iteration. The number of multiplications required for a cycle of iteration, for the simple methods, is essentially N^2 except in sparse systems where the majority of the coefficients are zero. For example, in systems where each variable is coupled to only one or two other variables the number of operations required per cycle of iteration may be less than $3N$.

In view of the above remarks we shall compare the iterative methods on the basis of the number of iterations required to achieve a specified accuracy. Let the iterative method $u^{(k)} = Gu^{(k-1)} + r$ be consistent, linear, and stationary so that $e_k = u^{(k)} - u = Ge_{k-1}$, as obtained in Section 3-2. Consequently, $e_k = G^k e_0$. For practical purposes we assert that an iterative method has converged when the norm of the error e_k is less than some predetermined number ρ of the norm of the error e_0. To determine a *meaningful* and *useful bound for* ρ we shall briefly review some concepts of linear algebra (see, for example, Varga [25]).

Let u be a vector with N complex components, $A = (\alpha_{ij})$ be an $N \times N$ complex matrix. Let u^* and A^* be the conjugate transposes of u and A respectively. The *Euclidean norm* of u is defined to be

$$\|u\| = [u^*u]^{1/2} = \left[\sum_{i=1}^{N} |u_i|^2 \right]^{1/2} \tag{3-78}$$

and the *Spectral norm* of A is defined to be

$$\|A\| = [\lambda(A^*A)]^{1/2} \tag{3-79}$$

where $\lambda(A^*A)$ is the spectral radius of A^*A. The spectral norm and spectral radius are related in the following manner:

(i) If A is Hermitian (i.e. $A^* = A$), then $\|A\| = \lambda(A)$
(ii) If A is arbitrary, then $\|A\| \geq \lambda(A)$
(iii) If $\lambda(A) < 1$ and n is large, then $\|A^n\| \approx [\lambda(A)]^n$.

From the definitions [Eqns (3-78) and (3-79)] it can be shown that

$$\|A\| = \sup_{x \neq 0} \frac{\|Ax\|}{\|x\|}. \tag{3-80}$$

Employing Eqn (3-80) we have

$$\|e_k\| = \|G^k e_0\| \leq \|G^k\| \, \|e_0\|. \tag{3-81}$$

Consequently, if e_0 is not the zero vector, $\|G^k\|$ provides an *upper bound* for the ratio $\rho = \|e_k\|/\|e_0\|$. Equation (3-63) also implies that $\|e_k\|$ behaves roughly like $ck^{\bar{N}-1}\lambda^k$ as $k \to \infty$, where c is a constant. For large values of k, $\|e_{k+1}\|/\|e_k\|$ averages to λ. Therefore, on the average, the error decreases by the factor λ at each step in the iteration. If $\|e_{k+1}\|/\|e_k\|$ were exactly λ ($\lambda < 1$) for all k, then $-\log_{10} \lambda$ would be the number of decimal digits of accuracy gained in each iterative step. If $\|e_{k+1}\|/\|e_k\| \to \lambda$, as $k \to \infty$, then $-\log_{10} \lambda$ would be the asymptotic number of decimal digits gained. Actually $\|e_{k+1}\|/\|e_k\|$ averages to λ, so the number

$$R(G) = -\log \lambda(G) \tag{3-82}$$

has been introduced by Young [20] as the *rate of convergence* of the linear iteration $u^{(k)} = Gu^{(k-1)} + r$, characterized by the matrix G.

Upon replacing $\|G^k\|$ by $[\lambda(G)]^k$ in Eqn (3-81) we see that $\rho \approx [\lambda(G)]^k$. *Thus a good approximation to the number of iterations required to reduce the initial error by a factor ρ is*

$$k = \frac{-\log \rho}{-\log \lambda(G)} = \frac{-\log \rho}{R(G)}. \tag{3-83}$$

Comparison of various iterative processes will be facilitated by employing a simple example. Consider the square $R: 0 < x < \pi, 0 < y < \pi$ in which the solution of the Dirichlet problem for the Laplace equation is to be obtained by employing the five-point molecule of Eqn (3-14). With $h = \pi/n$, R is divided into n^2 square nets with $N = (n-1)^2$ interior points. Let A_h denote the square matrix formed from the five-point formula on the left-hand side of Eqn (3-14). Let the N eigenvectors of $-A_h$ be denoted by $X_{pq}(ih, jh)$ with corresponding positive eigenvalues μ_{pq}, $p, q = 1, 2, \ldots, n-1$. From the defining relation

$$-A_h X_{pq} = \mu_{pq} X_{pq} \tag{3-84}$$

we have

$$X_{pq} = \sin pih \sin qjh \tag{3-85}$$

$$\mu_{pq} = 4h^{-2} \left[\sin^2 \left(\frac{ph}{2} \right) + \sin^2 \left(\frac{qh}{2} \right) \right]. \tag{3-86}$$

Clearly

$$\min \mu_{pq} = \mu_{1,1} = 8h^{-2} \sin^2 \frac{h}{2} \to 2 \quad \text{as } h \to 0$$

$$\tag{3-87}$$

$$\max \mu_{pq} = \mu_{n-1,n-1} = 8h^{-2}\left(1 - \sin^2 \frac{h}{2}\right) \sim \frac{8}{h^2} \quad \text{as } h \to 0.$$

The convergence of the method of Jacobi is controlled by the eigenvalues of $G = -D^{-1}(R + S)$. Since $A = R + D + S$, we have

$$G = -D^{-1}(A - D) = I - D^{-1}A. \tag{3-88}$$

In our sample problem $A = -A_h$ and $D = (4/h^2)I$ so that

$$G = I - \frac{h^2}{4}(A_h).$$

Since the eigenvalues of A_h are given by Eqn (3-86), the eigenvalues λ_{pq} of G are

$$\lambda_{pq} = 1 - \sin^2 \left(\frac{ph}{2} \right) - \sin^2 \left(\frac{qh}{2} \right)$$

$$= \frac{\cos ph + \cos qh}{2}, \quad p, q = 1, 2, \ldots, n-1.$$

Consequently, the spectral radius is

$$\lambda(G) = \max |\lambda_{pq}| = \cos h \sim 1 - \frac{h^2}{2} \quad \text{as } h \to 0$$

and the rate of convergence is

$$R(G) = -\log(\cos h) \sim -\log\left(1 - \frac{h^2}{2}\right) = \frac{h^2}{2} + O(h^4). \tag{3-89}$$

Therefore the rate of convergence of the Jacobi iteration is approximately $h^2/2$, which is rather slow for small values of h.

The convergence of the Gauss–Seidel iteration is examined in a similar manner. The dominant eigenvalue is found to be (Problem ? 23)

$$\lambda(G) = \max |\lambda_{pq}| = \cos^2 h \sim 1 - h^2, \quad \text{as } h \to 0.$$

Consequently, the rate of convergence is

$$R(G) = -\log \cos^2 h \sim h^2 + O(h^4) \tag{3-90}$$

which is *twice* as fast as that for the Jacobi method.

In the next section we shall see that the Gauss–Seidel method, accelerated by SOR, has

$$\lambda_{\text{opt}}(G) \sim 1 - 2h, \quad \text{as } h \to 0$$

so that

$$R_{\text{opt}}(G) \sim 2h + O(h^2). \tag{3-91}$$

The rate of convergence is now asymptotically $2h$, larger than that of Gauss–Seidel by the factor $2/h$.

As a numerical example we take $n = 30$, corresponding to a Dirichlet problem with 851 interior nodes. For the solution of the problem we have the following spectral radii:

Iteration of Jacobi: $\qquad \lambda = \cos(\pi/30) = \cos 6° = 0.9945$
Iteration of Gauss–Seidel: $\lambda = \cos^2 \pi/30 = \cos^2 6° = 0.9890$
SOR (optimum): $\qquad\quad \lambda = 1 - 2(\pi/30) = 0.7906.$

Therefore the error in the approximate solution is reduced by 0.55, 1.1, and 21 per cent per iteration, respectively. Optimum SOR clearly pays very well with little additional computational effort.

The number of iterations required to reduce the initial error by a factor ρ is given in Eqn (3-83). For the individual cases of the preceding paragraph we have

$$-\frac{2 \log \rho}{h^2}, \qquad -\frac{\log \rho}{h^2} \quad \text{and} \quad -\frac{\log \rho}{2h}$$

respectively.

In studying the convergence of an actual iterative process for solving $Au = v$, it is usually easier to consider the *residuals*

$$R_k = v - Au^{(k)} \tag{3-92}$$

than to calculate the errors e_k. This occurs because the R_k can be computed *without knowing* the solution $A^{-1}v$, while the e_k cannot. From the definition of e_k [Eqn (3-31)] we have $e_k = u^{(k)} - A^{-1}v$. Consequently,

$$Ae_k = Au^{(k)} - v = -R_k \tag{3-93}$$

and therefore, since $e_k = G_k e_{k-1}$,

$$R_k = -AG_k e_{k-1} = -AG_k A^{-1} A e_{k-1}$$
$$= (AG_k A^{-1})R_{k-1}. \tag{3-94}$$

Upon repetitive application of Eqn (3-94) we finally obtain

$$R_k = (AG_k G_{k-1} \ldots G_1 A^{-1})R_0 \tag{3-95}$$

which becomes

$$R_k = AG^k A^{-1} R_0 = (AGA^{-1})^k R_0 \tag{3-96}$$

in the case of a linear stationary iteration.

It is an easy matter to show that AGA^{-1} has the same Jordan canonical form as G, whereupon Eqn (3-63) implies that

$$R_k \sim -\binom{k}{\overline{N}-1} \sum_i \lambda_i^{k-\overline{N}+1} A Y_i, \quad \text{as } k \to \infty. \tag{3-97}$$

Consequently, the convergence criteria for R_k are the same as those for e_k.

PROBLEMS

3-22   Carry out the details of the solution for the eigenvalue–eigenvector problem [Eqns (3-84), (3-85), and (3-86)].

3-23   For the Dirichlet problem of Laplace's equation in $0 < y < \pi, 0 < x < \pi$, show that the spectral radius of the Gauss–Seidel iteration is $\lambda = \cos^2 h$.

3-24   Consider the same problem as that of Problem 3-23 on the rectangle $0 < x < a = rh, 0 < y < b = sh$. Show that the spectral radius of the Gauss–Seidel iteration is $\lambda = [\frac{1}{2}(\cos \pi/r + \cos \pi/s)]^2$ (Young [20]).

3-25   Complete the details of all steps leading from Eqn (3-93) to Eqn (3-96).

3-26   Let the maximum change in any variable be denoted by $\|u^{(k+1)}\| = \max_j |u_j^{(k+1)} - u_j^{(k)}|$. In a Jacobi iteration show that this value is greater than or equal to $\|e_k\| - \|e_{k+1}\|$ and hence find an upper bound for the absolute error at each stage (Collatz [24]). (*Hint*: Use the triangle inequality and Eqn (3-69).)

$$\textit{Ans:} \quad \|e_{k+1}\| \leq \frac{\theta_{max}}{1 - \theta_{max}} \|u^{(k+1)}\|.$$

3-7 Accelerations—successive over-relaxation (SOR)

In real problems it is normal for iterative processes to be slowly convergent. That is to say, $\lambda(G)$ is only slightly less than 1 and the rate of convergence is almost zero. The success of most iterations depends upon the application of special techniques called *accelerations* to hasten the progress of e_k to zero. In this section we shall discuss a basic acceleration formula due to Lyusternik [33] and the SOR technique.

Lyusternik [33] studied the simplest iterations in which the unique largest eigenvalue is real, $|\lambda_i| = \lambda$, with the corresponding $\bar{N} = 1$. Consequently, from Eqn (3-63), $e_k \sim \lambda_i^k Y_i$, so that

$$e_{k+1} \sim \lambda_i e_k, \quad \text{as } k \to \infty. \tag{3-98}$$

Even though λ is near 1 we can still employ Eqn (3-98) to substantially reduce the error e_{k+1}. From the definition, $A^{-1}v = u^{(k)} + e_k$, therefore it follows, from Eqn (3-98), that

$$A^{-1}v = u^{(k+1)} + \lambda_i e_k + \epsilon_k, \quad \|\epsilon_k\| \text{ small.}$$

Upon eliminating e_k we find that the solution $A^{-1}v$ is related to $u^{(k)}$ and $u^{(k+1)}$ through the relation

$$A^{-1}v = \frac{u^{(k+1)} - \lambda_i u^{(k)}}{1 - \lambda_i} + \frac{\epsilon_k}{1 - \lambda_i}. \tag{3-99}$$

Thus the vector

$$(1 - \lambda_i)^{-1}[u^{(k+1)} - \lambda_i u^{(k)}] \tag{3-100}$$

is very close to the solution, provided only that $\|\epsilon_k\|$ is small compared to $1 - \lambda_i$.

In practice, the Lyusternik acceleration is employed in the following manner: Let $u^{(k)}$ be the accepted value from the kth step of the iteration and compute $\bar{u}^{(k+1)} = Gu^{(k)} + r_k$; calculate the new value $u^{(k+1)}$ by means of

$$u^{(k+1)} = \frac{1}{1 - \lambda_i} [\bar{u}^{(k+1)} - \lambda_i u^{(k)}] \tag{3-101}$$

and repeat. This is clearly of the one-parameter relaxation form

$$u^{(k+1)} = \alpha \bar{u}^{(k+1)} + (1 - \alpha)u^{(k)}, \quad \alpha = \frac{1}{1 - \lambda_i}$$

with α large. We have previously observed a similar form in SOR [see Eqn (3-55)].

The successive over-relaxation scheme of Young–Frankel, discussed in Section 3-4 has several advantages, one of which is its extreme simplicity. The great contribution of this idea, like that of Lyusternik, lies in its acceleration of the convergence of an already convergent process. When the matrix A

has certain properties the application of SOR is found to be very worthwhile. The success of SOR has stimulated many studies aimed at extending its applicability. In this section we shall sketch these studies and briefly indicate proofs.

Young [20] showed that SOR is highly useful if the matrix A possesses a condition he called *Property (A)*. The matrix A is said to have Property (A) if there exists a permutation matrix π (of its rows and corresponding columns) such that

$$\pi A \pi^T = \begin{bmatrix} D_1 & F \\ G & D_2 \end{bmatrix} \tag{3-102}$$

where D_1, D_2 are square diagonal matrices and F and G are rectangular matrices. The arithmetic of the Jacobi method, and therefore the eigenvalues and rate of convergence, is independent of the order in which the points of the mesh are scanned. However, the Gauss–Seidel method and its over-relaxation refinement clearly depend upon the order in which one solves for the unknowns u_i of $Au = v$. If A has Property (A), by rearrangement (if necessary) of the rows and corresponding columns of A we may obtain the form of Eqn (3-102). More generally, we say that any ordering of the rows and columns of A is *consistent* if, starting from that ordering, we can permute the rows and corresponding columns in such a way that if $\alpha_{ij} \neq 0$, the ordering relation between the ith row and jth column is unchanged so that the matrix has the form (not unique)

$$\begin{bmatrix} D_1 & F_1 & 0 & \ldots & 0 & 0 \\ G_1 & D_2 & F_2 & \ldots & 0 & 0 \\ 0 & G_2 & D_3 & \ldots & 0 & 0 \\ & & \ddots & & 0 & 0 \\ 0 & 0 & \ldots & G_{p-2} & D_{p-1} & F_{p-1} \\ 0 & 0 & \ldots & 0 & G_{p-1} & D_p \end{bmatrix} \tag{3-103}$$

Here D_i are square diagonal $r_i \times r_i$ matrices and the F_i and G_i are $r_i \times r_{i+1}$ and $r_{i+1} \times r_i$ rectangular matrices, respectively.

We consider the five-point formula [Eqn (3-14)] for Laplace's equation to provide examples of the form of Eqn (3-103). Let us number the points as shown in Fig. 3-4(a). If we take the ordering 1; 4, 2; 7, 5, 3; 8, 6; 9—that is, along the diagonals—the matrix A would have the form shown in Fig. 3-4(b). The crosses indicate those positions which have nonzero coefficients. The matrices D_1, D_2, D_3, D_4, and D_5 are 1×1, 2×2, 3×3, 2×2, and 1×1 respectively. F_1 is 1×2 and G_1 is 2×1, etc. This ordering is clearly of the form (3-103)—that is, the matrix for the five-point molecule has Property (A). Other orderings are left for Problems 3-27 and 3-28.†

---

† The situation when A does not have Property (A) will be discussed later.

It is *not necessary* to take the equations in the order shown in Fig. 3-4(b) or in the Problems or in any order of form (3-103), provided that we take them in an order that is consistent with one of these acceptable forms. The consistency requirement ensures that the chosen arithmetic is *exactly* the same as that of an acceptable choice. Thus, the consistent ordering of Fig. 3-4(b) with that of Problem 3-28 implies identical arithmetic and hence the same eigenvalues. Property (A) enables Fig. 3-4(b) and Problem 3-27, involving quite different arithmetic, to give rise to the same eigenvalues.

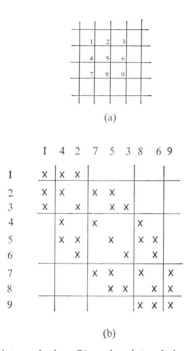

Fig. 3-4 **(a)** Mesh point numbering; **(b)** mesh point ordering along the diagonals

Young's results can be broken down into a sequence. In this discussion we write G_J, G_S, and G_ω as the iteration matrices for Jacobi, Gauss–Seidel, and SOR methods respectively.

(a) Let A have Property (A), then the eigenvalues of G_J are either zero or occur in pairs $\pm\ \mu_i$.

(b) Let A have Property (A) and be consistently ordered. Let $\omega \neq 0$ be any real number. If $\lambda \neq 0$, then λ is an eigenvalue of G_ω if and only if there exists an eigenvalue μ_i of G_J such that

$$\frac{(\lambda + \omega - 1)^2}{\lambda} = \omega^2 \mu_i^2. \tag{3-104}$$

This result can be developed by writing the eigenvalue problem for G_ω—that is,

$$\lambda(I - \omega R)Y = \{(1 - \omega)I + \omega S\}Y$$

in difference equation form—and solving. Since the eigenvalues of G_J are independent of the ordering of the matrix, the *eigenvalues of G_ω are the same for all consistent orderings.*

If $\omega = 1$, then SOR becomes the Gauss–Seidel method of successive displacement and we find $\lambda_i = \mu_i^2$. While some of the roots of G_S may be zero, the others are the squares of the corresponding roots for the Jacobi process. Thus we have the following spectral radius relation:

$$\lambda(G_S) = [\lambda(G_J)]^2 \tag{3-105}$$

from which it follows that

$$R(G_S) = 2R(G_J) \tag{3-106}$$

that is to say, the rate of convergence of the Gauss–Seidel method is just twice that of the Jacobi method when A has Property (A) and is consistently ordered.

However, we are more interested in selecting the best value ω of ω_b in the sense that the largest of the λ_i is a minimum with respect to ω. If A is symmetric, then G_J, while not necessarily symmetric, is similar to a symmetric matrix. With this condition and result (a) above, the eigenvalues of G_J are real and occur in pairs. Let $\lambda(G_J)$ be the spectral radius of the Jacobi method. By examining the mapping defined by Eqn (3-104), between the μ and λ planes, we can obtain our result. In fact, Eqn (3-104) shows that the λ_i are the roots λ of the equation†

$$\lambda - \omega\mu_i\lambda^{1/2} + \omega - 1 = 0, \quad \mu_i \text{ real}. \tag{3-107}$$

From this we get two values of $\lambda^{1/2}$

$$\lambda^{1/2} = \tfrac{1}{2}\{\omega\mu_i \pm \sqrt{[\omega^2\mu_i^2 - 4(\omega - 1)]}\}. \tag{3-108}$$

If $\mu_i = 0$ we obtain $\lambda = -(\omega - 1)$. Since the eigenvalues of G_J occur in (real)‡ pairs $\pm\mu_i$, take $\mu_i > 0$ and suppose $0 < \mu_i < 1$ for all i. For real roots $\lambda^{1/2}$, the larger is

$$\lambda^{1/2} = \tfrac{1}{2}\{\omega\mu_i + \sqrt{[\omega^2\mu_i^2 - 4(\omega - 1)]}\}.$$

By differentiation we find that

$$\frac{d\lambda^{1/2}}{d\omega} = -\frac{1}{2}\left[\frac{1 - \mu_i\lambda^{1/2}}{\lambda^{1/2} - \omega\mu_i/2}\right]. \tag{3-109}$$

---

† If the negative square root is selected it will lead to two other values of $\lambda^{1/2}$, but the same values of λ are obtained on squaring Eqn (3-108).

‡ A theory exists for the case of complex μ_i but, owing to its complications, it is omitted here (see Varga [25]).

At $\omega = 1$, $\lambda^{1/2} = \mu_i$. As ω increases from 1 the quantity $\lambda^{1/2}$ decreases provided $\lambda^{1/2} > \omega\mu_i/2$ as long as $\omega^2\mu_i^2 - 4(\omega - 1) > 0$. However, since $\mu_i < 1$, ω will eventually take a value, say $\omega = \omega_i$, such that

$$\omega_i^2\mu_i^2 - 4(\omega_i - 1) = 0, \quad 1 < \omega_i < 2. \tag{3-110}$$

For $\omega = \omega_i$, Eqn (3-107) has a double root $\lambda^{1/2} = (\omega_i - 1)^{1/2}$. For $\omega > \omega_i$, the product of the two roots is $\omega - 1 > \omega_i - 1$. Consequently, the absolute value of *at least* one of the two roots is greater than $(\omega_i - 1)^{1/2}$.

Consequently, if $0 < \mu_i < 1$ the minimum value (for $1 \leq \omega$) attained by the larger root $\lambda^{1/2}$ of Eqn (3-107) is found when Eqn (3-110) holds—that is, when

$$\omega = \omega_i = 2\mu_i^{-2}[1 - \sqrt{(1 - \mu_i^2)}] = \frac{2}{1 + \sqrt{(1 - \mu_i^2)}}. \tag{3-111}$$

For this value of ω_i the minimum value of $\lambda^{1/2}$ is $(\omega_i - 1)^{1/2}$, and the minimum value of λ is

$$\omega_i - 1 = \frac{1 - \sqrt{(1 - \mu_i^2)}}{1 + \sqrt{(1 + \mu_i^2)}}. \tag{3-112}$$

We leave it for Problem 3-31 to show that this value [Eqn (3-112)] is the least obtainable value for all ω $(-\infty < \omega < \infty)$. Therefore for *fixed* μ_i the quantity in Eqn (3-111) is the best choice for ω.

The whole set of eigenvalues occurs in pairs. Thus, as before, we need consider only the positive members. The *largest root* λ, equal to $\mu_1^2 = \max_i \mu_i^2$ when $\omega = 1$, is the critical value. As ω increases from 1 it decreases as long as $\omega < \omega_1$. *The optimum value ω_b of ω in the sense that $\lambda(G_\omega)$ is a minimum is*

$$\omega_b = \omega_1 = \frac{2}{1 + \sqrt{(1 - \mu_1^2)}} = \frac{2}{1 + \sqrt{[1 - \lambda^2(G_J)]}}. \tag{3-113}$$

This value of ω is larger than $\omega_2, \omega_3, \ldots$ so that the roots λ_i corresponding to *all other* root pairs $\pm \mu_i$ are complex with absolute value $\omega_b - 1$. That is, with $\omega = \omega_b$ *all* eigenvalues of G_{ω_b} have modulus $\omega_b - 1$.

(c) Assume that all μ_i of G_J are real with $|\mu_i| < 1$. The optimum value of ω for which $\lambda(G_\omega)$ is minimized is the value given in Eqn (3-113). The corresponding value of $\lambda(G_{\omega_b})$ is

$$\omega_b - 1 = \frac{1 - \sqrt{[1 - \lambda^2(G_J)]}}{1 + \sqrt{[1 - \lambda^2(G_J)]}}. \tag{3-114}$$

If ω is on the range $\omega_b \leq \omega < 2$, all eigenvalues of G_ω have the same modulus $\omega - 1$; that is (Problem 3-32),

$$\lambda(G_\omega) = \omega - 1, \quad \omega_b \leq \omega < 2 \tag{3-115}$$

In the example problem of Section 3-6 we found $\lambda(G_J) = \cos h \sim 1 - (h^2/2)$ as $h \to 0$. Consequently, for optimum SOR, we get the spectral radius as

$$\lambda(G_{\omega_b}) = \frac{1 - \sin h}{1 + \sin h} \sim 1 - 2h, \quad \text{as } h \to 0. \tag{3-116}$$

The rate of convergence is approximately $2h$. This is larger than that for the Gauss–Seidel method by $2h^{-1}$. *Computing time will therefore be reduced by the factor $2h^{-1}$ (asymptotically)—a very substantial amount!*

If $\lambda(G_J) > 1$, however, it can be shown that $\lambda(G_{\omega_b}) > 1$ and we do not obtain convergence with this method.

In the practical application of SOR, determination of the optimal ω is perhaps the most important problem. We add here a few useful remarks.

Let A be a symmetric matrix having Property (A).

(i) It is better to overestimate ω_b than to underestimate it by the same amount. This was first observed, and proved, by Young [20]. Overestimation of the true ω_b has a smaller adverse effect on the rate of convergence than an underestimation since the curve of $\lambda(G_\omega)$ versus ω has a slope of 1 for $\omega > \omega_b$ but an infinite slope as $\omega \to \omega_b^-$ (Problem 3-34). Carré [35], for example, found that if $\omega_b = 1.9$ then $\lambda(G_J) = 0.9986$ while $\lambda(G_S) = 0.9972$ and $\lambda(G_{\omega_b}) = 0.9$. But if we take $\omega = 1.875$, then $\lambda(G_\omega) = 0.9498$.

(ii) Nevertheless, we can estimate ω_b only by approaching it from below, for if $\omega > \omega_b$ the corresponding roots μ are complex. In practice one carries out a few iterations with $\omega = 1$ (Gauss–Seidel) in order to obtain an estimate for $\lambda(G_S)$ by means of Eqn (3-83). Then, since $\lambda(G_S) = \lambda^2(G_J)$ [Eqn (3-105)] ω_b can be estimated by Eqn (3-113). The accurate determination of the optimum value is slow. Carré suggests replacing Eqn (3-113) by

$$\omega_b = 2\left\{1 + \sqrt{\left[1 - \frac{(\lambda + \omega - 1)^2}{\lambda\omega^2}\right]}\right\}^{-1}$$

where λ is the largest eigenvalue of the iteration matrix corresponding to that particular ω.

(iii) One could carry out several iterations with various $1 \leq \omega < 2$ and observe the number of iterations for convergence. That value of ω yielding the minimum number is taken as ω_b. This is a useful practice if a problem is to be repeated a large number of times, changing only, say, boundary conditions. Bellman, Juncosa, and Kalaba [36] found this to be a useful device in a nonlinear problem which will be discussed subsequently.

(iv) Determination of the optimum over-relaxation parameter ω_b is an important and often difficult part of the problem. Two basic approaches are used. One is to carry out a number of SOR iterations with some $\omega < \omega_b$ and then, on the basis of numerical results, obtain a new estimate of ω_b. This procedure has been utilized by many, as discussed previously. New versions are due to Kulsrud [79] and Reid [80]. A second approach is to obtain an estimate using the 'power' method prior to carrying out the SOR iterations. Recent studies include those of Rigler [81], Wachspress [82], and Hageman and Kellogg [83].

PROBLEMS

3-27 Let the ordering of the points of Fig. 3-4(a) be 1, 3, 5, 7, 9; 2, 4, 6, 8. Set up the matrix corresponding to Fig. 3-4(b). Is it of the form (3-103)? What inference can you draw about the uniqueness of form (3-103)?

3-28 Repeat Problem 3-27 for the natural ordering 1, 2, 3; 4, 5, 6; 7, 8, 9.

3-29 Let the unique largest eigenvalue of the iteration matrix G be real. Employing Eqn (3-98) develop the Aitken [34] acceleration formula

$$u^{(k+2)} = \frac{u^{(k)}\bar{u}^{(k+2)} - \{u^{(k+1)}\}^2}{u^{(k)} - 2u^{(k+1)} + \bar{u}^{(k+2)}}.$$

Here $u^{(k)}$, $u^{(k+1)}$ are accepted results from the k and $(k+1)$ steps and $\bar{u}^{(k+2)} = Gu^{(k+1)} + r_k$.

3-30 Can the Lyusternik or Aitken methods be used if the eigenvalue of largest absolute value is complex? Why? (*Hint*: In Eqn (3-63) there are two terms of equal magnitude.)

3-31 Show from Eqn (3-109) that the value in Eqn (3-112) is the least that can be obtained for all $\omega(-\infty < \omega < \infty)$.

3-32 Verify the truth of Eqn (3-115).

3-33 By employing Eqn (3-114) show that as $\lambda(G_J)$ tends to zero, $R(G_{\omega_b})$ becomes asymptotic to $2\sqrt{[R(G_S)]}$.

3-34 Show that the graph of $\lambda(G_\omega)$ versus ω has a slope of 1 for $\omega > \omega_b$ but an infinite slope as $\omega \to \omega_b^-$. Plot the quantity $\lambda(G_\omega)$ versus ω.

3-8 Extensions of SOR

Method of Garabedian

An alternative approach, which does not depend upon Property (A), was conceived by Garabedian [37]. In this heuristic interpretation the SOR solution of the difference equation approximating the Dirichlet problem for the Laplace equation is thought of as the solution, by difference methods, of a *'time' dependent problem* for $(\partial u/\partial t) = L_h(u)$. $L_h(u)$ is an operator *different* from that of Laplace. The great value of this contribution is its applicability to various finite difference analogs of Laplace's equation regardless of whether or not they possess Property (A). We shall therefore briefly describe the idea.

Consider the five-point molecule [Eqn (3-14)] for solving Laplace's equation, in a region R, possessing linear boundary conditions of the form $au + bu_n = 0$. Using the ordering of Eqn (3-54) we have

$$U^{(k+1)}(x, y) = (1 - \omega)U^{(k)}(x, y) + \frac{\omega}{4}[U^{(k+1)}(x - h, y)$$

$$+ U^{(k+1)}(x, y - h) + U^{(k)}(x + h, y) + U^{(k)}(x, y + h)]. \quad (3\text{-}117)$$

We now suppose that $U^{(k)}$ represents the value at $t = k \Delta t = kp$ of some time-dependent function $U(x, y, t)$. Thus Eqn (3-117) can be rewritten as

$$U(x, y, t + p) = (1 - \omega) U(x, y, t) + \frac{\omega}{4}[U(x - h, y, t + p)$$

$$+ U(x, y - h, t + p) + U(x + h, y, t) + U(x, y + h, t)]. \quad (3\text{-}118)$$

When the leading terms of the formal Taylor series expansion of Eqn (3-118) about (x, y, t) are retained, we have

$$U + pU_t + \cdots = (1 - \omega)U + \frac{\omega}{4}[U - hU_x + \frac{h^2}{2} U_{xx} + pU_t$$

$$- hpU_{xt} + \cdots + U - hU_y + \frac{h^2}{2} U_{yy} + pU_t - hpU_{yt} + \cdots]. \quad (3\text{-}119)$$

On carrying out the algebra and discarding terms of degree higher than 2 in h, p, and $(2 - \omega)$, we have

$$p\left(1 - \frac{\omega}{2}\right)U_t = \frac{\omega}{4}[h^2(U_{xx} + U_{yy}) - hp(U_{xt} + U_{yt})]. \quad (3\text{-}120)$$

Now divide Eqn (3-120) by $\omega hp/4$ and set

$$\alpha = \frac{1}{2}\frac{1 - \omega/2}{(\omega/4)h} = \frac{2 - \omega}{\omega h}, \qquad \beta = \frac{h}{p}$$

whereupon Eqn (3-120) becomes

$$2\alpha U_t = \beta(U_{xx} + U_{yy}) - (U_{xt} + U_{yt}). \quad (3\text{-}121)$$

We can interpret this formalism to mean that Eqn (3-118) is a difference equation which approximates the 'time' dependent *hyperbolic* equation (3-121). Garabedian assumes that h and p are so small and that an appropriate ratio β (see Chapter 4) is so selected that the solution of the difference equation [Eqn (3-118)] behaves approximately like the exact solution of Eqn (3-121) in the region R of the (x, y) plane. The problem is therefore to choose $\alpha = \alpha(\omega)$ so that the true solution of Eqn (3-121) converges most rapidly, as $t \to \infty$, to the steady-state solution of $u_{xx} + u_{yy} = 0$ in R with the same boundary conditions.

Equation (3-121) is solved by introducing the new variable $s = \beta t + x/2 + y/2$ and separating the resulting equation

$$U_{xx} + U_{yy} - \tfrac{1}{2}U_{ss} - 2\alpha U_s = 0. \tag{3-122}$$

The series solution is

$$U = U_0(x, y) + \sum_{n=1}^{\infty} (a_n e^{-p_n s} + b_n e^{-q_n s})U_n(x, y) \tag{3-123}$$

where the U_n are the eigenfunctions of $-\nabla^2 U_n = \mu_n U_n$, $0 < \mu_1 \le \mu_2 \le \cdots$ with $aU + bU_n = 0$ on the boundary of R. The quantities p_n and q_n are

$$p_n = 2\alpha - \sqrt{(4\alpha^2 - 2\mu_n)}, \qquad q_n = 2\alpha + \sqrt{(4\alpha^2 - 2\mu_n)}.$$

As $t \to \infty$, $s \to \infty$ and we see from Eqn (3-123) that

$$U - U_0 \sim a_1 e^{-p_1 s} U_1(x, y) = a_1 e^{-p_1 \beta t} e^{-p_1(x/2 + y/2)} U_1(x, y)$$

since p_1 is the coefficient possessing the smallest real part of all the numbers p_i, q_i. *For the most rapid convergence* we must choose α so that the real part of p_1 is as large as possible. The real part of p_1 is maximized when the radical is zero—that is, when

$$\alpha = \alpha_{\text{opt}} = \left(\frac{\mu_1}{2}\right)^{1/2}. \tag{3-124}$$

From the definition of α we find that

$$\omega_{\text{opt}} = \frac{2}{1 + h\alpha_{\text{opt}}} = 2(1 - h\alpha_{\text{opt}}) + O[(h)^2]. \tag{3-125}$$

Some examples are demonstrative:

(a) Five-point molecule, Dirichlet problem, $R: 0 < x < \pi$, $0 < y < \pi$

$$\omega_{\text{opt}} = \frac{2}{1 + h} = 2 - 2h + O(h^2).$$

Five-point molecule, Dirichlet problem, $R: 0 < x < 1$, $0 < y < 1$

$$\omega_{\text{opt}} = \frac{2}{1 + \pi h} = 2 - 2\pi h + O(h^2).$$

(b) Nine-point molecule (see Fig. 3-5), Dirichlet problem, $R: 0 < x < \pi$, $0 < y < \pi$.

The nine-point approximation for Laplace's equation is

$$4[U(x - h, y) + U(x, y - h) + U(x + h, y) + U(x, y + h)]$$
$$+ U(x - h, y - h) + U(x - h, y + h) + U(x + h, y - h)$$
$$+ U(x + h, y + h) - 20U(x, y) = 0 \tag{3-126}$$

whose local discretization error is $O(h^6)$. This is, therefore, an extremely accurate approximation to use when solving Laplace's equation. Clearly, the matrix obtained from Eqn (3-126) is of positive type and diagonally dominant.

With ω as defined by $\omega = 2(1 + \alpha h)^{-1}$, Garabedian finds that the analog of Eqn (3-121) for Eqn (3-126), with $h = p$, is

$$2U_{xt} + 3U_{yt} + 5\alpha U_t = 3(U_{xx} + U_{yy}).\tag{3-127}$$

This leads to $\alpha_{\mathrm{opt}} = (13\mu_1)^{1/2}/5$, so that

$$\omega_{\mathrm{opt}} = \frac{2}{1 + (\tfrac{13}{25}\mu_1)^{1/2}h} = 2 - \frac{2}{5}(13\mu_1)^{1/2}h + O(h^2).$$

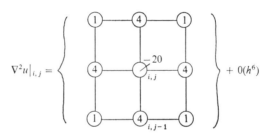

Fig. 3-5   Nine-point computational molecule for the Laplace operator

For the square $0 < x < \pi,\, 0 < y < \pi,\, \mu_1 = 2$, so that

$$\omega_{\mathrm{opt}} = 2 - 1.442\,\sqrt{(2)}h + O(h^2) = 2 - 2.04h + O(h^2)$$

very close to the value for the five-point formula. Thus the number of cycles of iteration required *will be approximately* the same for SOR solutions of the five- and nine-point formulae.

The nine-point formula for Laplace's equation has been examined in great detail by van de Vooren and Vliegenthart [38]. They compare the convergence rates for the Jacobi, Gauss–Seidel, and SOR methods for the nine-point formula. Although the matrix of the nine-point formula *does not possess* Property (A), the optimum relaxation factor for a rectangular region can be obtained by separation of variables. Their results differ slightly from Garabedian's (he overestimates ω_{opt}). For a square of side a, with equal mesh sizes, they obtain

$$\omega_{\mathrm{opt}} = 2 - 2.116\frac{\pi h}{a} + 2.24\left(\frac{\pi h}{a}\right)^2 + O(h^3)$$

$$\lambda(G_{\omega 9})\dagger = 1 - 1.791\frac{\pi h}{a} + 1.60\left(\frac{\pi h}{a}\right)^2 + O(h^3)\tag{3-128}$$

$$R(G_{\omega 9}) = 1.791\frac{\pi h}{a} + O(h^3).$$

† To distinguish the two molecules we write $\lambda(G_{\omega 9})$ or $\lambda(G_{\omega 5})$.

(c) Unequal mesh spacing

By employing the methods given in the previous sections one can, with considerable computational labor, develop appropriate results for a rectangular region $0 < x < a$, $0 < y < b$ with unequal mesh sizes $\Delta x = h$ and $k = \Delta y$. We record here those results when the five- and nine-point molecules are applied to the Dirichlet problem for Laplace's equation.

Jacobi five-point:

$$U_{i,j}^{(n+1)} = \frac{1}{2(h^2 + k^2)} \{k^2(U_{i+1,j}^{(n)} + U_{i-1,j}^{(n)}) + h^2(U_{i,j+1}^{(n)} + U_{i,j-1}^{(n)})\} \quad (3\text{-}129)$$

Jacobi nine-point (Problem 3-39):

$$\lambda(G_{J5}) = \frac{1}{h^2 + k^2} \left\{ k^2 \cos \frac{\pi h}{a} + h^2 \cos \frac{\pi k}{b} \right\}$$

$$R(G_{J5}) = \frac{1}{2} \frac{h^2 k^2}{h^2 + k^2} \pi^2 \left(\frac{1}{a^2} + \frac{1}{b^2} \right)$$

$$\lambda(G_{J9}) \quad \text{same as } \lambda(G_{J5}) \text{ when } \frac{1}{\sqrt{5}} < \frac{h}{k} < \sqrt{5}\dagger$$

$$R(G_{J9}) = \frac{3}{5} \frac{h^2 k^2}{h^2 + k^2} \pi^2 \left(\frac{1}{a^2} + \frac{1}{b^2} \right)$$

Gauss–Seidel (ordering as before) nine-point:

$$U_{i,j}^{(n)} = \frac{1}{20} \{U_{i+1,j+1}^{(n)} + U_{i-1,j+1}^{(n)} + U_{i+1,j-1}^{(n+1)} + U_{i-1,j-1}^{(n+1)}\}$$

$$- \frac{1}{10} \frac{h^2 - 5k^2}{h^2 + k^2} \{U_{i+1,j}^{(n)} + U_{i-1,j}^{(n+1)}\} + \frac{1}{10} \frac{5h^2 - k^2}{h^2 + k^2} \{U_{i,j+1}^{(n)} + U_{i,j-1}^{(n+1)}\}$$
$$(3\text{-}130)$$

Gauss–Seidel five-point (Problem 3-40):

$$\lambda(G_{S5}) = \lambda^2(G_{J5}); \qquad R(G_{S5}) = 2R(G_{J5})$$
$$\lambda(G_{S9}) = \lambda^2(G_{J9}); \qquad R(G_{S9}) = 2R(G_{J9})$$

Method of Kahan

Kahan [39] considered the case of real nonsingular matrices $A = (\alpha_{ij})$ for which

$$\alpha_{ij} = \alpha_{ji} \leq 0, \quad i \neq j, \qquad \alpha_{ii} > 0, \quad i = 1, 2, \ldots, N \quad (3\text{-}131)$$

and A is positive-definite. These conditions are satisfied when A is obtained from self-adjoint elliptic differential equations such as those described in Section 3-3 (Problem 3-41). Kahan develops the following result: *If the*

† See van de Vooren and Vliegenthart [38] for h/k outside this interval.

matrix A has properties (3-131) *and* ω_b *is defined by Eqn* (3-113), *then* $\lambda(G_{\omega_b}) \geq \omega_b - 1, 0 < \omega_b < 2$ *and* $\omega_b - 1 \leq \lambda(G_{\omega_b}) < \sqrt{(\omega_b - 1)}$.

The last condition of Kahan's theorem implies that the best convergence factor $\lambda(G_{\omega_{opt}})$ is fairly close to that obtainable with ω_b, when ω_b is near 2. The actual optimal value ω_{opt} of ω, which minimizes $\lambda(G_\omega)$, is not known for all matrices which satisfy conditions (3-131). The method of van de Vooren and Vliegenthart [38] may apply to this problem. Young and Frank [12] assert that the use of $\omega = \omega_b$ appears to give convergence rates like those for SOR where the matrix A has Property (A).

PROBLEMS

3-35 Perform the transformation carrying Eqn (3-121) into Eqn (3-122). Verify that Eqn (3-122) is hyperbolic.

3-36 Find a nine-point formula for Laplace's equation if $h = \Delta x \neq \Delta y = k$. What is the local error?

3-37 Develop Eqn (3-127) and find the solution by separation. Determine α_{opt} for that case.

3-38 When Eqn (3-126) is employed to solve Laplace's equation $\Delta^2 u = u_{xx} + u_{yy} = 0$, on a square grid, the local discretization error is $O(h^6)$, as $h \to 0$. Does the same result hold for the Poisson equation $\Delta^2 u = g$ where g is harmonic (i.e. $\Delta^2 g = 0$)? Is the nine-point formula any better than the five-point if g is not harmonic? (*Hint*: If $\Delta^2 u = 0$ what can be said about $\Delta^4 u, \Delta^6 u$?).

Ans: 'Yes' for the first part and 'No' for the second.

3-39 Write the nine-point formula analogous to Eqn (3-129) for Jacobi's method.

3-40 Write the five-point formula analogous to Eqn (3-130) for the Gauss–Seidel iteration.

3-41 Let Eqn (3-34) be self-adjoint. Show that the conditions of Eqn (3-131) hold if the approximations of Eqns (3-42) are employed.

3-9   Qualitative examples of over-relaxation

Young [20] investigated several methods for approximating ω_b for regions other than rectangles. We discuss two of these. In the first he chose ω_b to be that value obtained for a circumscribing rectangle. An alternative method was to use ω_b for a square of the same area as the actual region. In *both* cases it was found that ω_b so chosen was always larger than the actual best value ω_{opt}. When values of ω_b determined by these methods were used, instead of the best values, the number of extra iterations needed were of the order of 15–30 per cent.

Applications of SOR to a wide variety of problems has proved its usefulness as an accelerator, even when the conditions of the above partial list are not satisfied. We briefly discuss two examples—both of which are nonlinear.

The first example, by Padmanabhan, Ames, and Kennedy [40], concerns the early stages of collapse of a cylindrical homogeneous fluid mass in a density stratified medium. In the early stages of collapse, accelerative forces are much more significant than the viscous forces. A computational model based upon the inviscid, irrotational theory is developed. In terms of dimensionless variables the problem to be solved is

$$\phi_{xx} + \phi_{yy} = 0, \qquad u = -\phi_x, \qquad v = -\phi_y$$

$$\phi_{t|_{\eta^+}} = \left[\left(\frac{g}{\rho_0} \right) \left| \frac{\partial \rho}{\partial y} \right| \frac{y^2}{2} + \frac{1}{2} (u^2 + v^2) \right]\bigg|_{y=\eta} \qquad (3\text{-}132)$$

$$\left[v - u \left(\frac{\partial \eta}{\partial x} \right) \right]_{y=\eta} - \frac{\partial \eta}{\partial t} = 0.$$

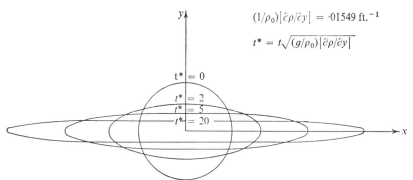

$(1/\rho_0)|\partial\rho/\partial y| = \cdot01549 \text{ ft.}^{-1}$

$t^* = t\sqrt{(g/\rho_0)|\partial\rho/\partial y|}$

Fig. 3-6   Wake profiles

The first equation is Laplace's equation for the potential ϕ. The second is the unsteady Bernoulli equation evaluated at the moving boundary's interface $y = \eta(x, t)$, thereby constituting a *dynamic boundary condition at the interface.* Since $y = \eta(x, t)$ is the equation of the interface, the third equation represents the *kinematic boundary condition at the interface.* At times $t < 0$, the initial velocities are zero ($u = v = \phi = \phi_t = 0$) and the initial interface is circular (Fig. 3-6). By application of symmetry conditions only the quarter circle $0 \le x \le a_0$, $-a_0 \le y \le 0$ is required. Along the straight boundaries the symmetry conditions generate the boundary conditions $\partial\phi/\partial y = 0$ and $\partial\phi/\partial x = 0$.

Beginning with the initial state, just discussed, the computational algorithm proceeds as follows:

(i) Calculate the velocity potential on the *boundary* (small Δt);

$$\phi(x, y, t + \Delta t)|_{y=\eta} = \phi(x, y, t)|_{y=\eta}$$

$$+ \Delta t \left[\frac{g}{\rho_0} \left| \frac{\partial \rho}{\partial y} \right| \frac{y^2}{2} + \frac{u^2 + v^2}{2} \right] \Bigg|_{\substack{y=\eta \\ \text{time } t}} + O(\Delta t) \quad (3\text{-}133)$$

where $\quad \phi(x, y, \Delta t)|_{y=\eta} = \Delta t \left[\frac{g}{\rho_0} \left| \frac{\partial \rho}{\partial y} \right| \frac{y^2}{2} \right] \Bigg|_{\substack{y=\eta \\ t=0}}$

(ii) Calculate the velocity potential in the *interior* at time $t + \Delta t$ employing the boundary shape at time t and the boundary potential calculated in Eqn (3-133).

Solve $\nabla^2 \phi = 0$ by over-relaxation with $\Delta x = h$, $\Delta y = k$, and ω_b calculated by Young's first method. This generates $\phi(ih, jk, t + \Delta t)$.

(iii) Calculate the gradient of ϕ

at interior points

$$-u|_{i,j} = \phi_x|_{i,j} = (2h)^{-1}[\phi_{i+1,j} - \phi_{i-1,j}] + O(h^2)$$

$$-v|_{i,j} = \phi_y|_{i,j} = (2k)^{-1}[\phi_{i,j+1} - \phi_{i,j-1}] + O(k^2)$$

at the boundary

backward differences employed to extrapolate from interior points.

(iv) Predict the new shape of the boundary employing a first-order difference in time of the kinematic boundary condition.

Conservation of matter was checked by monitoring the area of the wake. These agreed with the original value to within 5 per cent, and the results of the computation are shown in Fig. 3-6. The application of over-relaxation accelerated the convergence greatly over the Gauss–Seidel method. We examine an extension of this example to include viscous effects in Chapter 5.

As a second example we consider the numerical solution of

$$Lu = f(x, t, u) \quad (3\text{-}134)$$

by employing a Newton second-order process and over-relaxation. With L a linear parabolic operator, a comparison of the Newton and Picard approximations was made in Section 2-15. Here we sketch the results of Bellman, Juncosa, and Kalaba [41] when the second-order Newton process is applied to the equation

$$u_{xx} + u_{yy} = e^u \quad (3\text{-}135)$$

in the region $0 \le x \le \frac{1}{2}$, $0 \le y \le \frac{1}{4}$. Two sets of boundary conditions are considered, $u \equiv 0$ in the first case and $u \equiv 10$ in the second. This equation is of physical interest in diffusion–reaction problems, vortex problems, and electric space charge considerations. These are discussed in Ames [9].

Here, we compare the Picard linearization of Eqn (3-135) with the Newton linearization. These are respectively *outer iterations* of the form

$$u_{xx}^{(k+1)} + u_{yy}^{(k+1)} = \exp u^{(k)}$$

and [see Eqn (2-181)]

$$u_{xx}^{(k+1)} + u_{yy}^{(k+1)} = \exp u^{(k)} + [u^{(k+1)} - u^{(k)}] \exp u^{(k)}$$

where the latter is more conveniently written as

$$u_{xx}^{(k+1)} + u_{yy}^{(k+1)} - [\exp u^{(k)}]u^{(k+1)} = [1 - u^{(k)}] \exp u^{(k)}. \quad (3\text{-}136)$$

In each case the standard five-point molecule is employed for the Laplacian, that is

$$h^{-2}[U_{i+1,j} + U_{i,j+1} + U_{i-1,j} + U_{i,j-1} - 4U_{i,j}]$$

with $\Delta x = \Delta y = h$. As has been previously discussed, both equations have finite difference systems which require the solution of linear algebraic equations whose matrix form is†

$$(I - L_k - V_k)U^{(k+1)} = b^{(k)} \quad (3\text{-}137)$$

where I is the identity matrix and L_k and V_k are, respectively, lower triangular and upper triangular matrices.

The inner iteration—that is, the solution of Eqn (3-137)—is carried out by over-relaxation. The combined process is of the form

$$(I - \omega L_k)U_{r+1}^{(k+1)} = [\omega V_k - (\omega - 1)I]U_r^{(k+1)} + b^{(k)}, \quad r = 0, 1, 2, \ldots$$
$$(3\text{-}138)$$

where k is the *outer* iteration index, r is the *inner* iteration index, and ω is the relaxation factor. The iteration is carried out in the order imposed. It is clear that very little effort should be expended in obtaining an accurate solution to Eqn (3-137) if $U^{(k+1)}$ is not a very good approximation to the solution of Eqn (3-135). Similarly, there is little point in iterating on the index k (outer iteration) if the approximations given by Eqn (3-138) are too rough as a consequence of terminating the inner iteration on r too soon. Thus, there exists the open question of 'When should iteration cease on k or r at each step?' (The results of Douglas [42] and Young and Wheeler [43] are important here. Although their problems are different, they find that only *one cycle* of the Peaceman–Rachford alternating direction iteration (i.e. only m inner iterations) were performed for each outer iteration. Additional inner iterations did not hasten the convergence.)

Since the authors' goal was not to answer the question raised in the preceding paragraph, they used a simultaneous iteration based on a *blend* formula

$$(I - \omega L_k)U^{(k+1)} = [\omega V_k - (\omega - 1)I]U^{(k)} + \omega b^{(k)}. \quad (3\text{-}139)$$

Following Young's suggestion a value of ω for Laplace's equation was chosen

---

† An ordering has been introduced and the equations have been divided by the main diagonal coefficient.

by using the linear theory, which gave $\omega = 1.732$. Hence, the range of $\omega = 1.70$–1.74 was used initially for the boundary condition $u \equiv 0$. In this case no advantage was obtained from the use of the quadratically convergent Newton procedure over the results of the first-order Picard method. In fact, the same value of $\omega = 1.74$ gave the fastest convergence in both instances.

In the case of the boundary condition $u \equiv 10$, a number of runs for different values of ω were made to *determine experimentally an optimal value of ω.* The results of this experiment are given in Table 3–1 (note that the same accuracy is required in both cases).

Table 3–1

Picard Method		Newton Method	
Value of ω	Number of iterations	Value of ω	Number of iterations
1.00	134	1.00	148
1.10	106	1.40	66
1.20	78	1.50	50
1.30	76	1.55	41
1.35	71	1.60	42
1.40	66	1.65	46
1.50	53	1.70	51
1.55	63	1.74	58

With boundary conditions of 10, rapid changes near the boundary are suggested. Here the Newton method was better, requiring 41 iterations compared to 53 for the Picard approach. The optimal value of ω is slightly smaller for the Picard method.

As a byproduct of this analysis we observe that the Gauss–Seidel method ($\omega = 1$) compares rather poorly with over-relaxation, requiring from $2\frac{1}{2}$ to 4 times as many iterations. In the case of the boundary condition $u \equiv 0$ (not shown), where $\omega = 1.74$ is optimal, the GS scheme takes up to seven times as many iterations as SOR.

As in the parabolic case, the advantage accruing to the Newton linearization becomes significant when steep gradients occur. We also note that the Newton linearization is easily generalized to the case where L is quasilinear. A number of additional nonlinear examples are given by Ames [9], together with their physical implication.

PROBLEMS

3-42 Consider the over-relaxation iteration

$$u_{xx}^{(k+1)} + u_{yy}^{(k+1)} = bu_k[cu_{k+1} + (1 - c)u_k]$$

of $u_{xx} + u_{yy} = bu^2$ where c is the relaxation parameter. Compare this general form with that obtained by application of the Picard and Newton procedures.

3-43 Set up a computational algorithm for the equation of Problem 3-42 in the cases $c = 2$ and $c = 1$.

3-10 Other point iterative methods†

Other point iterative methods for solving $Au = v$ have been developed and successfully applied. In this section we will briefly discuss them, together with their advantages and disadvantages. Sufficient references will be provided so that additional details may easily be found. One of these methods may be better suited for a given problem than those previously discussed.

(a) Gradient method

When A is symmetric and positive-definite the gradient procedure is one of the important and interesting methods of solving $Au = v$. One form is developed by considering the real-valued quadratic form

$$E(u) = u^T Au - 2v^T u. \qquad (3-140)$$

Equation (3-140) can be written alternatively as

$$\begin{aligned} E(u) &= (u^T A - v^T)u - v^T u \\ &= (u^T A - v^T)u - u^T v + v^T A^{-1}v - v^T A^{-1}v \\ &= (u^T A - v^T)(u - A^{-1}v) - v^T A^{-1}v \\ &= (u - A^{-1}v)^T A(u - A^{-1}v) - v^T A^{-1}v. \qquad (3-141) \end{aligned}$$

Consequently, $E(u)$ attains its minimum value $-v^T A^{-1}v$ when $u = A^{-1}v$, the solution. For all other values of u, $E(u) > -v^T A^{-1}v$. Clearly, solving the system $Au = v$ is equivalent to developing a u which minimizes the quadratic form $E(u)$.

For an arbitrary $u^{(o)}$ we proceed from $u^{(k-1)}$ to $u^{(k)}$ by first computing the direction of steepest descent (downhill gradient direction) of $E(u)$ at $u^{(k-1)}$. This is the direction

$$-\operatorname{grad} E(u)|_{u=u^{(k-1)}} = -2(Au^{(k-1)} - v) = 2R_{k-1} \qquad (3-142)$$

where R_{k-1} is the vector residual error, previously employed.

Now $E(u)$ decreases in the direction of R_{k-1} so it seems reasonable to introduce a positive number ν_{k-1}, to be determined, such that the iteration takes the form

$$u^{(k)} = u^{(k-1)} + \nu_{k-1}R_{k-1}. \qquad (3-143)$$

Forsythe [44], and references therein, contain a complete discussion of the

---

† Like their predecessors these matrix problems *do not* have to arise from partial differential equations.

optimum gradient method. One way to choose the ν_{k-1} is such that $E(u^{(k-1)} + \nu R_{k-1})$ is made a minimum among all real ν. This quantity is minimized when ν is given the optimum value

$$R_{k-1}^T R_{k-1}/R_{k-1}^T A R_{k-1} \tag{3-144}$$

but this choice requires the storage of R_k as well as $u^{(k)}$. The additional computation of AR doubles the magnitude of the computation. Stiefel [45] points out that the use of the optimum value of ν may not only create extra work but is usually a *short-sighted strategy anyway*!

What then? We might select all the ν_k to have the same value ν whereupon, by Eqn (3-143),

$$u^{(k)} = u^{(k-1)} + \nu(v - Au^{(k-1)}).$$

Consequently, the error $e_k = u^{(k)} - A^{-1}v$ satisfies

$$e_k = e_{k-1} - \nu A e_{k-1} = (I - \nu A)e_{k-1} \tag{3-145}$$

which is a linear stationary iteration with $G = I - \nu A$. If $\mu_i > 0$ are the eigenvalues of A and λ_i those of G, it follows that $\lambda_i = 1 - \nu\mu_i$ $(i = 1, 2, \ldots, N)$ Therefore, for the spectral radius of G to be less than 1 we must have (as a very minimum for convergence)

$$0 < \nu < \frac{2}{\max_i \mu_i}. \tag{3-146}$$

But if we wish the *fastest possible convergence*, we must minimize max $|\lambda_i|$—that is, choose ν so that

$$\max_i |1 - \nu\mu_i| = \min. \tag{3-147}$$

One practical method of approximately choosing ν, so that Eqn (3-147) holds, revolves about the fact that we very often know the μ_i lie in an interval $a \le \mu_i \le b$ for all i $(0 < a < b < \infty)$. For every ν the function $|1 - \nu\mu|$ takes its maximum at one of the end points $\mu = a$ or $\mu = b$. The best choice of ν is that one for which

$$\max (|1 - \nu a|, |1 - \nu b|)$$

is smallest. This is easily seen to be the one for which $1 - \nu a = -(1 - \nu b)$; that is, $\nu = 2/(a + b)$. With this value of ν we have, for all i,

$$|1 - \nu\mu_i| \le 1 - \nu a = \frac{b - a}{b + a} = \frac{b/a - 1}{b/a + 1} < 1. \tag{3-148}$$

(b) Richardson's method

When A is a positive-definite symmetric matrix, the relaxation factor ν_{k-1} of Eqn (3-143) may depend upon the iteration number k. This idea dates to 1910 when Richardson [46] conceived it.

Since $e_k = (I - v_{k-1}A)e_{k-1}$ we have, upon successive application, the expression

$$e_k = \left[\prod_{i=0}^{k-1}(I - v_iA)\right]e_o$$

$$= P_k(A)e_o \qquad (3\text{-}149)$$

where $P_k(x)$ is a polynomial of degree k with zeros $1/v_i$ such that $P_k(0) = 1$. If the eigenvalues of A are μ_i, then those of $P_k(A)$ are $P_k(\mu_i)$. Suppose,† again, that all we know about the μ_i is that they lie in a certain *real* interval $0 < a \leq \mu_i \leq b < \infty$ for all i.

Richardson [46] suggested sprinkling the k zeros v_i^{-1} of $P_k(x)$ fairly uniformly over $a \leq x \leq b$. This *may be* effective if k is not too large.

Most authors (e.g. Shortley [47] and Young [48]) employ the more sophisticated idea of finding the $P_k(x)$ whose values for $a \leq x \leq b$ are in absolute value as small as possible in some sense. The usual one used is max $|P_k(x)|$. Thus one seeks $P_k(x)$ with $P_k(0) = 1$, such that

$$\max_{a \leq x \leq b} |P_k(x)| = \text{min.} \qquad (3\text{-}150)$$

The answer to the problem was given by Markhoff in 1892 (see [49]) as

$$P_k(x) = [T_k(y_0)]^{-1} T_k\left[\frac{b + a - 2x}{b - a}\right], \qquad y_0 = \frac{b + a}{b - a} > 1 \quad (3\text{-}151)$$

where $T_k(y) = \cos(k \cos^{-1} y)$ are the ordinary Chebyshev polynomials (see Problem 3-45) on $-1 \leq y \leq 1$.

As $k \to \infty$ it can be shown that

$$\max_{a \leq x \leq b} |P_k(x)| \leq 2[y_0 - \sqrt{(y_0^2 - 1)}]^k$$

from which it follows that the *average convergence factor* for the first k steps, $(\|e_k\|/\|e_0\|)^{1/k}$, is bounded by

$$y_0 - \sqrt{(y_0^2 - 1)}$$

and the *average rate of convergence*, $-\log(\|e_k\|/\|e_0\|)^{1/k}$, is bounded by

$$\log[y_0 + \sqrt{(y_0^2 - 1)}].$$

Since $y_0 = (b/a + 1)/(b/a - 1)$ and a, b are chosen optimally, the convergence factor could never be smaller than the value obtained by setting $y_0 = (P + 1)/(P - 1)$ where P is Todd's *P-condition number*

$$P = \frac{\max_i \mu_i}{\min_i \mu_i} \leq \frac{b}{a}$$

---

† Eigenvalue problems will be examined later in this volume, Section 5-3.

for positive-definite matrices. This follows since

$$y_0 - \sqrt{(y_0^2 - 1)} \geq (P + 1)(P - 1)^{-1} - \sqrt{[(P + 1)^2(P - 1)^{-2} - 1]}.$$

If $P \gg 1$, $y_0 = 1 + 2(a/b) + O[(a/b)^2]$, so that

$$y_0 - \sqrt{(y_0^2 - 1)} = 1 - 2\left(\frac{a}{b}\right)^{1/2} + O\left(\frac{a}{b}\right) \geq 1 - 2P^{-1/2} \qquad (3\text{-}152)$$

and the corresponding relation for the rate of convergence is

$$\log\left[y_0 + \sqrt{(y_0^2 - 1)}\right] = 2\left(\frac{a}{b}\right)^{1/2} \leq 2P^{-1/2}. \qquad (3\text{-}153)$$

As an example of Richardson's method we return to the Dirichlet problem for Laplace's equation which we have used previously (Section 3-6). In that problem we found [Eqns (3-87)]

$$\min \mu_{p,q} = \mu_{1,1} = 8h^{-2} \sin^2 \frac{h}{2} \sim 2$$

and

$$\max \mu_{p,q} = \mu_{n-1,n-1} = 8h^{-2}\left(1 - \sin^2 \frac{h}{2}\right) \sim 8h^{-2}$$

(as $h \to 0$). Selecting $a = \mu_{1,1}$ and $b = \mu_{n-1,n-1}$ we find $P = b/a \sim 4/h^2$ (as $h \to 0$) and $y_0 = 1 + h^2/2 + O(h^4)$. Consequently, the average convergence factor is

$$y_0 - \sqrt{(y_0^2 - 1)} = 1 - h + O(h^2) \qquad (3\text{-}154)$$

while the corresponding rate of convergence is approximately h. Thus the number of iterations varies as h^{-1} as $h \to 0$. The number of iterations is therefore of the same order of magnitude as for SOR. However, SOR converges approximately twice as fast as Richardson's method, and since it is simpler and requires less storage, its use is recommended instead of Richardson's method.

Young and Warlick [50] and Young [51] found that with Richardson's method there is difficulty in controlling rounding errors. The reason for this lies at the heart of the process. One cannot let $k \to \infty$ in Richardson's procedure. For as long as one actually uses Eqn (3-143) for each computation step, the ν_i must be known at the start of the computation. Since the ν_i^{-1} are the zeros of $P_k(x)$, one must have a fixed k, say K, in mind from the start. Once k is fixed one determines the ν_i, $i = 1, 2, \ldots, K$, from the tabulated zeros of the Chebyshev polynomials (see, for example, Snyder [52]) and then computes $u^{(1)}, \ldots, u^{(K)}$ from Eqn (3-143). If e_K is not sufficiently small, one uses the same values of ν_i to construct another cycle of K steps. Large values of K are desirable for good convergence but they introduce a possible loss of accuracy due to round-off within a single Richardson cycle.

Such processes are called *semi-iterative* by Varga [53]. We shall now examine these in more detail.

(c) Semi-iterative methods

Let the linear stationary iterative method

$$u^{(k+1)} = Gu^{(k)} + v \tag{3-155}$$

be consistent and suppose that there is given a set of coefficients $\beta_{n,k}$, $k = 0, 1, 2, \ldots, n$ — that is,

$$\beta_{0,0}$$

$$\beta_{1,0}, \ \beta_{1,1}$$

$$\beta_{2,0}, \ \beta_{2,1}, \ \beta_{2,2}$$

$$\cdots$$

such that

$$\sum_{k=0}^{n} \beta_{n,k} = 1. \tag{3-156}$$

Then the relation

$$v^{(n)} = \sum_{k=0}^{n} \beta_{n,k} u^{(k)} \tag{3-157}$$

defines a *semi-iterative process with respect to* the iterative method of Eqn (3-155). If $\beta_{n,k} = 0$ for all $k < n$ and $\beta_{n,n} = 1$, the semi-iterative method reduces to the original method. An optimum choice of the $\beta_{n,k}$ will therefore generate a semi-iterative method which ought to be as good as the original.

One such choice is described in Young and Frank [12]. Let the error vector of the nth iterant be

$$f_n = v^{(n)} - u = v^{(n)} - A^{-1}v$$

so that

$$f_n = \sum_{k=0}^{n} \beta_{n,k} u^{(k)} - \left(\sum_{k=0}^{n} \beta_{n,k} \right) u = \sum_{k=0}^{n} \beta_{n,k} e_k \tag{3-158}$$

where e_k is the error vector of the original method. From Eqn (3-32), $e_k = G^k e_0$, whereupon Eqn (3-158) becomes

$$f_n = \sum_{k=0}^{n} \beta_{n,k} G^k e_0. \tag{3-159}$$

As in Richardson's method we now let $P_n(x)$ be defined by

$$P_n(x) = \sum_{k=0}^{n} \beta_{n,k} x^k$$

so that Eqn (3-159) is expressible as

$$f_n = P_n(G) e_0 \tag{3-160}$$

in analogy with Eqn (3-149). Thus, we are led to the problem of minimizing the spectral radius of the polynomial $P_n(G)$. In our problem we assume G to be similar to a symmetric matrix and $P_n(x)$ to have real coefficients. Further, we suppose that G has real eigenvalues λ_i and the iteration equation [Eqn (3-155)] is convergent. Then the λ_i lie in a certain real interval

$$-1 < a \leq \lambda_i \leq b < 1 \tag{3-161}$$

for some a and b. Since

$$\lambda(P_n(G)) = \max_{1 \leq i \leq n} |P_n(\lambda_i)| \leq \max_{a \leq x \leq b} |P_n(x)| \tag{3-162}$$

it follows that our problem becomes that of minimizing $\max_{-1 < a \leq x \leq b < 1} |P_n(x)|$ such that $P_n(1) = 1$, for all n. The answer to this problem is, as before, supplied by the work of Markhoff [49]. The unique solution to this minimizing problem is the function

$$\bar{P}_n(x) = \frac{T_n\left[\dfrac{2x - (a + b)}{b - a}\right]}{T_n\left[\dfrac{2 - (a + b)}{b - a}\right]} \tag{3-163}$$

where T_n is the Chebyshev polynomial defined by

$$T_n(x) = \begin{cases} \cos(n \cos^{-1} x), & |x| \leq 1 \\ \cosh(n \cosh^{-1} x), & |x| > 1 \end{cases} \tag{3-164}$$

These polynomials satisfy the well-known three-term relation

$$T_{n+1}(x) - 2xT_n(x) + T_{n-1}(x) = 0, \quad n \geq 1$$
$$T_0(x) = 1, \qquad T_1(x) = x. \tag{3-165}$$

This recurrence relation provides the foundation for an *iterative method of second degree*. All of the previous procedures have been of first degree.

In general, $b = \lambda(G) > -a$, but the special case $b = \lambda(G) = -a$† holds if the original method, Eqn (3-155), is that of Jacobi and the matrix A is symmetric and has Property (A). (See Problem 3-48 for the more general case.)

Under this assumption Eqn (3-163) becomes $T_n(1/\lambda)\bar{P}_n(x) = T_n(x/\lambda)$. Thus the recurrence relation, Eqn (3-165), takes the form

$$T_{n+1}(1/\lambda)\bar{P}_{n+1}(x) = \frac{2x}{\lambda} T_n(1/\lambda)\bar{P}_n(x) - T_{n-1}(1/\lambda)\bar{P}_{n-1}(x) \quad \text{(3-166)}$$

for $n \geq 1$ and $\bar{P}_0(x) = 1$, $\bar{P}_1(x) = x$. Upon replacing x by G and employing Eqn (3-160), we have

$$T_{n+1}(1/\lambda)f_{n+1} = \frac{2}{\lambda} T_n(1/\lambda)Gf_n - T_{n-1}(1/\lambda)f_{n-1}, \quad n \geq 1. \tag{3-167}$$

† This assumption may also be used when one knows only that the eigenvalues of G are real and $\lambda(G)$ is to be estimated.

Upon division by $T_{n+1}(1/\lambda)$ and using the definition of f_n, Eqn (3-167) may be written in the form

$$v^{(n+1)} = \gamma_{n+1}\{Gv^{(n)} + v - v^{(n-1)}\} + v^{(n-1)}, \quad n \geq 1 \qquad (3\text{-}168)$$

where

$$\gamma_{n+1} = \frac{2T_n(1/\lambda)}{\lambda T_{n+1}(1/\lambda)}, \quad n \geq 1 \quad \text{and} \quad \gamma_1 = 1.$$

The recurrence relation [Eqn (3-168)], of degree two, provides for the computation of successive iterants of the semi-iterative method without intermediate computation of the iterants $u^{(n)}$ of the original method. The vectors $v^{(0)}$ and $v^{(1)}$ are, respectively, equal to the quantities $u^{(0)}$ and $u^{(1)}$ of the basic method.

Application of a semi-iterative method is similar to SOR with a *variable* relaxation factor. The additional storage of a vector ($v^{(n)}$ and $v^{(n-1)}$) could be an important consideration limiting the utility of this method in the solution of large problems.

The definition for rate of convergence, given by Eqn (3-82), is no longer meaningful for semi-iterative methods since the coefficients in Eqn (3-168), for example, change with each iteration step. Instead, as in Richardson's method, an average rate of convergence is employed. This is defined by means of the expression

$$R_n = R[\bar{P}_n(G)] = -\frac{1}{n}\log \lambda[\bar{P}_n(G)]. \qquad (3\text{-}169)$$

For the general result, given in Problem 3-48, this average rate is computed as follows. From Eqns (3-161), (3-162), and (3-163) we have

$$\lambda(\bar{P}_n(G)) = 1/T_n(\epsilon). \qquad (3\text{-}170)$$

From Snyder [52] we have

$$T_n(x) = \tfrac{1}{2}\{[x + \sqrt{(x^2 - 1)}]^n + [x + \sqrt{(x^2 - 1)}]^{-n}\}$$

so that with $\sigma = 1/\epsilon$ and $\alpha = \sigma/[1 + \sqrt{(1 - \sigma^2)}]$ we find

$$R_n = -\frac{1}{n}\log \frac{2\alpha^n}{1 + \alpha^{2n}}. \qquad (3\text{-}171)$$

Moreover,

$$\lim_{n \to \infty} R_n = -\log \alpha. \qquad (3\text{-}172)$$

Golub and Varga [54] observed that semi-iterative methods can be used with considerable computational advantage if the matrix A, of $Au = v$, has the properties

(a) A is symmetric and positive-definite;

(b) A has Property (A), and

(c) A is consistently ordered and has the form

$$A = \begin{bmatrix} D_1 & F \\ G & D_2 \end{bmatrix} \qquad (3\text{-}173)$$

where D_1 and D_2 are *square diagonal* matrices. The procedure to be described will be called the *cyclic Chebyshev semi-iterative method*.

Let u and v be partitioned to conform with the partitioning of A so that the considered problem has the form

$$\begin{bmatrix} D_1 & F \\ G & D_2 \end{bmatrix} \begin{bmatrix} u_1 \\ u_2 \end{bmatrix} = \begin{bmatrix} v_1 \\ v_2 \end{bmatrix} \tag{3-174}$$

When the Jacobi process is applied, one has

$$\begin{bmatrix} u_1^{(n+1)} \\ u_2^{(n+1)} \end{bmatrix} = \begin{bmatrix} 0 & H \\ K & 0 \end{bmatrix} \begin{bmatrix} u_1^{(n)} \\ u_2^{(n)} \end{bmatrix} + \begin{bmatrix} c_1 \\ c_2 \end{bmatrix}$$

where $H = -D_1^{-1}F$, $K = -D_2^{-1}G$, $c_1 = D_1^{-1}v_1$, $c_2 = D_2^{-1}v_2$.

Conditions (3-173) imply that the eigenvalues λ_i of the matrix

$$G = \begin{bmatrix} 0 & H \\ K & 0 \end{bmatrix}$$

are real and satisfy $-\lambda \le \lambda_i \le \lambda$, where λ is the spectral radius of G.

The optimum semi-iterative method specified by Eqn (3-168) is

$$\begin{bmatrix} v_1^{(n+1)} \\ v_2^{(n+1)} \end{bmatrix} = \gamma_{n+1} \left\{ \begin{bmatrix} 0 & H \\ K & 0 \end{bmatrix} \begin{bmatrix} v_1^{(n)} \\ v_2^{(n)} \end{bmatrix} + \begin{bmatrix} c_1 \\ c_2 \end{bmatrix} - \begin{bmatrix} v_1^{(n-1)} \\ v_2^{(n-1)} \end{bmatrix} \right\} + \begin{bmatrix} v_1^{(n-1)} \\ v_2^{(n-1)} \end{bmatrix}, \quad n \ge 1 \tag{3-175}$$

with γ_{n+1} as before. Additionally,

$$\begin{bmatrix} v_1^{(0)} \\ v_2^{(0)} \end{bmatrix} \quad \text{and} \quad \begin{bmatrix} v_1^{(1)} \\ v_2^{(1)} \end{bmatrix}$$

are equal, respectively, to the corresponding vectors from the base iteration, Eqn (3-155). Equation (3-175) can be written as the pair of equations

$$v_1^{(n+1)} = \gamma_{n+1}[Hv_2^{(n)} + c_1 - v_1^{(n-1)}] + v_1^{(n-1)} \tag{3-176a}$$

$$v_2^{(n+1)} = \gamma_{n+1}[Kv_1^{(n)} + c_2 - v_2^{(n-1)}] + v_2^{(n-1)} \tag{3-176b}$$

for $n \ge 1$. Equations (3-176) determine the vector sequences $\{v_1^{(n)}\}_0^\infty$ and $\{v_2^{(n)}\}_0^\infty$ but require essentially *double* the storage requirements of the SOR method. However, if we begin with the single vector guess $v_2^{(0)}$ and recall that $\gamma_1 = 1$, then Eqn (3-176a), for $n = 0$, becomes

$$v_1^{(1)} = Hv_2^{(0)} + c_1 \tag{3-177}$$

and Eqn (3-176b), for $n = 1$, becomes

$$v_2^{(2)} = \gamma_2[Kv_1^{(1)} + c_2 - v_2^{(0)}] + v_2^{(0)}.$$

Thus, only the subsequences $\{v_1^{(2n+1)}\}_0^\infty$ and $\{v_2^{(2n)}\}_0^\infty$ need be calculated from

$$\begin{aligned}
v_1^{(2n+1)} &= \gamma_{2n+1}[Hv_2^{(2n)} + c_1 - v_1^{(2n-1)}] + v_1^{(2n-1)}, & n \geq 1 \\
v_2^{(2n+2)} &= \gamma_{2n+2}[Kv_1^{(2n+1)} + c_2 - v_2^{(2n)}] + v_2^{(2n)}, & n \geq 0
\end{aligned} \tag{3-178}$$

where $v_1^{(1)}$ is given by Eqn (3-177).

The cyclic Chebyshev semi-iterative method described by Eqns (3-177) and (3-178) requires only the *single* vector guess $v_2^{(0)}$ and only *half* of the components of $v^{(n)}$ need be computed for each iteration. No additional storage requirements over those of Gauss–Seidel or SOR are necessary. Using the spectral radius of the associated matrix G as an approximation of the spectral norm, one finds that the convergence rate of the cyclic Chebyshev method is the same as that of SOR. However, for a finite number of iterations, the spectral norm of the associated matrix of SOR is slightly larger than that of the cyclic Chebyshev method. Golub and Varga [54] compare the average rate of convergence of the cyclic Chebyshev process with the rates of convergence of several variants of the SOR method. In all these numerical experiments the number of required iterations using the cyclic Chebyshev method were essentially the same, but slightly less than the number of iterations required using the SOR procedure.

If the matrix A satisfies the conditions necessary for SOR to apply (Section 3-7), Varga [53] proves that the optimum semi-iterative method relative to SOR is simply the SOR method itself. Sheldon [55] presents a modification of SOR which is subject to acceleration as a semi-iterative method. Known as *symmetric-successive-over-relaxation* (SSOR), a complete iteration consists of two half-iterations. The first is simply the SOR method while the second half-iteration is the SOR method applied to the equations in the reverse order. The net result of a complete iteration is that the matrix G of SSOR is symmetric and therefore has real eigenvalues, whereupon the aforementioned semi-iterative methods apply.

Analysis of SSOR is complicated, but investigations by Sheldon [56] indicate that the average convergence rate of SSOR with semi-iteration for Laplace's equation in a rectangle is greater than or equal to $[R(L_\omega)]^{1/2}$ where $R(L_\omega)$ is the convergence rate of SOR. However, the work per iteration of SSOR with semi-iteration is approximately two-and-one-half times that of SOR.

Some of the point iterative methods are compared below in respect of the approximate average rate of convergence for the solution of *Laplace's equation with Dirichlet boundary conditions* for the square with a square net of side h and a five-point molecule.

	Method	*Convergence rate*†
(a)	Jacobi	$\frac{1}{2}h^2$
(b)	Gauss–Seidel	h^2
(c)	Optimum SOR	$2h$
(d)	Semi-iterative	
	(i) General	h
	(ii) Cyclic Chebyshev	$2h$
	(iii) SSOR	$\sim h^{1/2}$
(e)	Richardson	$\alpha h,\ 0.59 \leq \alpha \leq 1$
		(α depends upon period)

PROBLEMS

3-44 Consider the Dirichlet problem for Laplace's equation employing the five-point molecule. Find the best value of ν by means of the analysis leading to Eqn (3-148). What familiar method results? *Ans:* $\nu = \frac{1}{4}$

3-45 Adjust the Chebyshev polynomials to the interval $a \leq x \leq b$ and scale them to obtain Eqn (3-151).

3-46 Verify Eqns (3-152) and (3-153).

3-47 Develop Eqn (3-168) from Eqns (3-167) and (3-165).

3-48 When the eigenvalues of G (Section 3-10(c)) are real and the minimum eigenvalue is *greater* than $-\lambda(G)$ and can be estimated, the quantity a in Eqn (3-163) is set equal to this eigenvalue. Develop the recursive relation corresponding to Eqn (3-168) by employing a method similar to that of the text.

Ans: (Young and Frank [12])

$$v^{(n+1)} = v^{(n-1)} + \frac{4}{b-a}\,\Gamma_{n+1}[Gv^{(n)} + v - v^{(n)}]$$

$$+ 2\frac{b+a}{b-a}\,\Gamma_{n+1}[v^{(n-1)} - v^{(n)}], \quad n \geq 2 \tag{3-179}$$

$$\Gamma_{n+1} = T_n(\epsilon)/T_{n+1}(\epsilon), \quad \epsilon = [2 - (b+a)]/(b-a);$$

$$v^{(0)} = u^{(0)}, \quad v^{(1)} = \frac{2}{2-(b+a)}[Gv^{(0)} + v] - \frac{b+a}{2-(b+a)}v^{(0)}$$

3-49 Apply the semi-iterative method, based on the Jacobi process, to Poisson's equation $u_{xx} + u_{yy} = -2$ to obtain that function u which vanishes on the boundaries, $x = \pm 1$, $y = \pm 1$, of the unit square. Use the five-point molecule for Laplace's equation.

3-11 Block iterative methods

All of the previously discussed iterative methods belong to the 'point iterative' category—that is, at each step the approximate solution is modified at a

---

† This approximate convergence rate is only the lowest order term in an asymptotic series in powers of h.

single point of the domain. Furthermore, the value of each component in the kth iteration, $u^{(k)}$, is determined by an *explicit* formula. By explicit we mean that the kth approximation of the ith component can be determined, by itself, employing already computed approximate values of the other unknowns. It is natural to call such processes *explicit iterative* methods.

Explicit methods have natural extensions to *block iterative processes* in which groups of components of $u^{(k)}$ are modified simultaneously. This will involve the simultaneous solution of a system of equations. Consequently, individual components are implicitly defined in terms of other components of the same group or block. Such a method is called an *implicit iterative* or *block iterative* method.† The redefinition of an explicit method so that it becomes implicit often leads to an increase in the convergence rate at the cost of some complication in the computational algorithm. The blocks may be single rows, two rows, etc. The entire set of point iterative methods, previously discussed, is convertible into block iterative methods. This development for the Jacobi, Gauss–Seidel, and SOR procedures has been carried out by Arms, Gates, and Zondek [23], Friedman [21], Keller [22], Cuthill and Varga [57], and Varga [58] (see also Varga [25]). Herein we shall give the details for the *single* line methods without extensive proofs. Advantages and disadvantages will be discussed while additional details are left for the literature.

Let R be a two-dimensional rectangular domain of mesh points with m rows. Let the general linear elliptic equation, Eqn (3-34), be discretized to obtain the five-point difference scheme, Eqn (3-35), which we write in the form

$$U_{i,j} = \theta_1 U_{i+1,j} + \theta_2 U_{i-1,j} + \theta_3 U_{i,j+1} + \theta_4 U_{i,j-1} + T_{i,j} \quad (3\text{-}180)$$

where

$$\theta_i = \beta_i/\beta_0 \quad \text{and} \quad T_{i,j} = -h^2 g_{ij}/\beta_0$$

With line iterative methods one improves the values of the approximate solution simultaneously on an entire line of points. Therefore, the iterative formula for *Jacobi row iteration* is

$$U_{i,j}^{(k+1)} = \theta_1 U_{i+1,j}^{(k+1)} + \theta_2 U_{i-1,j}^{(k+1)} + \theta_3 U_{i,j+1}^{(k)} + \theta_4 U_{i,j-1}^{(k)} + T_{i,j} \quad (3\text{-}181)$$

and for the *Gauss–Seidel row iteration* is

$$U_{i,j}^{(k+1)} = \theta_1 U_{i+1,j}^{(k+1)} + \theta_2 U_{i-1,j}^{(k+1)} + \theta_3 U_{i,j+1}^{(k)} + \theta_4 U_{i,j-1}^{(k+1)} + T_{i,j}. \quad (3\text{-}182)$$

As indicated, in the Gauss–Seidel row iteration the rows are improved in the order of increasing y.

For each y, determination of the improved values requires the solution of a system of N linear equations in N unknowns, where N is the number of interior mesh points on the given row. As previously observed, for formulae

† The term *group iterative* method is also used.

such as Eqn (3-180), the desired solution is coupled only with two other points on a given row and column. Consequently, the $N \times N$ matrix will be tridiagonal and its solution is found *directly* by means of the Thomas algorithm described in Section 2-3 (see also Problem 3-52).

Successive line over-relaxation is defined in an obvious way as

$$U_{i,j}^{(k+1)} = U_{i,j}^{(k)} + \omega[\overline{U}_{i,j}^{(k+1)} - U_{i,j}^{(k)}] \tag{3-183}$$

where

$$\overline{U}_{i,j}^{(k+1)} = \theta_1 \overline{U}_{i+1,j}^{(k+1)} + \theta_2 \overline{U}_{i-1,j}^{(k+1)} + \theta_3 U_{i,j+1}^{(k)} + \theta_4 \overline{U}_{i,j-1}^{(k+1)} + T_{i,j}. \tag{3-184}$$

Additionally, the whole theory of over-relaxation applies equally well to other blocks of points, as discussed by Forsythe and Wasow [14, p. 269].

As an example of this implicit method we consider the Dirichlet problem for Laplace's equation in the square $0 < x < \pi$, $0 < y < \pi$, previously treated in Section 3-6. Let the $(n-1)^2$ interior mesh points be $x = ih$, $y = jh$, $i, j = 1, 2, \ldots, n-1$, with $nh = \pi$. Using Jacobi's method by lines, we determine the $U^{(k+1)}$ by solving the system

$$U_{i+1,j}^{(k+1)} + U_{i,j+1}^{(k)} + U_{i-1,j}^{(k+1)} + U_{i,j-1}^{(k)} - 4U_{i,j}^{(k+1)} = 0. \tag{3-185}$$

The eigenvalues λ^L of the operator Q^L corresponding to Eqn (3-185), with zero boundary conditions, together with the functions $X(x, y) = X(ih, jh)$, are determined such that

$$\lambda X(x + h, y) + X(x, y + h) + \lambda X(x - h, y)$$
$$+ X(x, y - h) - 4\lambda X(x, y) = 0 \tag{3-186}$$

holds. With $X(x, y) = \sin px \sin qy$ we have

$$(2\lambda \cos ph + 2 \cos qh - 4\lambda) \sin px \sin qy = 0$$

which is satisfied if

$$\lambda_{p,q}^L = \frac{\cos qh}{2 - \cos ph}, \qquad p, q = 1, 2, \ldots, n - 1. \tag{3-187}$$

Since $h = \pi/n$ it follows that these $(n-1)^2$ eigenvalues, $\lambda_{p,q}^L$, are less than 1 in absolute value for all p and q. Further, the spectral radius

$$\lambda^L = \max_{p,q} \lambda_{p,q}^L = \frac{\cos h}{2 - \cos h} \sim 1 - h^2, \quad \text{as } h \to 0 \tag{3-188}$$

which governs the rate of convergence of this method. The rate of convergence is h^2 while the rate of convergence for the *point* Jacobi method was $h^2/2$. Thus the Jacobi method by lines converges at the same rate as the Gauss–Seidel method by points and twice as fast as the Jacobi method by points.

It is now a simple matter to examine *SOR by lines*. Since over-relaxation can be applied without altering the previously developed theory, Eqn (3-114) now takes the form

$$\omega_{opt} - 1 = \frac{1 - \sqrt{[1 - (\lambda^L)^2]}}{1 + \sqrt{[1 - (\lambda^L)^2]}} \sim 1 - 2h\sqrt{2}, \quad \text{as } h \to 0. \quad (3\text{-}189)$$

Consequently, the convergence rate is $2\sqrt{(2)}h$, an improvement by the factor $\sqrt{2}$ over the corresponding optimum SOR by points. Whether this gain compensates for the extra time needed in solving the implicit equations is a matter that must be studied for each problem. The modest convergence rate advantage of SOR by lines is supplemented by one considerable advantage. Less restrictions are required on the difference equations to obtain a block tridiagonal form (see Problem 3-52) than are needed to have Property (A). For example, the nine-point difference approximation to Laplace's equation (Fig. 3-5, Section 3-8) does not have Property (A), but it does take the block tridiagonal form of Problem 3-52 when the lines are used as blocks.

Varga [58] examined SOR by lines in the case that two lines of mesh points are included in each block. The rate of convergence of this method is found to be approximately $\sqrt{2}$ times greater than that of SOR by single line calculation. He and Cuthill and Varga [57] show that it is possible in many cases to perform both single and double line SOR in approximately the same number of arithmetic operations per mesh point as required by point SOR. In these cases line iterative methods result in a decrease in total computational effort over the point methods.

Many other possible blocks can be formed, but for computational ease it is important that the block tridiagonal or block five-diagonal form be retained —that is, the unknowns of each block should be coupled only among themselves (Problem 3-52) or with unknowns of adjacent blocks or with blocks adjacent to the adjacent blocks.

Some of the line iterative methods are compared below in respect of approximate average rates of convergence for the solution of Laplace's equation with Dirichlet boundary conditions for the square with a square net of side h.

Method	*Convergence rate*
(a) Single line (five-point molecule)	
(i) Jacobi	h^2
(ii) Gauss–Seidel (Problem 3-54)	$2h^2$
(iii) SOR	$2\sqrt{(2)}h$
(b) Single line (nine-point molecule, improved accuracy)	
(i) Jacobi	h^2
(ii) Gauss–Seidel	$2h^2$
(iii) SOR	$2\sqrt{(2)}h$
(c) Two-line SOR	$4h$

PROBLEMS

3-50 Write the algorithm for a *Gauss–Seidel* single column iteration.

3-51 Write the algorithm for a Jacobi two (adjacent) row iteration.

3-52 Let $AU = V$ where A has the block tridiagonal form

$$
\begin{bmatrix}
D_1 & S_1 & 0 & 0 & & \cdots & & 0 \\
R_1 & D_2 & S_2 & 0 & & \cdots & & 0 \\
\vdots & \vdots & \vdots & \vdots & & & & \vdots \\
0 & 0 & 0 & \cdots & R_{m-2} & D_{m-1} & S_{m-1} \\
0 & 0 & 0 & \cdots & 0 & R_{m-1} & D_m
\end{bmatrix}
\begin{bmatrix}
U_1 \\ U_2 \\ \vdots \\ U_{m-1} \\ U_m
\end{bmatrix}
=
\begin{bmatrix}
V_1 \\ V_2 \\ \vdots \\ V_{m-1} \\ V_m
\end{bmatrix}
$$

and the square matrices D_i are tridiagonal. Express the Jacobi line iteration and the Gauss–Seidel line iteration in matrix form.

$$Ans:\ \text{Jacobi};\ R_{i-1}U_{i-1}^{(k)} + D_iU_i^{(k+1)} + S_iU_{i+1}^{(k)} = V_i$$

3-53 Consider $(au_x)_x + (cu_y)_y + fu = g$; that is, the self-adjoint form of Eqn (3-34). Use difference representations of the form

$$
(au_x)_x = \frac{a(x + \tfrac{1}{2}s_1h, y)[(u_1 - u_0)/s_1h] - a(x - \tfrac{1}{2}s_2h, y)[(u_0 - u_2)/s_2h]}{[s_1 + s_2)/2]h}
\tag{3-190}
$$

where $u_1 = u(x + s_1h, y)$, $u_0 = u(x, y)$, $u_2 = u(x - s_2h, y)$, and a corresponding approximation for $(cu_y)_y$. Derive a difference equation and write the point Gauss–Seidel iteration. Define the line SOR and write it in matrix form.

3-54 Carry out a comparable analysis to that described in Eqns (3-185)–(3-188) for the Gauss–Seidel method by lines. What is the convergence rate?

$$Ans:\ 2h^2$$

3-55 Develop a comparable analysis to that described in Eqns (3-185)–(3-188) using the nine-point molecule (Section 3-8) instead of the five-point molecule. Is the rate of convergence changed?

$$Ans:\ \text{No};$$

$$\lambda_9^L = \frac{(\cos h)(2 + \cos h)}{5 - 2\cos h} \sim 1 - h^2 \quad \text{as } h \to 0$$

3-12 Alternating direction methods

The SOR method by lines proceeds by taking all the lines in the same direction. Thus in Fig. 3-4(a), for example, we first solve for the values at 1, 2, 3, then for 4, 5, 6, and finally for 7, 8, 9. We then begin again with 1, 2, 3, and so forth. Convergence is often improved by following the first sequence with a second in the column direction. Thus a complete iteration consists of a first half in the row direction followed by a second half in the column direction.

Such methods are aptly designated *alternating direction implicit* methods or ADI *methods* for short. The first of these, developed by Peaceman and Rachford [59] (PRADI for short), is related to a procedure developed by Douglas [60] for solving the equation $u_t = u_{xx} + u_{yy}$ (see Chapter 5). Douglas and Rachford [61] presented a method similar to that of PRADI characterized by its ease of generalization to three dimensions. Birkhoff, Varga, and Young [62]† summarize the state of knowledge to 1962 in a long survey paper complete with a lengthy series of computations and numerous references. Tateyama, Umoto, and Hayashi [63] present a variant of the classical PRADI. Their double interlocking variant proceeds on every other line using old values and on the remainder using the new values.

The two basic processes, PRADI and DRADI, are similar. We begin our discussion with the process of Peaceman and Rachford for Laplace's equation approximated by a five-point molecule in a rectangular domain with equal mesh sizes h. This formula, when centered at (i, j), is

$$U_{i+1,j} + U_{i-1,j} + U_{i,j+1} + U_{i,j-1} - 4U_{i,j} = 0. \qquad (3\text{-}191)$$

The iteration proceeds from $U_{i,j}^{(k)}$ to the determination of

$$U_{i,j}^{(k+1/2)} = U_{i,j}^{(k)} + \rho_k[U_{i+1,j}^{(k+1/2)} + U_{i-1,j}^{(k+1/2)}$$
$$- 2U_{i,j}^{(k+1/2)}] + \rho_k[U_{i,j+1}^{(k)} + U_{i,j-1}^{(k)} - 2U_{i,j}^{(k)}] \quad (3\text{-}192)$$

by a single row (line) iteration followed by a single column iteration determined from

$$U_{i,j}^{(k+1)} = U_{i,j}^{(k+1/2)} + \rho_k[U_{i+1,j}^{(k+1/2)} + U_{i-1,j}^{(k+1/2)} - 2U_{i,j}^{(k+1/2)}]$$
$$+ \rho_k[U_{i,j+1}^{(k+1)} + U_{i,j-1}^{(k+1)} - 2U_{i,j}^{(k+1)}]. \quad (3\text{-}193)$$

Equation (3-192), with $\rho_k = \frac{1}{4}$, defines a method which is similar to, but is *not* the same as, the Jacobi row iteration. The quantities ρ_k, called iteration parameters, may depend upon k. In any event it is important that the same values be employed for both parts of the iterative step.

More generally let us suppose that $G(x, y)$ is nonnegative while A and C are positive functions. Let the numerical solution of the self-adjoint equation

$$G(x, y)u - \frac{\partial}{\partial x}\left[A(x, y)\frac{\partial u}{\partial x}\right] - \frac{\partial}{\partial y}\left[C(x, y)\frac{\partial u}{\partial y}\right] = S(x, y) \quad (3\text{-}194)$$

be sought in the interior of a bounded plane region R which takes specified values u on the boundary B of R. A rectangular mesh is chosen in R of mesh lengths h, k. On $R(h, k)$ we approximate $-hk[Au_x]_x$ by Hu and $-hk[Cu_y]_y$ by Vu where H and V are finite difference operators of the form

$$HU_{i,j} = -a_{i,j}U_{i+1,j} + 2b_{i,j}U_{i,j} - c_{i,j}U_{i-1,j} \qquad (3\text{-}195)$$

$$VU_{i,j} = -\alpha_{i,j}U_{i,j+1} + 2\beta_{i,j}U_{i,j} - \gamma_{i,j}U_{i,j-1}. \qquad (3\text{-}196)$$

† A rational explanation, with any generality, of the effectiveness of ADI methods is still lacking.

Several choices for a, b, c, α, β, and γ are possible. Since symmetric matrices are highly desirable, the most common choice for these coefficients is

$$a_{i,j} = kA_{i+1/2,j}/h, \qquad c_{i,j} = kA_{i-1/2,j}/h, \qquad 2b_{i,j} = a_{i,j} + c_{i,j}$$

$$\alpha_{i,j} = hC_{i,j+1/2}/k, \qquad \gamma_{i,j} = hC_{i,j-1/2}/k, \qquad 2\beta_{i,j} = \alpha_{i,j} + \gamma_{i,j}. \tag{3-197}$$

These choices make H and V symmetric matrices. Other possibilities are presented in Birkhoff and Varga [64]. Here, we shall examine only the case for $h = k$. The reader may consult Birkhoff, Varga, and Young [62] for the general case.

The aforementioned discretization defines an approximate solution of the Dirichlet problem for Eqn (3-194) as the algebraic solution of

$$(H + V + \Sigma) U = K. \tag{3-198}$$

The matrix Σ is the *nonnegative diagonal* matrix formed from $h^2G_{i,j}$, while K is the vector formed by adding $h^2S_{i,j}$ to those values from Eqns (3-195) and (3-196) determined from known boundary values of B. H and V have *positive diagonal entries* and *nonpositive off-diagonal elements*. Both H and V are *diagonal dominant* and *positive-definite*† (Varga [25]). If the network $R(h, k)$ is connected, then the matrices $H + V$ and $H + V + \Sigma$ are irreducible. If a Stieltjes matrix is irreducible, then its matrix *inverse has all positive elements* (Varga [25]). By ordering the mesh points by rows one can make H tridiagonal; by ordering them by columns one can make V tridiagonal. Even though H and V are *similar to tridiagonal matrices*, they cannot be made tridiagonal simultaneously!

The approximate solution of Eqn (3-194) for mixed boundary conditions of the form

$$\frac{\partial u}{\partial n} + d(x, y)u = s(x, y), \quad d > 0 \text{ on } B \tag{3-199}$$

can be reduced to a matrix problem of the form of Eqn (3-198) having the same properties. This is also true if the mesh lengths are different or variable (see Varga [25] and Frankel [65]).

The boundary value problem has now been reduced to the solution of an algebraic equation, Eqn (3-198). Rapid solution for large networks is desired.

Equation (3-198) is obviously equivalent, for any matrices D and E, to each of

$$(H + \Sigma + D)U = K - (V - D)U \tag{3-200}$$

$$(V + \Sigma + E)U = K - (H - E)U \tag{3-201}$$

provided $(H + \Sigma + D)$ and $(V + \Sigma + E)$ are nonsingular. Peaceman and Rachford [59] first employed these forms for the case $\Sigma = 0$, $D = E = \rho I$.

---

† Real symmetric positive-definite matrices with nonpositive off-diagonal elements are called *Stieltjes matrices*.

The generalization to $\Sigma \neq 0$ and arbitrary $D = E$ was done by Wachspress and Habetler [66].

Peaceman and Rachford [59] proposed solving Eqn (3-198) in the case $\Sigma = 0$,† $D = E = \rho I$ by choosing an appropriate set of positive numbers ρ_k, called iteration parameters, and calculating the sequence of vectors $U^{(k+1/2)}$, $U^{(k+1)}$ defined by

$$
\begin{align}
(H + \Sigma + \rho_k I)U^{(k+1/2)} &= K - (V - \rho_k I)U^{(k)} \\
(V + \Sigma + \rho_k I)U^{(k+1)} &= K - (H - \rho_k I)U^{(k+1/2)}.
\end{align} \tag{3-202}
$$

The set, Eqns (3-202), formed from Eqns (3-200) and (3-201) with $D_k = E_k = \rho_k I$ can be extended to the case $D_k = \rho_k I$, $E_k = \bar{\rho}_k I$, thereby defining the actual *Peaceman–Rachford method*

$$
\begin{align}
U^{(k+1/2)} &= (H + \Sigma + \rho_k I)^{-1}[K - (V - \rho_k I)U^{(k)}] \\
U^{(k+1)} &= (V + \Sigma + \bar{\rho}_k I)^{-1}[K - (H - \bar{\rho}_k I)U^{(k+1/2)}].
\end{align} \tag{3-203}
$$

If the matrices to be inverted are similar to positive-definite (hence nonsingular) well-conditioned *tridiagonal* matrices, Eqns (3-203) can be rapidly solved by employing the Thomas algorithm. Our main object is to select the initial vector $U^{(0)}$ and the iteration parameters ρ_k and $\bar{\rho}_k$ in order to make the process converge rapidly.

The Douglas–Rachford [61] variant, defined originally for $\Sigma = 0$, can be defined for general Σ by

$$
\begin{align}
U^{(k+1/2)} &= (H_1 + \rho_k I)^{-1}[K - (V_1 - \rho_k I)U^{(k)}] \\
U^{(k+1)} &= (V_1 + \rho_k I)^{-1}[V_1 U^{(k)} - \rho_k U^{(k+1/2)}]
\end{align} \tag{3-204}
$$

where $H_1 = H + \frac{1}{2}\Sigma$ and $V_1 = V + \frac{1}{2}\Sigma$. This amounts to setting $D_k = E_k = \rho_k I - \frac{1}{2}\Sigma$ in Eqns (3-202) and carrying out some elementary manipulations (Problem 3-57). Since H_1 and V_1 are, by suitable rearrangements of their rows and corresponding columns, tridiagonal matrices, the iterative method [Eqns (3-204)] can be carried out directly by the Thomas algorithm. In this variant, as well as Eqns (3-203), the vector $U^{(k+1/2)}$ is treated as an auxiliary vector and is discarded immediately after $U^{(k+1)}$ is calculated.

PROBLEMS

3-56 Let R be the unit square $0 < x < 1$, $0 < y < 1$. Consider the Dirichlet problem for

$$
xy - \frac{\partial}{\partial y}\left[(x + y)^2 \frac{\partial u}{\partial x}\right] - \frac{\partial}{\partial y}\left[(x + y)^2 \frac{\partial u}{\partial y}\right] = 1
$$

with $u(x, 1) = u(x, 0) = 1$, $u(1, y) = 0$, $u(0, y) = 2$. With $h = k = \frac{1}{3}$, write the discretization in the form of Eqn (3-198). Identify H, V, and Σ.

† This is not essential. The case for general Σ is discussed.

3-57 With the aftermentioned assumptions, develop the Douglas–Rachford equations [Eqns (3-204)].

3-58 Eliminate the vector $U^{(k+1/2)}$ between the two equations of (3-203) and hence obtain a linear iterative method of the form $U^{(k+1)} = T_k U^{(k)} + G_k$. Is the method stationary?

3-59 Formulate the Peaceman–Rachford method for Problem 3-56.

3-13 Summary of ADI results

At this writing the study of ADI methods continues. An intensive discourse on these works goes beyond the scope of this volume. Nevertheless, we wish to record here some of the more useful results together with the background references necessary for additional study. The associated equation for each theorem is Eqn (3-194).

(1) Any stationary ADI process with all $D_k = D$ and all $E_k = E$ is convergent provided $\Sigma + D + E$ is symmetric and positive-definite and that $2H + \Sigma + D - E$ and $2V + \Sigma + E - D$ are positive-definite (Birkhoff *et al.* [62]).

(2) If $\rho, \bar{\rho} > 0, 0 \le \theta, \bar{\theta} \le 2$, then the stationary ADI method defined with $\theta' = 2 - \theta$ by

$$(H + \tfrac{1}{2}\theta\Sigma + \rho I)U^{(k+1/2)} = K - (V + \tfrac{1}{2}\theta'\Sigma - \rho I)U^{(k)}$$

$$(V + \tfrac{1}{2}\bar{\theta}\Sigma + \bar{\rho}I)U^{(k+1)} = K - (H + \tfrac{1}{2}\bar{\theta}'\Sigma - \bar{\rho}I)U^{(k+1/2)} \tag{3-205}$$

is convergent.† Consequently the Douglas–Rachford method is convergent for any fixed $\rho > 0$.

(3) Let a be the smallest and b the largest eigenvalues of H_1 and α be the least and β the largest eigenvalues of V_1. Then the spectral radius of the DRADI method is less than or equal to

$$F = \min\left\{ \left(\frac{b - \sqrt{(ab)}}{b + \sqrt{(ab)}}\right)\left(\frac{\beta - \sqrt{(ab)}}{\beta + \sqrt{(ab)}}\right), \left(\frac{\sqrt{(\alpha\beta)} - a}{\sqrt{(\alpha\beta)} + a}\right)\left(\frac{\beta - \sqrt{(\alpha\beta)}}{\beta + \sqrt{(\alpha\beta)}}\right) \right\} \tag{3-206}$$

with the optimum ρ (corresponding to maximum rate of convergence) equal to $\sqrt{(ab)}$ if the first term (in braces) of Eqn (3-206) is smaller than or equal to the second, and $\rho = \sqrt{(\alpha\beta)}$ otherwise. The *convergence rate*, for a single fixed value of ρ, is then at least $-\ln F$.

(4) The rate of convergence of the ADI methods can be appreciably increased by the application of several iteration parameters. These are used successively in a cyclic order. The theory of convergence, and of the selection of good iteration parameters when more than one cycle is used, has not been

---

† This is a generalization of the Douglas–Rachford idea. Here $D = \rho I - \theta\Sigma$, $E = \bar{\rho}I - \bar{\theta}\Sigma$.

fully developed. For those cases which generate matrices H, V, and Σ which are pairwise commutative, a satisfactory theory does exist (Birkhoff and Varga [64]). The asymptotic convergence rates of Douglas and Rachford [61] were shown to apply to the self-adjoint elliptic difference equations [Eqns (3-200) and (3-201)] in a connected plane network if and only if the symmetric matrices H, V, and Σ are pairwise commutative—that is, if and only if

$$HV = VH, \qquad H\Sigma = \Sigma H, \qquad V\Sigma = \Sigma V. \tag{3-207}$$

An equivalent set of assumptions are

$$HV = VH, \qquad \Sigma = \sigma I, \tag{3-208}$$

and H and V are each *similar* to *nonnegative diagonal matrices*.

Wachspress and Habetler [66] observed that one can obtain matrices H, V, and Σ satisfying Eqn (3-208) from differential equations of the form

$$\nu E_2(x)F_1(y)u - F_1(y)\frac{\partial}{\partial x}\left[E_1(x)\frac{\partial u}{\partial x}\right] - E_2(x)\frac{\partial}{\partial y}\left[F_2(y)\frac{\partial u}{\partial y}\right] = S(x, y) \tag{3-209}$$

in the rectangle $R: 0 \leq x \leq X, 0 \leq y \leq Y$. The functions E_1, E_2, F_1, and F_2 are assumed to be continuous and positive in R and $\nu \geq 0$. This is a special case of Eqn (3-194) with $A(x, y) = E_1(x)F_1(y)$, $C(x, y) = E_2(x)F_2(y)$, $G(x, y) = \nu E_2(x)F_1(y)$.

In order to obtain pairwise commutative matrices H, V, and Σ, we first choose mesh sizes h and k such that X/h and Y/k are integers. We then divide Eqn (3-209) by $E_2(x)F_1(y)$, obtaining

$$\nu u - \frac{1}{E_2(x)}\frac{\partial}{\partial x}\left[E_1(x)\frac{\partial u}{\partial x}\right] - \frac{1}{F_1(y)}\frac{\partial}{\partial y}\left[F_2(y)\frac{\partial u}{\partial y}\right] = \frac{S(x, y)}{E_2(x)F_1(y)}. \tag{3-210}$$

Replacing $-hk[E_1u_x]_x$ and $-hk[F_2u_y]_y$ by the expressions given in Eqns (3-195) and (3-196), we get

$$(H + V + \Sigma)U(x, y) = t(x, y)$$

where

$$HU(x, y) = A_0(x)U(x, y) - A_1(x)U(x + h, y) - A_3(x)U(x - h, y)$$

$$VU(x, y) = C_0(y)U(x, y) - C_2(y)U(x, y + k) - C_4(y)U(x, y - k) \tag{3-211}$$

$$\Sigma = hk\nu$$

and

$$t(x, y) = hkS(x, y)/E_2(x)F_1(y)$$

$$A_1(x) = kE_1(x + \tfrac{1}{2}h)/hE_2(x)$$

$$C_2(y) = hF_2(y + \tfrac{1}{2}k)/kF_1(y), \text{ etc.}$$

(5) The Peaceman–Rachford method for solving Eqn (3-198) is defined by

$$(H_1 + \rho_k I)U^{(k+1/2)} = K - (V_1 - \rho_k I)U^{(k)}$$
$$(V_1 + \rho_k I)U^{(k+1)} = K - (H_1 - \rho_k I)U^{(k+1/2)}$$

(3-212)

where $H_1 = H + \frac{1}{2}\Sigma$, $V_1 = V + \frac{1}{2}\Sigma$ and the conditions of Eqn (3-208) hold. Two choices of the iteration parameters are in present usage. One choice was given by Peaceman and Rachford [59] and the other by Wachspress [66, 67]. Let $\bar{a}$ and $\bar{b}$ be such that for all eigenvalues μ of H and ν of V, $\bar{a} \le \mu, \nu \le \bar{b}$.

The parameters of Peaceman and Rachford are

$$\rho_j^{(P)} = \bar{b}(\bar{a}/\bar{b})^{(2j-1)/2m}, \quad j = 1, 2, \ldots, m$$

(3-213)

and those of Wachspress are

$$\rho_j^{(W)} = \bar{b}(\bar{a}/\bar{b})^{(j-1)/(m-1)}, \quad m \ge 2, \quad j = 1, 2, \ldots, m.$$

(3-214)

Neither of these parameter sets is optimum. However, their use makes the Peaceman–Rachford method effective. Estimates of the average rate of convergence and of the optimum choice of the number m of parameters have been developed in Birkhoff *et al.* [62].

For fixed m, let

$$\delta = \frac{1-z}{1+z}, \quad z = c^{1/(2m)}, \quad c = \frac{\bar{a}}{\bar{b}}.$$

(3-215)

It then follows that the average rate of convergence is

$$\bar{R}_m^{(P)} = -\frac{2}{m} \log \delta$$

(3-216)

when the Peaceman–Rachford parameters are used. The optimum value of m, relative to the Peaceman–Rachford parameters, is found by studying the behavior of $\bar{R}_m^{(P)}$ as a function of m, where m is assumed to be a continuous variable (Problem 3-62). The function $\bar{R}_m^{(P)}$ is maximized when

$$\delta = \bar{\delta} = \sqrt{2} - 1 \doteq 0.414$$

(3-217)

and the corresponding value of $\bar{R}_m^{(P)}$ is

$$\bar{R}_m^{(P)} = \frac{4(\log \bar{\delta})^2}{-\log c} \doteq \frac{3.11}{-\log c}.$$

(3-218)

This value $\bar{\delta}$ will generally correspond to a nonintegral value of m, and the actual value of $\bar{R}_m^{(P)}$, for integral m, would generally be less than that given by Eqn (3-218).

In practice, the following procedure is useful:

(i) Estimate $\bar{a}$, $\bar{b}$, and compute $c = \bar{a}/\bar{b}$.
(ii) Find the smallest integer m such that

$$(\bar{\delta})^{2m} \le c, \quad \bar{\delta} = \sqrt{2} - 1.$$

(iii) Determine the iteration parameters by Eqns (3-213).

(iv) Average rate of convergence is

$$\bar{R}_m^{(P)} = -\frac{2}{m} \log \delta, \qquad \delta = \frac{1 - c^{1/(2m)}}{1 + c^{1/(2m)}}.$$

Corresponding results for the Wachspress parameters are determined in a manner similar to that for Peaceman–Rachford parameters. For a fixed number m of Wachspress parameters, let

$$\epsilon = \left(\frac{1 - y}{1 + y}\right)^2, \qquad y = c^{1/2(m-1)} \tag{3-219}$$

whereupon

$$\bar{R}_m^{(W)} = -\frac{2}{m} \log \epsilon. \tag{3-220}$$

The optimum value of m, relative to the Wachspress parameters, is found from the approximate optimum value of

$$\epsilon = \bar{\epsilon} = \delta^2 = (\sqrt{2} - 1)^2 \doteq 0.172$$

by the following practical scheme:

(i) Estimate $\bar{a}, \bar{b}$, and compute $c = \bar{a}/\bar{b}$.

(ii) Find the smallest integer m such that

$$(\bar{\epsilon})^{m-1} \le c.$$

(iii) Determine the iteration parameters by Eqn (3.214).

(iv) $\bar{R}_m^{(W)} = -\dfrac{2}{m} \log \epsilon$,

$$\epsilon = \left[\frac{1 - c^{1/2(m-1)}}{1 + c^{1/2(m-1)}}\right]^2$$

(6) If the optimum m is chosen relative to the Peaceman–Rachford parameters, then

$$\liminf_{c \to 0} \bar{R}_m^{(P)} (-\log c) \ge 3.11$$

and

$$\limsup_{c \to 0} \bar{R}_m^{(P)} (-\log c) \le 4.57.$$

If the optimum m is chosen relative to the Wachspress parameters then

$$\liminf_{c \to 0} \bar{R}_m^{(W)} (-\log c) \ge 6.22$$

and

$$\limsup_{c \to 0} \bar{R}_m^{(W)} (-\log c) \le 7.66.$$

These results suggest that the *Wachspress parameters are superior by a factor of approximately* 2. Extensive numerical experiments tend to confirm this observation, provided one chooses the optimum values of m.

Of course, neither of the above methods results in strictly optimum parameters. Wachspress [68] has devised an algorithm for calculating optimum parameters when the number of parameters, m, is a power of 2. However, for each of five regions† the five-point finite difference analog of the Dirichlet problem for Laplace's equation was solved by Peaceman–Rachford ADI in Birkhoff *et al.* [62]. The results confirm the superiority of the Wachspress parameters over those of Peaceman–Rachford for the unit square, provided one chooses good values of m by the foregoing procedures. But, for the other regions, there was little information to allow a choice between the two sets. The optimum parameters were not appreciably better than those of Wachspress. Because of the theoretical superiority of the Wachspress parameters over those of Peaceman and Rachford, and because the Wachspress parameters are easy to compute as compared with the optimum ones, the authors recommend their use.‡

(7) Research still continues on ADI methods. Some of the results include those of Hubbard [69], who examines alternating direction schemes for the heat equation in general domains. Widlund [70] discusses convergence rates of ADI methods for linear equations

$$-\frac{\partial}{\partial x}\left(a(x)\frac{\partial u}{\partial x}\right) - \frac{\partial}{\partial y}\left(b(y)\frac{\partial u}{\partial y}\right) + c(x, y)u = f(x, y) \qquad (3\text{-}221)$$

with Dirichlet data on compact domains.

Douglas *et al.* [71] describe a *multistage* ADI procedure for solving equations of the form

$$(A_1 + A_2 + A_3 + A_4)U = K \qquad (3\text{-}222)$$

where each A_i is a Hermitian, positive semi-definite operator which can be represented as a tridiagonal matrix. If the domain of integration is bounded, the elliptic equation

$$\nabla \cdot [a(x, y, z)\nabla u] = f(x, y, z) \qquad (3\text{-}223)$$

leads to Eqn (3.222).

Fairweather and Mitchell [72] discuss an alternative computational procedure for ADI methods. Their research was motivated by the results of D'Yakonov [73] who found in the two-dimensional parabolic case that the PRADI and DRADI methods lose accuracy if the boundary conditions are

---

† The regions were the unit square, unit square with a $\frac{4}{10} \times \frac{4}{10}$ square removed from the center, unit square with $\frac{1}{5} \times \frac{1}{5}$ square removed from each corner, L-shaped region, and right isosceles triangle.

‡ Numerical experiments by Tateyama *et al.* [63] on the Dirichlet problem for Laplace's equation in the rectangle, triangle, and L-shaped region (but employing interlaced ADI) show a preference for the Peaceman–Rachford parameters.

time-dependent. (This loss of accuracy was independently found by Fairweather [74].) The Mitchell-Fairweather computational method overcomes this difficulty and must also be adopted to take full advantage of the high-accuracy ADI method of Mitchell and Fairweather [75].

Lynch and Rice [76] investigate the PRADI method for solving elliptic partial difference equations with parameters chosen in such a way that they exploit smoothness properties of the initial error.

Application of ADI methods to hyperbolic problems has been investigated by Lees [77] and Fairweather and Mitchell [78]. Applications to fluid mechanics will be discussed in Chapter 5.

(8) Some of the alternating direction methods are compared below in respect of approximate average rates of convergence for the solution of Laplace's equation with Dirichlet boundary conditions for the square with a square net of side h.

Method	*Convergence rate*
Alternating Direction (Peaceman–Rachford)	
(a) Peaceman–Rachford parameters	
(1) Fixed number m	$\dfrac{4}{m}\left(\dfrac{h}{2}\right)^{1/m}$
(2) Variable number	$-\dfrac{1.55}{\log\,(h/2)}$
(b) Wachspress parameters	
(1) Fixed number m	$\dfrac{8}{m}\left(\dfrac{h}{2}\right)^{1/(m-1)}$
(2) Variable number	$-\dfrac{3.11}{\log\,(h/2)}$

PROBLEMS

3-60 If Eqns (3-195) and (3-196) are applied directly to Eqn (3-209) before division by E_2F_1, do the matrices H and V commute?

Ans: Though symmetric they do not, in general, commute.

3-61 Complete the development of Eqns (3-211) by giving A_0, C_0, A_3, and C_4.

3-62 Solve Eqns (3-215) for m as a function of c and δ and hence obtain an expression for $\bar{R}_m^{(P)}(\delta, c)$. Show that this function is maximized (with respect to δ) when $\delta = \sqrt{2} - 1$. Find the corresponding value of $\bar{R}_m^{(P)}$.

3-63 Consider the Laplace equation in the unit square with $h = \frac{1}{20}$. Using the results of Section 3-6, find the optimum value of m and the corresponding Peaceman–Rachford and Wachspress iteration parameters.

3-14 Some nonlinear examples

(a) Mildly nonlinear elliptic equations

Douglas [84] considers the applicability of alternating direction methods to the Dirichlet problem for the 'mildly' nonlinear elliptic equation

$$u_{xx} + u_{yy} = Q(x, y, u) \tag{3-224}$$

in a rectangle R, under the assumption that $0 < m \leq \partial Q/\partial u \leq M < \infty$ and $u = g(x, y)$ on the boundary of R.

The procedure used by Douglas involves a two-level iteration similar to the inner–outer iteration discussed in Section 3-9. The outer is a modified Picard iteration and the inner is an alternating direction (Peaceman–Rachford) method. Specifically, the five-point computational molecule given by Eqn (3-15) is used for $u_{xx} + u_{yy}$, which we label (after Douglas) $\Delta_x^2 + \Delta_y^2$. Thus the finite difference analog of Eqn (3-224) is

$$
\begin{aligned}
(\Delta_x^2 + \Delta_y^2)u_{i,j} &= Q(x_i, y_j, u_{i,j}) \text{ in } R, \\
u_{i,j} &= g_{i,j} \text{ on the boundary of } R
\end{aligned}
\tag{3-225}
$$

which are nonlinear algebraic equations.

The solution of Eqns (3-225) is done by the Picard type *outer iteration*

$$(\Delta_x^2 + \Delta_y^2)u_{i,j}^{(n+1)} - Au_{i,j}^{(n+1)} = Q(x_i, y_j, u_{i,j}^{(n)}) - Au_{i,j}^{(n)} \tag{3-226}$$

and on the boundary of R, $u_{i,j}^{(n+1)} = g_{i,j}$. The solution of this *linear* system is accomplished by an *inner iteration* which is carried out by means of an alternating direction method. The optimum value of A, in Eqn (3-226), is shown to be

$$A = \tfrac{1}{2}(M + m). \tag{3-227}$$

The choice of such a procedure as Eqn (3-226) improves the operation of the process. The number of sweeps of the alternating direction process required for each outer iteration is $O(-\ln h)$, which leads to an estimate of the total number of calculations required to obtain a uniformly good approximation as $O[h^{-2}(\ln h)^2]$.

A very important result of this analysis is that the number of outer iterations is independent of h!

Direct generalization of the Douglas approach to

$$\nabla \cdot [a(x, y)\nabla u] = Q(x, y, u) \tag{3-228}$$

is difficult using ADI for the inner iterations. Douglas discusses the use of SOR for the inner iterations, thereby extending his scheme to Eqn (3-228) at the expense of additional computation. This method can be generalized to three dimensions.

(b) The equation $\nabla \cdot [F \nabla u] = 0$

The case of interest to us is when

$$F = F[|\nabla u|] = F[(u_x^2 + u_y^2)^{1/2}]. \tag{3-229}$$

Problems leading to such equations occur in heat conduction where the thermal conductivity depends upon $|\nabla u|$, as discussed by Slattery [85] and Serrin [86]. Vertical heat transfer from a horizontal surface by turbulent free convection (see Priestley [87]) and turbulent flow of a liquid with a free surface over a plane (see Philip [88]) have a steady-state mathematical model of the form $\nabla \cdot [F(u) \nabla u] = 0$ or Eqn (3-229).

The partial differential equations in magnetostatics, as they apply in highly saturated rotating machinery, have the form $\nabla \cdot [\mu \nabla V] = 0$ where $V =$ scalar potential, $\mathbf{H} = -\nabla V$. The magnetic permeability $\mu = \mu(H)$† establishes the relation between B and H. Determination of μ has proceeded by means of fitting experimental data with mathematical expressions of the form

$$B = \mu(H)H, \qquad H = [V_x^2 + V_y^2]^{1/2}. \tag{3-230}$$

Fischer and Moser [89] have investigated the fitting of the magnetization curve and tabulate fifteen different fitting functions, some of which are

$$(a + bH)^{-1}, \qquad aH^{-1} \tanh bH, \qquad H^{-1} \exp [H/(a + bH)].$$

Several numerical studies in rectangular and curvilinear geometries have been undertaken by Trutt et al. [90] to obtain solutions of Eqn (3-229) with $\mu = (a + bH)^{-1}$. The calculation uses the standard five-point molecule and the SOR algorithm. The choice of the relaxation parameter, ω, is based upon that for Laplace's equation. Large gradients are anticipated in this problem and, consequently, the evidence of the Bellman et al. [41] calculation suggests that numerical experimentation would be helpful in problems of this type if the optimal value of ω is desired.

The calculation proceeded by selecting initial guesses $\mu^{(0)}$ and $V^{(0)}$ over the domain. At the kth step $V^{(k)}$ and $\mu^{(k)}$ are known. $V^{(k+1)}$ is calculated using SOR with $\mu = \mu^{(k)}$, then $\mu^{(k+1)}$ is calculated using $V = V^{(k+1)}$ with SOR. The domain of integration is the region of Fig. 3-7 containing an air gap between the two iron 'fields.' In addition, the domain contains corners. The interior corners have singularities at which special treatment becomes necessary. These were treated by the technique of mesh refinement—a procedure which will be discussed later in Chapter 5.

(c) Laminar flow of non-Newtonian fluids

As a last example we briefly discuss the problem of determining the laminar

---

† We write **H** for the vector magnetic field and H for its magnitude.

steady flow of a non-Newtonian fluid in a unit square duct. The dimension-less formulation is

$$\frac{\partial}{\partial x}\left[w\frac{\partial u}{\partial x}\right] + \frac{\partial}{\partial y}\left[w\frac{\partial u}{\partial y}\right] + \frac{f \cdot Re}{2} = 0 \qquad (3\text{-}231)$$

$$w = \left[\left(\frac{\partial u}{\partial x}\right)^2 + \left(\frac{\partial u}{\partial y}\right)^2\right]^{(n-1)/2} \qquad (3\text{-}232)$$

$$\int_0^1 \int_0^1 u(x, y)\, dx\, dy = 1 \qquad (3\text{-}233)$$

$u(x, y) = 0$ on the boundary Γ of the unit square $\qquad$ (3-234)

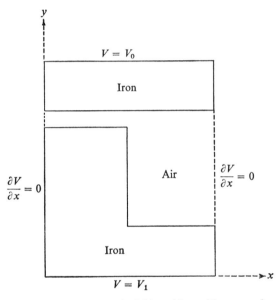

Fig. 3-7 Typical electromagnetic field problem with corner singularities

which is taken as $0 \le x \le 1, 0 \le y \le 1$. These equations involve the dimensionless variables u = velocity and w = viscosity given by the 'power law,' Eqn (3-232), with the non-Newtonian parameter $0 < n \le 1$. Here f is a friction factor and Re the Reynolds number. The ultimate objective is to find $f \cdot Re$ so that these equations hold. The problem is examined in detail by Young and Wheeler [43], and an alternative method has been described by Cryer [91] and Winslow [92].

A preliminary analysis reduces the problem's complexity; in fact it suffices to solve this problem with $f \cdot Re = 2$. Suppose that we solve

$$[WU_x]_x + [WU_y]_y + 1 = 0$$
$$W = [(U_x)^2 + (U_y)^2]^{(n-1)/2} \qquad (3\text{-}235)$$
$$U = 0 \text{ on } \Gamma.$$

Upon setting $u = cU$, Eqn (3-233) becomes

$$\int_0^1 \int_0^1 u \, dx \, dy = c \int_0^1 \int_0^1 U \, dx \, dy = 1$$

or

$$c = \left[\int_0^1 \int_0^1 U \, dx \, dy \right]^{-1}.$$

Since $u = cU$:

$$w = [(u_x)^2 + (u_y)^2]^{(n-1)/2} = c^{n-1}[(U_x)^2 + (U_y)^2]^{(n-1)/2}$$

$$= c^{n-1}W$$

so that

$$[wu_x]_x + [wu_y]_y = c^n\{[WU_x]_x + [WU_y]_y\}$$

$$= c^n(-1) = -\frac{f \cdot Re}{2}.$$

From this relation it follows that after the calculation of U in Eqn (3-235) we may calculate $f \cdot Re$ by

$$\frac{f \cdot Re}{2} = c^n = \left[\int_0^1 \int_0^1 U \, dx \, dy \right]^{-n}. \tag{3-236}$$

The problem's symmetry means that it suffices to solve the problem in one of the four quarters of the square, say $\frac{1}{2} \le x \le 1, \frac{1}{2} \le y \le 1$.

REFERENCES

1. Garabedian, P. R. *Partial Differential Equations.* Wiley, New York, 1964.
2. Hadamard, J. *Lectures on Cauchy's Problem in Linear Partial Differential Equations.* Dover, New York, 1952.
3. Sneddon, I. N. *Elements of Partial Differential Equations.* McGraw-Hill, New York, 1957.
4. Prandtl, L. *Phys. Z.,* **4**, 758, 1903.
5. Timoshenko, S. and Goodier, J. N. *Theory of Elasticity,* 2nd edit. McGraw-Hill, New York, 1951.
6. Householder, A. S. *Principles of Numerical Analysis.* McGraw-Hill, New York, 1953.
7. Bodewig, E. *Matrix Calculus.* Wiley (Interscience), New York, 1956.
8. Gustavson, F., Liniger, W., and Willoughby, R. *Symbolic Generation of an Optimal Crout Algorithm for Sparse Systems of Linear Equations.* IBM Watson Research Center, Yorktown Heights, New York, 1968.
9. Ames, W. F. *Nonlinear Partial Differential Equations in Engineering.* Academic Press, New York, 1965.
10. Taussky, O. *Am. Math. Mon.,* **56**, 672, 1949.
11. Geiringer, H. *Reissner Anniversary Volume,* p. 365. University of Michigan Press, Ann Arbor, 1949.

12. Young, D. M. and Frank, T. G. *A Survey of Computer Methods for Solving Elliptic and Parabolic Partial Differential Equations*, vol. 2, no. 1. Rome Centre Int. de Calcul, 1963.
13. Jacobi, C. G. J. *Gessamelte Werke*, vol. 3, p. 467. Berlin, 1884.
14. Forsythe, G. E. and Wasow, W. R. *Finite Difference Methods for Partial Differential Equations*. Wiley, New York, 1960.
15. Gerling, C. L. *Die Ausgleichs-Rechnungen der Practischen Geometrie*. Hamburg and Gotha, 1843.
16. Seidel, L. *Abh. bayer Akad. Wiss.*, **11**, 81, 1874.
17. Southwell, R. V. *Relaxation Methods in Engineering Science*. Clarendon Press, Oxford, 1940.
18. Southwell, R. V. *Relaxation Methods in Theoretical Physics*. Clarendon Press, Oxford, 1946.
19. Frankel, S. P. *Mathl. Tabl. natn. Res. Coun., Wash.*, **4**, 65, 1950.
20. Young, D. M. *Trans. Am. Math. Soc.*, **76**, 92, 1954.
21. Friedman, B. The iterative solution of elliptic difference equations, *Rept No. NYO*-7698. Courant Institute of Mathematical Sciences, New York University, New York, 1957.
22. Keller, H. B. *Q. appl. Math.*, **16**, 209, 1958.
23. Arms, R. J., Gates, L. D., and Zondek, B. *J. Soc. ind. appl. Math.*, **4**, 220, 1956.
24. Collatz, L. *Z. angew. Math. Mech.*, **22**, 357, 1942.
25. Varga, R. S. *Matrix Iterative Numerical Analysis*. Wiley, New York, 1962.
26. Householder, A. S. *J. Ass. comput. Mach.*, **5**, 205, 1958.
27. Stein, P. and Rosenberg, R. L. *J. London math. Soc.*, **23**, 111, 1948.
28. Todd, J. (ed.). *Survey of Numerical Analysis*, chap. 6. McGraw-Hill, New York, 1962.
29. Crandall, S. H. *Engineering Analysis*. McGraw-Hill, New York, 1956.
30. Reich, E. *Ann. math. Statist.*, **20**, 448, 1949.
31. Ostrowski, A. M. *Univ. Roma Ist. naz. alta Mat. rend. Mat. Appl.*, **13**, 140, 1954.
32. Guillemin, E. A. *The Mathematics of Circuit Analyses*, p. 154. Wiley, New York, 1949.
33. Lyusternik, L. A. *Trudy mat. Inst. V. A. Steklova*, **20**, 49, 1947.
34. Aitken, A. C. *Proc. R. Soc. Edinb.*, **62**, 269, 1937.
35. Carré, B. A. *Comput. J.*, **4**, 73, 1961.
36. Bellman, R., Juncosa, M., and Kalaba, R. Some numerical experiments using Newton's method for nonlinear parabolic and elliptic boundary value problems, *Rept. No. P*-2200. Rand Corp., Santa Monica, California, 1961.
37. Garabedian, P. R. *Mathl. Tabl. natn. Res. Coun., Wash.*, **10**, 183, 1956. *See also* The mathematical theory of three-dimensional cavities and jets, *Bull. Am. math. Soc.*, **62**, 219, 1956.
38. van de Vooren, A. I. and Vliegenthart, A. C. *J. engng. Math.*, **1**, 187, 1967.
39. Kahan, W. Gauss–Seidel method of solving large systems of linear equations. Ph.D. dissertation, University of Toronto, Canada, 1958.
40. Padmanabhan, H. I., Ames, W. F., and Kennedy, J. F. Wake deformation in density stratified fluids. Paper presented at October 1968 Soc. Ind. Appl. Math., Philadelphia Meeting.
41. Bellman, R., Juncosa, M. and Kalaba, R. Some numerical experiments using Newton's method for nonlinear parabolic and elliptic boundary value problems, *Rept. No. P*-2200. Rand Corp., Santa Monica, California, 1961.

42. Douglas, J., Jr. *Num. Math.*, **3**, 92, 1961.
43. Young, D. M. and Wheeler, M. F. Alternating direction methods for solving partial difference equations, in *Nonlinear Problems of Engineering* (W. F. Ames, ed.), p. 220. Academic Press, New York, 1964.
44. Forsythe, G. E. *Bull. Am. math. Soc.*, **59**, 299, 1953.
45. Stiefel, E. *Comment. math. Helvet.*, **29**, 157, 1955.
46. Richardson, L. F. *Phil. Trans. R. Soc.*, **A210**, 307, 1910.
47. Shortley, G. *J. appl. Phys.*, **24**, 392, 1953.
48. Young, D. *J. Math. Phys.*, **32**, 254, 1954.
49. Markhoff, W. *Math. Ann.*, **77**, 213, 1916 (translated by J. Grossmann from Russian original).
50. Young, D. and Warlick, C. H. On the use of Richardson's method for the numerical solution of Laplace's equation on the ORDVAC, *BRL Mem. Rept.* 707. Aberdeen Proving Ground, Aberdeen, Maryland, 1953.
51. Young, D. *J. Math. Phys.*, **32**, 243, 1954.
52. Snyder, M. A. *Chebyshev Methods in Numerical Approximation.*, Prentice-Hall Inc., Englewood Cliffs, N.J., 1966.
53. Varga, R. S. *J. Soc. ind. appl. Math.*, **5**, 39, 1957.
54. Golub, G. H. and Varga, R. S. *Num. Math.*, **3**, 147, 1961.
55. Sheldon, J. *Mathl. Tabl. natn. Res. Coun., Wash.*, **9**, 101, 1955.
56. Sheldon, J. *J. Ass. comput. Mach.*, **6**, 494, 1959.
57. Cuthill, E. H. and Varga, R. S. *J. Ass. comput. Mach.*, **6**, 236, 1959.
58. Varga, R. S. Factorization and normalized iterative methods, in *Boundary Problems in Differential Equations* (R. E. Langer, ed.). University of Wisconsin Press, Madison, Wisconsin, 1960.
59. Peaceman, D. W. and Rachford, H. H., Jr. *J. Soc. ind. appl. Math.*, **3**, 28, 1955.
60. Douglas, J., Jr. *J. Soc. ind. appl. Math.*, **3**, 42, 1955.
61. Douglas, J., Jr and Rachford, H. H., Jr. *Trans. Am. math. Soc.*, **82**, 421, 1956.
62. Birkhoff, G., Varga, R. S., and Young, D. Alternating direction implicit methods, in *Advances in Computers* (F. L. Alt and M. Rubinoff, eds), pp. 189–273. Academic Press, New York, 1962.
63. Tateyama, N., Umoto, J., and Hayashi, S. *Mem. Fac. Engng Kyoto Univ.*, **29**, 149, 1967.
64. Birkhoff, G. and Varga, R. S. *Trans. Am. math. Soc.*, **92**, 13, 1959.
65. Frankel, S. *Mathl. Tabl. natn. Res. Coun., Wash.*, **4**, 65, 1950.
66. Wachspress, E. L. and Habetler, G. J. *J. Soc. ind. appl. Math.*, **8**, 403, 1960.
67. Wachspress, E. L. CURE: A generalized two-space dimension multigroup coding for the IBM 704, *Knolls Atomic Power Lab. Rept No. KAPL1724.* General Electric Co., Schenectady, New York, 1957.
68. Wachspress, E. L. *J. Soc. ind. appl. Math.*, **10**, 339, 1962.
69. Hubbard, B. E. *Soc. ind. appl. Math. J. Num. Analysis*, **2**, 448, 1965.
70. Widlund, O. B. *Maths Comput.*, **21**, 500, 1966.
71. Douglas, J., Jr, Garder, A. O., and Pearcy, C. *Soc. ind. appl. Math., J. Num. Analysis*, **3**, 570, 1966.
72. Fairweather, G. and Mitchell, A. R. *Soc. ind. appl. Math., J. Num. Analysis*, **4**, 163, 1967.
73. D'Yakonov, Ye. G. *Zh. vychisl. Mat. Mat. Fiz.*, **2**, 549, 1962; **3**, 385, 1963.
74. Fairweather, G. Doctoral thesis, University of St Andrews, St Andrews, Scotland, 1965.
75. Mitchell, A. R. and Fairweather, G. *Num. Math.*, **6**, 285, 1964.

76. Lynch, R. E. and Rice, J. R. *Maths Comput.*, **22**, 311, 1968.
77. Lees, M. *J. Soc. ind. appl. Math.*, **10**, 610, 1962.
78. Fairweather, G. and Mitchell, A. R. *J. Inst. math. Appl.*, **1**, 309, 1965.
79. Kulsrud, H. E. *Communs Ass. comput. Mach.*, **4**, 184, 1961.
80. Reid, J. K. *Comput. J.*, **9**, 200, 1966.
81. Rigler, A. K. *Maths. Comput.*, **19**, 302, 1965.
82. Wachspress, E. *Iterative Solution of Elliptic Systems and Applications to the Neutron Diffusion Equations of Reactor Physics.* Prentice-Hall, Englewood Cliffs, N.J., 1966.
83. Hageman, L. A. and Kellogg, R. B. *Maths. Comput.*, **22**, 60, 1968.
84. Douglas, J., Jr. *Num. Math.*, **3**, 92, 1961.
85. Slattery, J. C. *Appl. sci. Res.*, **A12**, 51, 1963.
86. Serrin, J. In *Handbuch der Physik* (S. Flugge, ed.), vol. 8, pt I, p. 255. Springer, Berlin, 1959.
87. Priestly, C. H. B. *Aust. J. Phys.*, **1**, 176, 1954.
88. Philip, J. R. *Aust. J. Phys.*, **9**, 570, 1956.
89. Fischer, J. and Moser, H. *Arch. Elektrotech*, **42**, 286, 1956.
90. Trutt, F. C., Erdelyi, E. A., and Jackson, R. F. *I.E.E.E. Trans. Aerospace*, **1**, 430, 1963.
91. Cryer, C. W. *J. Ass. comput. Mach.*, **14**, 363, 1967.
92. Winslow, A. M. *J. comput. Phys.*, **1**, 149, 1966.

4

Hyperbolic equations

4-0 Introduction

Initial value or propagation problems are described by parabolic and hyperbolic equations. The former was the subject of our discussion in Chapter 2. Hyperbolic equations arise in transport (neutron diffusion and radiation transfer), wave mechanics, gas dynamics, vibrations, and other areas.

A convenient vehicle for these introductory remarks will be the simple wave equation

$$u_{tt} - u_{xx} = 0 \tag{4-1}$$

whose general solution, obtained by D'Alembert, can be calculated. If u_{tt} and u_{xx} are continuous, the change to (characteristic) variables

$$\theta = x + t, \quad \psi = x - t, \quad u(x, t) = v(\theta, \psi)$$

changes Eqn (4-1) into $v_{\theta\psi} = 0$, whose solution is

$$v = f(\theta) + g(\psi)$$

where f and g are arbitrary differentiable functions. If we insist that $v_{\theta\psi} = v_{\psi\theta}$, there are no other solutions. Consequently,

$$u(x, t) = f(x + t) + g(x - t) \tag{4-2}$$

is a solution of Eqn (4-1) if f and g are twice differentiable but otherwise arbitrary.

For the pure initial value problem, we now prescribe the initial conditions

$$u(x, 0) = F(x), \quad u_t(x, 0) = G(x). \tag{4-3}$$

A solution of the form of Eqn (4-2) will satisfy these if

$$f(x) + g(x) = u(x, 0) = F(x)$$

$$f'(x) - g'(x) = u_t(x, 0) = G(x).$$

Upon differentiating the first of these equations, two linear algebraic equations for f' and g' are obtained. Solving these and integrating leads to the two solutions

$$f(x) = \frac{1}{2}\left\{ F(x) + \int_0^x G(\eta)\, d\eta \right\} + C$$

$$g(x) = \frac{1}{2}\left\{ F(x) - \int_0^x G(\eta)\, d\eta \right\} + D$$

where C and D are constants of integration. By employing Eqn (4-2) we have

$$u(x, t) = \frac{1}{2}\left\{F(x + t) + F(x - t) + \int_{x-t}^{x+t} G(\eta)\,d\eta\right\} + E. \qquad (4\text{-}4)$$

However, $u(x, 0) = F(x) = F(x) + E$, so that $E = 0$. Consequently, the solution of the pure initial value problem defined by Eqns (4-1) and (4-3) is

$$u(x, t) = \frac{1}{2}\left\{F(x + t) + F(x - t) + \int_{x-t}^{x+t} G(\eta)\,d\eta\right\}. \qquad (4\text{-}5)$$

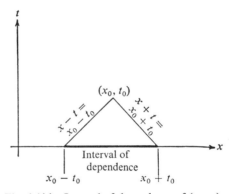

Fig. 4-1(a) Interval of dependence of (x_0, t_0)

Some important observations are immediately evident from Eqn (4-5). The value of the solution at a point (x_0, t_0) is

$$u(x_0, t_0) = \frac{1}{2}\left\{F(x_0 + t_0) + F(x_0 - t_0) + \int_{x_0-t_0}^{x_0+t_0} G(\eta)\,d\eta\right\}$$

see Fig. 4-1(a). Thus, the value of the solution at (x_0, t_0) *depends only upon the initial data* on that segment of the x-axis cut out by the lines $x - t = x_0 - t_0$ and $x + t = x_0 + t_0$. This segment is called the *interval of dependence* of the point (x_0, t_0). Conversely, the set of points (x, t) at which the solution is influenced by the initial data at a point $(x_0, 0)$ on the x-axis is the region bounded by the lines $x + t = x_0$ and $x - t = x_0$, as shown in Fig. 4-1(b). This region is called the *domain of influence*† of the point $(x_0, 0)$. Thus, we see that the characteristics ($x \pm t = $ constant of the equation $u_{xx} - u_{tt} = 0$) play a basic role in developing solutions for hyperbolic equations.

The concepts of domains of dependence and influence and of characteristics are fundamental to all hyperbolic equations. The extent to which initial and boundary conditions determine *unique* solutions can be deduced for a large number of cases from the following three theorems which are described for the

---

† There are corresponding domains for negative t since Eqn (4-5) holds in that case also.

two simultaneous first-order *quasilinear* equations [Eqns (1-8)] previously employed in Chapter 1,

$$a_1 u_x + b_1 u_y + c_1 v_x + d_1 v_y = f_1$$
$$a_2 u_x + b_2 u_y + c_2 v_x + d_2 v_y = f_2.$$

(4-6)

Proofs of these results and additional information may be found in Bernstein [1], Garabedian [2], Courant and Friedrichs [3], and Courant and Hilbert [4].

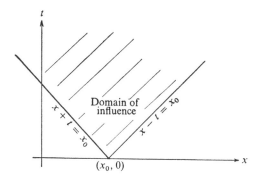

Fig. 4-1(b)   Domain of influence of $(x_0, 0)$

(1) Let us suppose that continuously differentiable values of u and v are specified on the *noncharacteristic* curve CD of Fig. 4-2(a). We also assume CD to be continuously differentiable. A solution to Eqns (4-6), assuming these prescribed values, is uniquely determined in the region CDE bounded by the initial curve CD, the β characteristic CE, and the α characteristic DE.† The direction of propagation is assumed to be upward, but if it were reversed there would be a corresponding 'triangle' of uniqueness below CD.

(2) Let us suppose that CD is a noncharacteristic curve which is continuously differentiable. A unique solution is determined in the region CDEF of Fig. 4-2(b) where DE and EF are characteristics, provided that u *and* v are known at C and continuously differentiable values of u *or* v are given along each of the segments CD and CF. The values at C must be compatible with the characteristics. A unique solution can sometimes be assured even when a discontinuity appears at C.

(3) In the case sketched in Fig. 4-2(c), with CE and CD characteristics, a unique solution is determined in the region CDFE where EF and FD are characteristics, when u and v are known at C and continuously differentiable values of u or v are given along CE and CD. The values at C must be compatible with the characteristics.

In the results just stated it is assumed that no boundary interference or other obstruction is present in the considered domain. It is quite possible that

† We 'label' the characteristics 'α' and 'β' for identification purposes only.

unanticipated boundaries, such as shock waves, flame fronts, and other discontinuities, may appear within the solution domain. Such discontinuities of properties are propagated by their own special laws and represent boundaries between regions where different equations must be solved. Usually these locations *are not* known in advance and must be determined by a simultaneous computation with the continuous solutions.

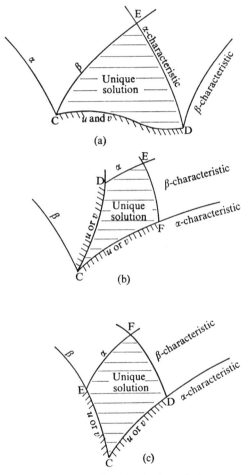

Fig. 4-2 Uniqueness domains for hyperbolic systems

As an illustration of the importance of the uniqueness question, we consider the simple first-order equation†

$$2u_x + u_y = 1. \tag{4-7}$$

† The characteristics for the quasilinear equation $au_x + bu_y = c$ are determined from $dx/a = dy/b$. Along the characteristics we have $dx/a = du/c$. We discuss the general case subsequently.

The characteristics are the straight lines $y = \frac{1}{2}x + e$, and along the characteristics we have $u = \frac{1}{2}x + g = y + g - e$ (Fig. 4-3). If the initial data $u(x_i, 0) = u_i$, for u, are specified on the noncharacteristic line segment $y = 0$, $0 < x < 1$, then the value of u for $y > 0$ is obtained by integrating along the characteristics drawn from the points x_i on the initial line segment. Thus $u(x, y)$, from the solution on the characteristics, is

$$u(x, y) = u_i + \frac{1}{2}(x - x_i) = u_i + y \qquad (4\text{-}8)$$

on the line $y = \frac{1}{2}(x - x_i)$ for each x_i, $0 < x_i < 1$. Further, we find from Bernstein [1] that the solution is unique in the region bounded by the terminal characteristics originating at $x = 0$ and $x = 1$ in Fig. 4-3.

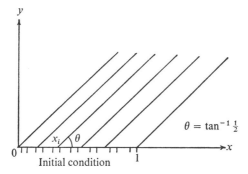

Fig. 4-3 Characteristics for the first-order equation [Eqn (4-7)]

On the other hand, if the initial curve is one of the characteristics, say the line $x/2 = y$, passing through the origin, the situation is quite different. The 'nonuniqueness condition' associated with the characteristics makes further investigation mandatory. There is a possibility of discontinuous first partial derivatives u_x and u_y on $y = x/2$. From the relations along the characteristics we see that there is a solution only if the initial data are $u = x/2 = y$ on $y = x/2$. Elsewhere the solution is not unique since we can take, for example,

$$u = x/2 + A(y - x/2)$$

which is a solution *for any value of A*. The nonuniqueness results from the fact that effectively the 'terminal characteristics' are coincidental.

The preceding arguments can be applied to second-order equations and first-order systems with qualitatively similar results. In general, if the initial curve is a characteristic we may have no solution at all (if the initial data are not properly chosen) and in any case *no unique* solution.

From Chapter 1 we see that the characteristic curves are loci of possible discontinuities in the derivatives of u and v. The possibility of propagation of discontinuous initial values into the field can be discussed employing the example previously given, Eqn (4-7). Suppose, in that example, that the data on the initial line, $y = 0$, are prescribed as

$$u = f(x), \quad 0 < x < x_1; \quad u = g(x), \quad x_1 < x < 1. \tag{4-9}$$

Further, suppose $f(x_1) \neq g(x_1)$ so that u is double valued at $x = x_1$. From the solution [Eqn (4-8)] along the characteristics, this double valued nature will persist all along the specific characteristic $y = \frac{1}{2}(x - x_1)$. The values to the left of this characteristic will be determined by $u = f(x)$ and to the right by $g(x)$.

If the initial data are such that there is a discontinuous slope, this will also propagate into the integration field. For example, if on $y = 0$,

$$u = x^2 \text{ for } 0 < x < \tfrac{1}{2}, \quad u = -x + \tfrac{1}{2} \text{ for } \tfrac{1}{2} < x < 1 \tag{4-10}$$

the solution of the differential equation is

$$u(x, y) = x_i^2 + y$$

on the line $y = \frac{1}{2}(x - x_i)$ to the left of $y = \frac{1}{2}(x - \frac{1}{2})$ and

$$u(x, y) = \tfrac{1}{2}[x - 3x_i + 1]$$

on the line $y = \frac{1}{2}(x - x_i)$ to the right of $y = \frac{1}{2}(x - \frac{1}{2})$. There are discontinuities in u_x and u_y along the characteristic $y = \frac{1}{2}(x - \frac{1}{2})$.

If, in more general problems, a discontinuity exists across a certain characteristic at one point, there will be a discontinuity across that characteristic along its entire length. We have also noted that characteristic curves are the natural boundaries for determining which portions of a solution domain are influenced by which boundary conditions. Thus we see that propagated discontinuities and the segmenting by the characteristics restricts the use of finite difference methods. *Care must be exercised to ensure that discrete models of the continuous system reflect all these facts.* The characteristics may not pass through many points of, say, a rectangular grid and any propagated discontinuities would give rise to difficult computational problems.

4-1   The quasilinear system

Knowledge of the characteristics is important in the development and understanding of numerical methods for hyperbolic systems. The so-called 'method of characteristics' is the *natural* numerical procedure for hyperbolic systems in two independent variables. The rationale underlying their use is that, by an appropriate choice of coordinates, the original system of hyperbolic first-order equations can be replaced by a system expressed in characteristic coordinates. Characteristic coordinates are the *natural* coordinates of the system in the sense that, in terms of these coordinates, differentiation is much simplified, as we shall see. This reduction becomes particularly simple when applied to two equations in two independent variables.

Consider the quasilinear hyperbolic system

$$\sum_{i=1}^{n} \{a_{ji}u_x^i + b_{ji}u_y^i\} + d_j = 0, \quad j = 1, 2, \ldots, n \tag{4-11}$$

of n coupled equations in n unknowns u^i, $i = 1, 2, \ldots, n$, where the coefficients a_{ji}, b_{ji}, and d_j are functions of x, y, and the u^i. If the coefficients are independent of the u^i, the system is linear. The method of the sequel applies equally well to linear or quasilinear systems.

The assumption of a quasilinear system of the form of Eqn (4-11) is no essential restriction. Many nonlinear initial value problems can be transformed into quasilinear form containing a large number of equations and unknowns (see Problems 4-1, 4-2).

We now adopt the matrix notation for Eqn (4-11) by setting

$$A = [a_{ji}], \qquad B = [b_{ji}], \qquad d = [d_j], \qquad u = [u^i]$$

where A and B are $n \times n$ matrices, and d and u are column vectors. The system takes the matrix form

$$Au_x + Bu_y + d = 0\dagger \tag{4-12}$$

Equation (4-12) is subjected to the linear transformation

$$v = Tu \tag{4-13}$$

where the determinant of $T \neq 0$, $T = [t_{ji}]$, and the t_{ji} may depend upon x, y, and the u^i but not upon the derivatives of the u^i. Such a transformation has the form

$$v^j = \sum_{i=1}^{n} t_{ji}u^i, \quad j = 1, 2, \ldots, n \tag{4-14}$$

and under such a transformation Eqn (4-12) takes a new but similar form. The new system is equivalent to the original in the sense that every solution of one is also a solution of the other.

The linear transformation of Eqn (4-12),

$$TAu_x + TBu_y + Td = 0 \tag{4-15}$$

is used to develop a *canonical form*. A convenient one is such that

$$TA = ETB \tag{4-16}$$

where E is a diagonal matrix, say,

$$E = \begin{bmatrix} e_1 & 0 & \ldots & 0 \\ 0 & e_2 & \ldots & 0 \\ & & \ddots & \\ 0 & 0 & \ldots & e_n \end{bmatrix} \tag{4-17}$$

† The notation u_x is used to mean the vector formed from the x derivatives of the u^i, $i = 1, 2, \ldots, n$.

Under the assumption of Eqn (4-16) we may rewrite Eqn (4-15) as

$$ETBu_x + TBu_y + Td = 0. \tag{4-18}$$

Let us examine the form of these equations by setting $TB = A^* = [a_{ji}^*]$, $Td = d^* = [d_j^*]$, so that Eqn (4-18) becomes

$$EA^*u_x + A^*u_y + d^* = 0. \tag{4-19}$$

The jth equation takes the form

$$\sum_{i=1}^{n} a_{ji}^* (e_j u_x^i + u_y^i) + d_j^* = 0. \tag{4-20}$$

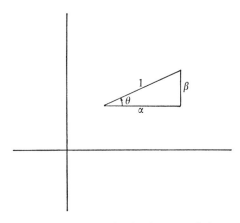

Fig. 4-4 Notation used in the development of characteristics

Now, if $\alpha\,\mathbf{i} + \beta\,\mathbf{j}$ is the unit vector for which

$$e_j = \frac{\alpha}{\beta} = \cot\theta$$

we can write (see Fig. 4-4)

$$e_j u_x^i + u_y^i = \frac{\alpha}{\beta} u_x^i + u_y^i$$

$$= \frac{1}{\beta} \{u_x^i \alpha + u_y^i \beta\}$$

$$= \frac{1}{\beta} \{u_x^i \cos\theta + u_y^i \sin\theta\} \tag{4-21}$$

which, except for the factor $1/\beta$, is the directional derivative in the direction defined by the vector $\alpha\mathbf{i} + \beta\,\mathbf{j}$ (Fig. 4-4). This, of course, depends upon e_j and hence on j. Thus every equation of the transformed system, Eqn (4-19), *contains differentiation in one direction only.* The removal of the complication of having more than one differential operator in each equation brings the theory closer to that of ordinary differential equations.

Calculation of the diagonal matrix E is accomplished from its definition $TA = ETB$. This is equivalent to the system of equations

$$\sum_{k=1}^{n} t_{jk} a_{ki} = \sum_{k=1}^{n} e_j t_{jk} b_{ki}; \quad i = 1, 2, \ldots, n$$

or   (4-22)

$$\sum_{k=1}^{n} (a_{ki} - e_j b_{ki}) t_{jk} = 0$$

which is a system of n homogeneous equations for the t_{jk}, $k = 1, 2, \ldots, n$. For a nontrivial solution to exist, the necessary and sufficient condition is

$$\det (A - e_j B) = 0 \qquad (4-23)$$

which is an algebraic equation generally involving x, y, and u^i, since A and B are functions of those variables.

We now consider the case in which

$$\det (A - \lambda B) = 0$$

has n *distinct real roots*.† In this case we choose the n roots (they are really eigenvalues) for e_j, $j = 1, 2, \ldots, n$, and determine the elements of the T matrix from Eqns (4-22). (We are tacitly assuming that $\det B \neq 0$, $\det T \neq 0$, and $\det A^* = \det TB \neq 0$ in all of this discussion.)

Equation (4-19) is called the *normal form*. The direction $\alpha_k \mathbf{i} + \beta_k \mathbf{j}$, for which $e_k = \alpha_k/\beta_k$, is called the kth characteristic direction. The n differential equations

$$\cot \theta_k = \frac{dx}{dy} = e_k \qquad (4-24)$$

are called the *characteristics* of the system. Their direction, at every point, is precisely the characteristic direction there. The 'method of characteristics' has the foregoing as its theoretical foundation.

Generally, the characteristic directions are not known until the solution of the problem is obtained. However, for a *system of two equations* the concept of characteristics can nevertheless be used to transform the given system into a simpler system. The simpler system, called the *canonical equations*, consists of four rather than two equations—this is acceptable since the new system is easier to compute.

The starting point for the development of the canonical system (for two equations) is the observation that the characteristics form two one-parameter families of curves that may be considered as (not necessarily orthogonal)

---

† This case is probably the most useful. However, hyperbolic systems may have multiple roots. The details may be found in Courant and Hilbert [4].

natural coordinates. Through every point there passes one curve of each family and these curves have different directions at each point since the e_j are usually functions of not only the independent but also the dependent variables.

If these curves are to be used as the coordinate axes for a new curvilinear coordinate system, then which of the infinitely many transformations

$$x = x(\alpha, \beta)$$
$$y = y(\alpha, \beta) \tag{4-25}$$

shall we use? A particularly useful one is to identify α with the e_1 family and β with the e_2 family so that

$$x_\alpha = \frac{dx}{dy} y_\alpha = e_1 y_\alpha \tag{4-26}$$

$$x_\beta = e_2 y_\beta.$$

The quantities $e_j u_x^i + u_y^i, j = 1, 2$, then become

$$e_1 u_x^i + u_y^i = \frac{x_\alpha}{y_\alpha} u_x^i + u_y^i$$

$$= \frac{1}{y_\alpha} \left\{ \frac{\partial u^i}{\partial x} \frac{\partial x}{\partial \alpha} + \frac{\partial u^i}{\partial y} \frac{\partial y}{\partial \alpha} \right\}$$

$$= \frac{u_\alpha^i}{y_\alpha} \tag{4-27}$$

and

$$e_2 u_x^i + u_y^i = \frac{u_\beta^i}{y_\beta}. \tag{4-28}$$

Whereupon the system $EA^* u_x + A^* u_y + d^* = 0$, which is equivalent to

$$a_{11}^*(e_1 u_x^1 + u_y^1) + a_{12}^*(e_1 u_x^2 + u_y^2) + d_1^* = 0$$

$$a_{21}^*(e_2 u_x^1 + u_y^1) + a_{22}^*(e_2 u_x^2 + u_y^2) + d_2^* = 0,$$

becomes

$$a_{11}^* u_\alpha^1 + a_{12}^* u_\alpha^2 + d_1^* y_\alpha = 0$$

$$a_{21}^* u_\beta^1 + a_{22}^* u_\beta^2 + d_2^* y_\beta = 0. \tag{4-29}$$

Equations (4-29), together with

$$x_\alpha = e_1 y_\alpha$$
$$x_\beta = e_2 y_\beta, \tag{4-30}$$

are the *canonical equations* of the system and have the dependent variables u^1, u^2, x, and y. They can be constructed from the original system without previously solving it.

The canonical hyperbolic system shares with the normal form the property that each equation has differentiations in one direction only. Moreover, in the canonical case, these directions coincide with the coordinate directions. This considerable improvement is supplemented by the fact that the system does not contain the independent variables, α and β, explicitly.

According to our derivation, every solution of the original set satisfies this canonical characteristic system. The converse—every solution of the canonical system satisfies the original system provided the Jacobian $x_\alpha y_\beta - x_\beta y_\alpha = (e_1 - e_2) y_\alpha y_\beta \neq 0$—is easily verified.

When the two differential equations are linear, e_1 and e_2 are known functions of x and y. Thus Eqns (4-30) are not coupled with Eqns (4-29). Consequently, Eqns (4-30) determine two families of characteristic curves independent of the solution.

In the reducible case (see Problem 1-8), where d_1^*, d_2^* are both zero and the other coefficients depend on u^1, u^2 only, the situation is similar to the linear case. Then e_1 and e_2 are known functions of u and v and Eqns (4-29) are independent of x and y and can thus be solved separately. If d_1^* and d_2^* do not vanish, but depend on u^1 and u^2, then these remarks are also valid.

For reducible equations, the characteristic curves in the (u, v) plane (the images of the characteristics in the (x, y) plane) are independent of the special solution u, v considered. They are obtainable directly from Eqns (4-29) as

$$a_{11}^* \frac{du^1}{du^2} = -a_{12}^*$$

$$a_{21}^* \frac{du^1}{du^2} = -a_{22}^*.$$

Before discussing the details of the numerical method, we will examine two problems in Section 4-2.

PROBLEMS

4-1 Transform $u_{tt} = (1 + \epsilon u_x)^\alpha u_{xx}$ into two first-order equations. Find the characteristics and the canonical equations.

4-2 Apply the quasilinear theory to the dimensionless equations for the longitudinal sound wave propagation in a one-dimensional gas (see Ames [5])

$$u_t + uu_x + c^2 \rho^{-1} \rho_x = 0, \qquad \rho_t + (\rho u)_x = 0, \qquad c = \rho^{(\gamma - 1)/2}$$

where γ is the ratio of specific heats. Find the characteristics and the canonical equations.

$$Ans: \quad \frac{dx}{dt}\bigg|_{\alpha, \beta} = u \pm c; \quad \rho u_\alpha + c\rho_\alpha = 0, \quad \rho u_\beta - c\rho_\beta = 0,$$

$$x_\alpha = (u + c)t_\alpha, \, x_\beta = (u - c)t_\beta$$

4-3   By setting $u_t = q$ and $u_x = p$ transform the equations for the constant tension moving threadline (see Ames *et al.* [6])

$$u_{tt} + 2Vu_{xt} + [V^2 - c^2(1 + u_x^2)^{-1}]u_{xx} = 0$$

$$V_t + VV_x = c^2(1 + u_x^2)^{-1/2}[(1 + u_x^2)^{-1/2}]_x \qquad (4\text{-}31)$$

$$[(1 + u_x^2)^{1/2}]_t + [V(1 + u_x^2)^{1/2}]_x = 0, \quad c \text{ constant}$$

into *four* first-order equations. The dependent variables are p, q, and V. Show that there are only *three* characteristics and that the system is not overdetermined.

Ans: $V \pm c(1 + p^2)^{-1/2}, V$

4-2   Introductory examples

(a) Direct calculation of the primitive variables

Characteristics have been employed by Swope and Ames [7] to investigate the linear transverse vibrations of a moving threadline forced at one end. This problem represents an example of the type permitting direct calculation of the primitive variable(s).

The linear dimensionless threadline equation is

$$y_{tt} + \alpha y_{xt} + \beta y_{xx} = 0 \qquad (4\text{-}32)$$

where $\alpha = 2V/c$ and $\beta + 1 = \alpha^2/4$. Equation (4-32) is reducible to two first-order equations, in the classical manner, by setting $p = y_x, q = y_t$, whereupon we find

$$q_t + \alpha q_x + \beta p_x = 0$$
$$q_x - p_t = 0. \qquad (4\text{-}33)$$

The solution of this system generates the first partial derivatives and a further investigation of, say, $dy = p\,dx + q\,dt$ is required to obtain the transverse displacement y.†

An alternative is sometimes possible, and, where practical, saves computation. Since our goal is the *direct* computation of y, we reduce Eqn (4-32) to first-order equations by choosing two such equations from a general class with arbitrary coefficients. The coefficients are then determined by recombining the lower order equations and matching them with the original form.

For the problem at hand let the two first-order equations be‡

$$a_1 y_x + b_1 u_t + d_1 y_t = 0$$
$$a_2 y_x + b_2 u_x + d_2 y_t = 0. \qquad (4\text{-}34)$$

Recombination is easily accomplished. The b_2-multiplied first equation,

---

† The transverse displacement y is the *primitive* variable here.
‡ This choice is not unique.

after differentiation with respect to x, has subtracted from it the b_1-multiplied second equation, after differentiation with respect to t. Thus we find

$$b_2 a_1 y_{xx} + (b_2 d_1 - b_1 a_2) y_{xt} - b_1 d_2 y_{tt} = 0. \qquad (4\text{-}35)$$

By comparing Eqns (4-32) and (4-35) we observe that they are identical if

$$b_2 a_1 = \beta$$
$$b_2 d_1 - b_1 a_2 = \alpha, \qquad -b_1 d_2 = 1. \qquad (4\text{-}36)$$

The solution of Eqns (4-36) is clearly not unique. One choice is $a_2 = 0$, $b_2 = 1$, $d_2 = 1$, $a_1 = \beta$, $b_1 = -1$, $d_1 = \alpha$, but alternatives are easily selected (Problem 4-4). With this choice, Eqns (4-34) become

$$\beta y_x - u_t + \alpha y_t = 0$$
$$u_x + y_t = 0. \qquad (4\text{-}37)$$

From the second of Eqns (4-37) we see that the new variable u is related to y through $u = -\int y_t \, dx + $ (arbitrary function of time), *but* the solution for u is of no immediate interest in this problem.

(b) Stress wave propagation

A typical application of the quasilinear theory is given by Ames and Vicario [8] in their study of the stress wave propagation in a moving medium. The dimensionless equations for a linear elastic medium are

$$m V_x + V m_x + m_t = 0$$
$$V V_x + V_t + m^{-3} m_x = 0. \qquad (4\text{-}38)$$

To employ the standard notation of Section 4-1 we set

$$u^1 = V, \qquad u^2 = m, \quad x \to x, \quad y \to t \qquad (4\text{-}39)$$

whereupon Eqn (4-12) becomes

$$\begin{bmatrix} u^2 & u^1 \\ u^1 & (u^2)^{-3} \end{bmatrix} \begin{bmatrix} u_x^1 \\ u_x^2 \end{bmatrix} + \begin{bmatrix} 0 & 1 \\ 1 & 0 \end{bmatrix} \begin{bmatrix} u_y^1 \\ u_y^2 \end{bmatrix} = 0. \qquad (4\text{-}40)$$

The matrices identified in Eqn (4-12) are

$$A = \begin{bmatrix} u^2 & u^1 \\ u^1 & (u^2)^{-3} \end{bmatrix}, \qquad B = \begin{bmatrix} 0 & 1 \\ 1 & 0 \end{bmatrix}, \qquad d = 0.$$

The computation of $E = \begin{bmatrix} e_1 & 0 \\ 0 & e_2 \end{bmatrix}$ is accomplished by means of Eqn (4-23).

Thus,

$$\det (A - \lambda B) = \det \begin{bmatrix} u^2 & u^1 - \lambda \\ u^1 - \lambda & (u^2)^{-3} \end{bmatrix} = 0$$

or

$$(u^2)^{-2} - (u^1 - \lambda)^2 = 0.$$

Thus we select

$$e_1 = u^1 + (u^2)^{-1}, \qquad e_2 = u^1 - (u^2)^{-1}$$

so that the characteristics are

$$\left.\frac{dx}{dy}\right|_1 = u^1 + (u^2)^{-1}, \qquad \left.\frac{dx}{dy}\right|_2 = u^1 - (u^2)^{-1} \tag{4-41}$$

or, in the original variables,

$$\left.\frac{dx}{dt}\right|_1 = V + m^{-1}, \qquad \left.\frac{dx}{dt}\right|_2 = V - m^{-1}.$$

The transformation matrix T, from Eqn (4-16), becomes

$$\begin{bmatrix} t_{11} & t_{12} \\ t_{21} & t_{22} \end{bmatrix}\begin{bmatrix} u^2 & u^1 \\ u^1 & (u^2)^{-3} \end{bmatrix} = \begin{bmatrix} u^1 + (u^2)^{-1} & 0 \\ 0 & u^1 - (u^2)^{-1} \end{bmatrix}\begin{bmatrix} t_{11} & t_{12} \\ t_{21} & t_{22} \end{bmatrix}\begin{bmatrix} 0 & 1 \\ 1 & 0 \end{bmatrix}$$

which, upon expansion, requires values of t_{ij} so that

$$\begin{bmatrix} u^2 t_{11} + u^1 t_{12}, & u^1 t_{11} + (u^2)^{-3} t_{12} \\ u^2 t_{21} + u^1 t_{22}, & u^1 t_{21} + (u^2)^{-3} t_{22} \end{bmatrix} = \begin{bmatrix} [u^1 + (u^2)^{-1}]t_{12}, & [u^1 + (u^2)^{-1}]t_{11} \\ [u^1 - (u^2)^{-1}]t_{22}, & [u^1 - (u^2)^{-1}]t_{21} \end{bmatrix}$$

Equating corresponding elements in the first row we find that both lead to

$$u^2 t_{11} - (u^2)^{-1} t_{12} = 0$$

which is satisfied (not unique) if $t_{11} = 1$ and $t_{12} = (u^2)^2$. Equating corresponding elements in the second row we find that both lead to

$$u^2 t_{21} + (u^2)^{-1} t_{22} = 0$$

which is satisfied if $t_{21} = 1$ and $t_{22} = -(u^2)^2$. Consequently,

$$T = \begin{bmatrix} 1 & (u^2)^2 \\ 1 & -(u^2)^2 \end{bmatrix}$$

and operating on Eqn (4-40) with T generates

$$\begin{aligned} (u^2)^2[e_1 u_x^1 + u_y^1] + [e_1 u_x^2 + u_y^2] = 0 \\ -(u^2)^2[e_2 u_x^1 + u_y^1] + [e_2 u_x^2 + u_y^2] = 0. \end{aligned} \tag{4-42}$$

From Eqns (4-27) and (4-28) it follows that Eqns (4-42) take the form

$$(u^2)^2 u_\alpha^1 + u_\alpha^2 = 0$$
$$-(u^2)^2 u_\beta^1 + u_\beta^2 = 0 \tag{4-43}$$

where α and β indicate the characteristic directions e_1, e_2 respectively.

Next, Eqns (4-41) become

$$x_\alpha = \{u^1 + (u^2)^{-1}\} y_\alpha$$
$$x_\beta = \{u^1 - (u^2)^{-1}\} y_\beta. \tag{4-44}$$

Equations (4-43) and (4-44) are the *canonical equations* of Section 4-1. In the original variables they read

$$m^2 V_\alpha + m_\alpha = 0$$
$$-m^2 V_\beta + m_\beta = 0 \tag{4-45}$$

and

$$x_\alpha = (V + m^{-1}) y_\alpha$$
$$x_\beta = (V - m^{-1}) y_\beta. \tag{4-46}$$

Clearly, each of Eqns (4-45) involves differentiations in one direction only, namely, the appropriate characteristic direction. In many cases, of which this is typical, a further step is often helpful. One may introduce new variables, termed the *Riemann invariants*, by integrating Eqns (4-45) in the α and β directions, respectively. Thus one obtains

$$V - m^{-1} = -2r(\beta) \tag{4-47}$$

$$V + m^{-1} = 2s(\alpha) \tag{4-48}$$

where $r(\beta)$ is invariant along the β characteristics and $s(\alpha)$ is invariant along the α characteristics. In terms of these Riemann invariants, r and s, we can write

$$V = s - r, \qquad m^{-1} = s + r. \tag{4-49}$$

The value of the Riemann invariants is especially useful for this case but not quite so helpful in others.† For more general details the reader is referred to Courant and Friedrichs [3], Ames [5], and Jeffrey and Taniuti [9].

In this example the characteristic equations [Eqns (4-41)] are expressible in r and s as

$$x_s = (V + m^{-1}) t_s = 2s t_s \tag{4-50}$$

$$x_r = (V - m^{-1}) t_r = -2r t_r. \tag{4-51}$$

Upon differentiating Eqn (4-50) with respect to r and Eqn (4-51) with respect

† This example is seen to correspond to an artificial gas with $\gamma = -1$, as discussed in Ames [5, p. 79].

to s, and equating the results, we have $(2st_s)_r + (2rt_r)_s = 0$ or $t_{rs} = 0$. On integrating, the general solution for t is

$$t = f[-\tfrac{1}{2}(V - m^{-1})] + g[\tfrac{1}{2}(V + m^{-1})]$$

where f and g are arbitrary but differentiable functions. The general solution, for x, is left for Problem 4-5.

PROBLEMS

4-4 Select an alternative solution of Eqns (4-36). With this new set write out the form of Eqns (4-34) and compare this with Eqns (4-37). Are there any advantages accruing to your set?

4-5 Find the general solution, for x, of Eqns (4-50) and (4-51).

4-6 Find the Riemann invariants for the equations of Problem 4-2.

$$\textit{Ans:} \quad \left. \begin{array}{l} u + l(\rho) = 2r(\beta) \\ u - l(\rho) = -2s(\alpha) \end{array} \right\} \text{ where } l(\rho) = \int_0^\rho c\,\frac{d\rho}{\rho}$$

4-7 For Problem 4-2 find the equations for t and x in terms of the Riemann invariants. (*Hint:* Use the invariants written in u and c rather than in u and ρ.)

Ans: The Poisson–Euler–Darboux equation

$$0 = t_{rs} + \frac{\mu}{r + s}(t_r + t_s), \qquad \mu = \tfrac{1}{2}(\gamma + 1)/(\gamma - 1)$$

4-8 If $\gamma = 3$, integrate the t-equation of Problem 4-7 and thus obtain the general solution for t.

4-3 Method of characteristics

The concept of characteristics was introduced in Section 1-2 as a vehicle for the classification of equations. In the previous sections of this chapter quasilinear hyperbolic equations were shown to be substantially simplified if characteristics, that is the *natural* coordinates of the system, were used. The basic rationale underlying the use of characteristics is that, by an appropriate choice of coordinates, the original system of hyperbolic equations can be replaced by a system whose coordinates are the characteristics. The simplifications are particularly useful when applied to problems involving one or two first-order (one second-order) equations in two independent variables.

Our main concern is the quasilinear second-order equation [Eqn (1-14) of Chapter 1]

$$au_{xx} + bu_{xy} + cu_{yy} = f \tag{4-52}$$

where a, b, c, and f are functions of x, y, u, u_x, and u_y but not u_{xx}, u_{xy}, or u_{yy}.

The analysis is performed as in Section 1-2 or Section 4-1. Using the former let us again ask when a knowledge of u, u_x, and u_y on the initial curve serves to determine u_{xx}, u_{xy}, and u_{yy} uniquely, so that the differential equation [Eqn (4-52)] is satisfied? If these derivatives exist, we have in all cases

$$d(u_x) = u_{xx}\, dx + u_{xy}\, dy$$
$$d(u_y) = u_{xy}\, dx + u_{yy}\, dy \tag{4-53}$$

and the differential equation provides the third relation.

Equations (4-52) and (4-53) comprise three equations for u_{xx}, u_{xy}, and u_{yy} whose solution exists, and is unique, unless the determinant

$$\begin{vmatrix} a & b & c \\ dx & dy & 0 \\ 0 & dx & dy \end{vmatrix} = 0$$

or

$$\frac{dy}{dx} = \frac{1}{2a}\{b \pm \sqrt{(b^2 - 4ac)}\}. \tag{4-54}$$

Equation (4-54) defines the characteristics. We suppose $b^2 - 4ac > 0$; that is, the system is hyperbolic. When Eqn (4-54) holds there is no solution at all unless the other determinants of the system also vanish, so that, necessarily, we have

$$\begin{vmatrix} a & f & c \\ dx & d(u_x) & 0 \\ 0 & d(u_y) & dy \end{vmatrix} = 0 \tag{4-55}$$

or

$$a\, d(u_x)\, dy - f\, dx\, dy + c\, dx\, d(u_y) = 0 \tag{4-56}$$

(see Problem 4-9). Upon dividing by dx and identifying the characteristics as $dy = \alpha\, dx$, $dy = \beta\, dx$, Eqn (4-56) becomes

$$a\alpha\, d(u_x) + c\, d(u_y) - f\, dy = 0$$

and

$$a\beta\, d(u_x) + c\, d(u_y) - f\, dy = 0. \tag{4-57}$$

Equations (4-57) specify the conditions which the solutions must satisfy along the characteristics.† With real characteristics, and if the initial curve is not a characteristic (Problem 4-13) (see Section 4-0), we shall be able to employ Eqns (4-54) and (4-57) in a step-by-step numerical procedure. This process simultaneously constructs the characteristic grid and the solution of the differential equation [Eqn (4-52)] at the grid points.‡

The numerical calculation§ is as follows: Suppose that u, u_x, and u_y are

---

† We shall call these 'the equations along the characteristics'.
‡ To be especially noted is the simplicity that is realized if, in Eqn (4-52), $f = 0$. Similar simplicity occurs if $f_1 = f_2 \equiv 0$ in Eqns (4-6)—that is, in the reducible case.
§ The method goes back to a monograph by Massau [10].

specified on the noncharacteristic initial curve Γ shown in Fig. 4-5. As in Fig. 4-5, we draw, from points on Γ, the two families of characteristics. For any two adjacent points P and Q on Γ, let the α characteristic from P intersect the β characteristic from Q at R. Since the characteristics generally depend upon the solution we must determine the (x, y) coordinates of R, as well as the values of u_x and u_y at this point, from the prescribed values on Γ. In principle, any numerical method for integrating ordinary differential equations can be employed to simultaneously solve Eqns (4-54) and (4-57).

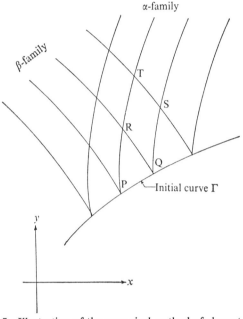

Fig. 4-5 Illustration of the numerical method of characteristics

If Euler's method (Problem 4-14) is employed, PR is approximated by a straight line and therefore yields an approximation which is $O(h)$. Here h is some number indicating the order of magnitude of the distances of the initial grid points on Γ—for instance, the maximum distance of two adjacent grid points. The discretization error can be reduced, at the price of increased computational labor, by employing arcs of parabolas in place of the straight lines of Euler's method. Thus we have (see, for example, Crandall [11, p. 174]):

$$\frac{y(R) - y(P)}{x(R) - x(P)} = \tfrac{1}{2}\{\alpha(R) + \alpha(P)\}$$

$$\frac{y(R) - y(Q)}{x(R) - x(Q)} = \tfrac{1}{2}\{\beta(R) + \beta(Q)\}$$

(4-58)

as second-order approximations for Eqns (4-54) and

$$\tfrac{1}{2}[a(R)\alpha(R) + a(P)\alpha(P)][u_x(R) - u_x(P)] + \tfrac{1}{2}[c(R) + c(P)][u_y(R) - u_y(P)]$$
$$- \tfrac{1}{2}[f(R) + f(P)][y(R) - y(P)] = 0 \quad (4\text{-}59)$$
$$\tfrac{1}{2}[a(R)\beta(R) + a(Q)\beta(Q)][u_x(R) - u_x(Q)] + \tfrac{1}{2}[c(R) + c(Q)][u_y(R) - u_y(Q)]$$
$$- \tfrac{1}{2}[f(R) + f(Q)][y(R) - y(Q)] = 0$$

as approximations for Eqns (4-57). Lastly, for the determination of u we need an approximation to

$$du = (u_x)\,dx + (u_y)\,dy. \quad (4\text{-}60)$$

This is given, again to second order, by either of the equations

$$u(R) - u(P) = \tfrac{1}{2}[u_x(R) + u_x(P)][x(R) - x(P)]$$
$$+ \tfrac{1}{2}[u_y(R) + u_y(P)][y(R) - y(P)]$$
$$u(R) - u(Q) = \tfrac{1}{2}[u_x(R) + u_x(Q)][x(R) - x(Q)] \quad (4\text{-}61)$$
$$+ \tfrac{1}{2}[u_y(R) + u_y(Q)][y(R) - y(Q)].$$

Equations (4-58), (4-59), and one of (4-61) constitute five nonlinear equations for the determination of the unknowns x, y, u_x, u_y, and u at the point R. They are usually solved by an iterative process such as the one we shall describe. In the absence of other specific information we would, for example, at the first step, identify $\alpha(R)$ with $\alpha(P)$ and $\beta(R)$ with $\beta(Q)$ in the first and second of Eqns (4-58). Improved values are then calculated and the process is repeated until convergence is achieved. By repetition of this basic computational step, the solution can be computed at other grid points adjacent to the initial curve such as S. From R and S we proceed to T, and so forth. At a boundary the process requires modification to insert the given boundary condition.

No restrictive conditions to ensure stability and convergence are to be expected for this method since it is based on a network which approximates that of characteristic curves. A proof of convergence, in the case of the Euler approximation, is given in Sauer [12].

In the important reducible case, $f = 0$ and α and β are independent of x and y. Consequently, we can solve Eqns (4-58) for $x(R)$ and $y(R)$ as

$$x(R) = \frac{y(Q) - y(P) + \tfrac{1}{2}[\alpha(P) + \alpha(R)]x(P) - \tfrac{1}{2}[\beta(Q) + \beta(R)]x(Q)}{\tfrac{1}{2}[\alpha(P) + \alpha(R)] - \tfrac{1}{2}[\beta(Q) + \beta(R)]}$$

$$y(R) = y(P) + \tfrac{1}{2}[\alpha(P) + \alpha(R)][x(R) - x(P)]$$
$$= y(Q) + \tfrac{1}{2}[\beta(Q) + \beta(R)][x(R) - x(Q)].$$

The calculation of $y(R)$, by both of the latter equations, provides a useful check on the numerical work.

Naturally, other numerical methods for simultaneous ordinary differential equations may be employed. Thomas [13] used several Adams' procedures for integrating our system including adapting them to solve hyperbolic systems in canonical form, Eqns (4-29) and (4-30). Thomas asserts that the complications inherent in these methods are more than offset by the larger step size allowed, but *even* the simpler methods require at least the solution of a linear system of fourth order at each point.

PROBLEMS

4-9 Let x and b be column vectors with n elements and A be an $n \times n$ matrix. If $Ax = b$ and $\det A = 0$ show that a necessary condition for the existence of finite solutions for x is that when b is substituted for *any* column of A the resulting determinant must also vanish. Interpret this result geometrically when $n = 2$.

4-10 Find the conditions analogous to Eqns (4-57) that the solutions to Eqns (4-6) must satisfy along the characteristics.

$$Ans: \quad [(b_1c_2 - b_2c_1)\frac{dy}{dx} - (b_1d_2 - b_2d_1)]\, du + \left[(d_1c_2 - d_2c_1)\frac{dy}{dx}\right] dv$$

$$+ \left[(c_1f_2 - c_2f_1)\frac{dy}{dx} - (d_1f_2 - d_2f_1)\right] dy = 0 \quad (4\text{-}62)$$

4-11 Find the equations along the characteristics for the equation of Problem 4-1.

4-12 Find the equations along the characteristics for the equations of Problem 4-2.

$$Ans: du \pm c\rho^{-1}\, d\rho = 0 \quad (4\text{-}63)$$

4-13 Suppose that the initial curve for $u_{xx} = u_{yy}$ is the characteristic $y = x$. Show that, in general, the equation has *no solution* and in any case *no unique solution*. (*Hint:* Find the equations along the characteristics $y \pm x =$ constant to be $u_y \pm u_x =$ constant.) If $u = 1$, $u_x = 1$, $u_y = -1$ on $y = x$, show that $u = 1 + (x - y) + A(x - y)^2$ is a solution for any value of A.

4-14 Employ a Euler approximation in preparing Eqns (4-54) and (4-57) for numerical computation. (*Hint:* PR in Fig. 4-5 is taken as a straight line having the slope $\alpha(P)$—that is, the value taken by $dy/dx = \alpha$ at the point P.)

4-15 In Section 1-2 the supersonic nozzle problem was discussed and the characteristics were developed in Eqns (1-23a) and (1-23b). Find the differential equations for u and v along the characteristics.

4-16 Let $\sin \mu = c(u^2 + v^2)^{-1/2}$ and show that both characteristic directions for Problem 4-15 make angles μ with the streamline direction given by $\theta = \tan^{-1}(v/u)$. Reformulate the results of Problem 4-15 in terms of the new variables q and θ, where $u = q \cos \theta$, $v = q \sin \theta$.

$$Ans: dy = \tan(\theta \pm \mu)\, dx; \quad q\, d\theta = \pm dq \cot \mu$$

4-17   Integrate the equations of Problem 4-16 along the characteristics.

$$Ans: \ \theta - \omega = \alpha, \qquad \theta + \omega = \beta,$$

$$\omega = \int_1^q \left\{ \frac{6(q^2 - 1)}{6 - q^2} \right\}^{1/2} \frac{dq}{q}$$

4-4   Constant states and simple waves

Before giving a detailed example, we pause to re-examine the quasilinear system, Eqns (4-6), with characteristics specified by Eqns (1-12) and the differential equations for u and v, along the characteristics, by Problem 4-10. The important reducible case has coefficients which depend only upon u and v *and not upon* the independent variables x and y. Consequently, the coefficient functions $a_1, a_2, b_1, b_2, \ldots$, even if they are complicated functions of

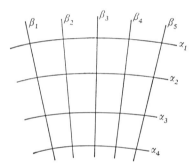

Fig. 4-6   Simple wave configuration

u and v, may still give parallel straight lines as characteristics. Such an event occurs in a region of the integration domain in which both u and v remain constant. Such a region is aptly called a *constant state*. At first glance this may appear a trivial circumstance, but it is physically important since wave speeds are always finite. Thus solutions, at least initially, often consist of several regions of constant state interconnected by regions in which u and v change. In a region of constant state, all characteristics of the same family are identified by the same constant, since they are all parallel.

Now suppose that u and v remain constant along *any one*, say β, characteristic but vary in value from one β characteristic to the next. Generally the slopes will be constant for each β characteristic, but will be different between β characteristics so that the β family will be nonparallel straight lines. They often resemble a fan, as shown in Fig. 4-6. Since u and v are constant along any specific β characteristic, each α characteristic will have the same slope as it crosses this β line. This slope varies from β line to β line so the α family consists of curves. In this case, the β characteristics carry a different label but all the α characteristics carry the same label. This results since, as the curves marked α_i cross (say) β_2, they take on the same value of u and v. Therefore, the integrated relation between u and v along any α_i must be identical with all others. The configuration just described is called a *simple wave*.

Apart from their relative simplicity, simple waves are important because of the following theorem:

In a solution containing constant state regions, the regions adjacent to constant states are always simple waves.

A proof of this theorem is found in Jeffrey and Taniuti [9, p. 69–71]. This character of simple waves plays an important role in the construction of solutions of hyperbolic systems.

4-5 Typical application of characteristics

A typical example where constant states and simple waves play a role is that of the one-dimensional isentropic flow of an inviscid gas expanding behind a piston. We have previously given the equations in Problem 4-2. Suppose that for $t' < 0$ the gas is at rest with density ρ_0, the piston is at $x' = a$, and the sound velocity of gas is c_0 in its initial state. A dimensionless formulation can be obtained by introducing the new variables

$$u = u'/c_0, \qquad c = c'/c_0, \qquad \rho = \rho'/\rho_0, \qquad x = x'/a, \qquad t = t'c_0/a$$

so that the equations become

$$u_t + uu_x + c^2\rho^{-1}\rho_x = 0$$
$$\rho_t + u\rho_x + \rho u_x = 0 \qquad (4\text{-}64)$$
$$c = \rho^{(\gamma-1)/2}.$$

At $t = 0$ the piston is withdrawn with (dimensionless) velocity ϵ $(0 < \epsilon < 1)$. The problem is therefore characterized by Eqns (4-64) with the *initial* conditions

$$u(x, 0) = 0, \qquad \rho(x, 0) = 1, \quad 0 < x < 1 \qquad (4\text{-}65)$$

and boundary conditions

$$u(0, t) = 0, \qquad u(1 + \epsilon t, t) = \epsilon. \qquad (4\text{-}66)$$

The last condition describes the piston motion; that is, its position at time t is given by $1 + \epsilon t$ and it is moving with velocity ϵ.

From Problem 4-2 the characteristics are

$$\left.\frac{dx}{dt}\right|_\alpha = u + c, \qquad \left.\frac{dx}{dt}\right|_\beta = u - c \qquad (4\text{-}67)$$

and the equations along the characteristics, from Problem 4-12, are

$$du + c\rho^{-1}\,d\rho = 0 \quad \text{(along α characteristics)}$$
$$du - c\rho^{-1}\,d\rho = 0 \quad \text{(along β characteristics)} \qquad (4\text{-}68)$$

We cannot obtain the characteristic curves in advance, since Eqns (4-67) depend upon u and c. However, we can integrate Eqns (4-68) when we recall that $c = \rho^{(\gamma-1)/2}$. Thus, we have (if air is employed, $\gamma = 7/5$, the value we use here),

$$\frac{2}{\gamma-1} c + u = 5c + u = \alpha$$

$$\frac{2}{\gamma-1} c - u = 5c - u = \beta$$

(4-69)

so that *after the characteristics are found* these relations, when solved for u and c, yield the complete solution as

$$c = \frac{\alpha + \beta}{10}, \qquad u = \frac{\alpha - \beta}{2}.$$

(4-70)

These equations yield the values of u and c at a point in terms of the labels on the two characteristics passing through the point.

We now examine the configuration of the characteristic curves in the integration domain of this problem, as sketched in Fig. 4-7. The initial data

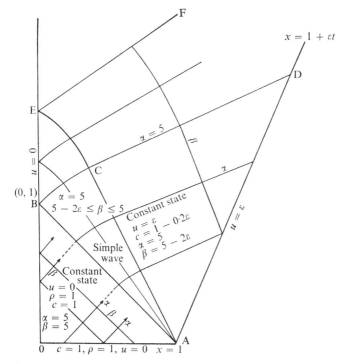

Fig. 4-7   Integration domain illustrating the solution by characteristics of a problem in isentropic compressible flow of a gas

specify that $c = 1$, $\rho = 1$, and $u = 0$ at $t = 0$. Consequently, from Eqns (4-69) we see that characteristic curves originating from the initial line OA *all* bear the labels $\alpha = 5$ or $\beta = 5$. Further, all these characteristic curves are straight lines with slopes $+1$ for the α characteristics and -1 for the β characteristics [Eqns (4-67)]. The boundary condition $u = 0$ on OB coupled with the second of Eqns (4-70) implies that the α characteristics reflected from OB carry the same label as the incident β characteristics. Therefore, the solution in the triangular-shaped region OAB is a region of *constant state* having $\alpha = 5$, $\beta = 5$ and, from Eqns (4-70), $u = 0$, $c = 1$. The boundary curve AB is the line $t = 1 - x$, so the coordinates of B are $(0, 1)$. The physical meaning of this constant state is that this domain consists of portions of the gas that have not yet been reached by the disturbance wave created by the moving piston. A unique solution in this triangle OAB is guaranteed by theorems (1) and (2) of Section 4-0. (The reader may wish to examine the remaining discussion in the light of the other theorems of Section 4-0.)

The constant state cannot continue beyond AB since the boundary condition $u = \epsilon$ on AD disagrees with $u = 0$ of the constant state. Adjacent to the constant state will be a *simple wave* whose extent will be determined by the condition of the solution on its right. The extent of the simple wave is determined by the requirement that along the last of the β characteristics, AC, the value of u must be ϵ. The β characteristics are straight lines whose slope and label vary from line to line. The α characteristics are all curved and bear the same label $\alpha = 5$ (compare Fig. 4-6), inherited from the region of constant state. In this *centered* simple wave region u and c, as functions of β, are

$$u = \frac{5 - \beta}{2}, \qquad c = \frac{5 + \beta}{10}. \tag{4-71}$$

From Eqns (4-67) the slope of the β characteristic is

$$\left.\frac{dt}{dx}\right|_\beta = \frac{1}{u - c} = \frac{5}{10 - 3\beta}. \tag{4-72}$$

Now the last β characteristic requires $u = \epsilon$. From Eqns (4-71) we find $\beta = 5 - 2\epsilon$ and hence $c = 1 - 0.2\epsilon$, and $(dt/dx)|_\beta = 5/(6\epsilon - 5)$. The results are shown in Fig. 4-7.

By analysis analogous to the above, we find the constant state $\alpha = 5$, $\beta = 5 - 2\epsilon$, $u = \epsilon$, $c = 1 - 0.2\epsilon$ to the right of CA in the triangular-shaped region ACD (Problem 4-18). The values of u and c can be obtained for any point in the simple wave region from Eqns (4-71) and (4-72). Upon fixing the value of β as β_a, $5 - 2\epsilon \le \beta_a \le 5$, the slope [Eqn (4-72)] of the β characteristic is fixed, and since all β characteristics pass through $x = 1$, $t = 0$, the line is uniquely determined. Along this characteristic $u = (5 - \beta_a)/2$, $c = (5 + \beta_a)/10$.

A more explicit solution is occasionally obtainable for the simple wave region. For this reason we seek to determine the coordinates of the point C and curve BC in analytic form. Let (x, t) be any point in the simple wave region BCA. From the analytic geometry of this region it follows that the slope of any β characteristic is expressible as

$$\left.\frac{dt}{dx}\right|_\beta = \frac{t}{x - 1}$$

which, when equated to the alternate expression [Eqn (4-72)], gives

$$\frac{t}{x - 1} = \frac{5}{10 - 3\beta}. \tag{4-73}$$

Along any α characteristic, in the centered simple wave, the quantity β appears to be a convenient parameter. With $x = x(\beta)$, $t = t(\beta)$, the β derivative of Eqn (4-73) becomes

$$(10 - 3\beta)\frac{dt}{d\beta} - 3t = 5\frac{dx}{d\beta}. \tag{4-74}$$

Now, from Eqns (4-67) and the fact that $\alpha = 5$ in the simple wave, we deduce that

$$\left.\frac{dt}{dx}\right|_\alpha = \frac{5}{15 - 2\beta}$$

which becomes

$$\frac{(15 - 2\beta)}{5}\frac{dt}{d\beta} = \frac{dx}{d\beta} \tag{4-75}$$

in parametric form (along α characteristics). Eliminating $dx/d\beta$ between Eqns (4-74) and (4-75), there results

$$\frac{dt}{d\beta} = -\frac{3t}{5 + \beta}$$

whose integral is

$$t = \left(\frac{10}{5 + \beta}\right)^3, \quad 5 - 2\epsilon \le \beta \le 5. \tag{4-76}$$

It is now easy to show (Problem 4-19) that

$$x = 1 - \left(\frac{3\beta}{5} - 2\right)\left(\frac{10}{5 + \beta}\right)^3, \quad 5 - 2\epsilon \le \beta \le 5. \tag{4-77}$$

Upon eliminating the parameter β, we obtain the explicit equation

$$x = 1 + 5t - 6t^{2/3}, \quad 1 \le t \le \left(\frac{10}{10 - 2\epsilon}\right)^3 \tag{4-78}$$

for the curve BC. The coordinates of the point C and the boundary CD of the constant state region adjacent to the moving boundary are now easily determined (Problem 4-20).

The analytic solution has been determined with moderate ease up to the curve BCD. To proceed further we must next examine the more difficult problem in region BCE. This domain has two families of curved characteristics and, therefore, the computational considerations will be similar to those of the general problem. While approximate procedures can be utilized, we employ the numerical method of characteristics to obtain the solution in

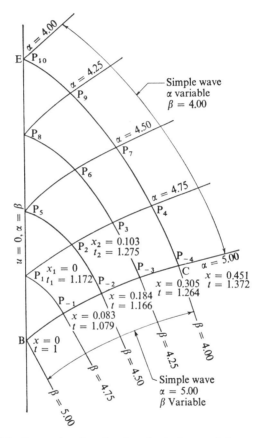

Fig. 4-8 Integration domain where both characteristics are curved

BCE. An exploded view of BCE is given in Fig. 4-8. For definiteness we have selected $\epsilon = \frac{1}{2}$ in the remainder of our discussion.

According to Eqns (4-70) we can evaluate u and c in terms of α and β as $u = \frac{1}{2}(\alpha - \beta)$ and $c = \frac{1}{10}(\alpha + \beta)$. All that remains is to calculate the coordinates (x, t) of the characteristic net P_i, $i = 1, 2, \ldots, 10$ of Fig. 4-8. Here we can label all our characteristics in advance and all the solution values at each P_i by using Eqns (4-70). The boundary condition, $u = 0$, on BE generates the condition $\alpha = \beta$ from Eqns (4-70). The (x, t) values along BC follow from Eqns (4-76) and (4-77). The precise location of the intersection points P_i are obtained by integrating Eqns (4-67), beginning with the specified conditions on BC. A more useful form of these equations is obtained by setting Eqns (4-70) into Eqns (4-67)—thus,

$$\frac{dt}{dx}\bigg|_\alpha = \frac{1}{u + c} = \frac{5}{3\alpha - 2\beta} \tag{4-79}$$

$$\frac{dt}{dx}\bigg|_\beta = \frac{1}{u - c} = \frac{5}{2\alpha - 3\beta}. \tag{4-80}$$

The solution is initiated by locating P_1 (clearly $x = 0$ here) followed by location of the P_i in sequence up to P_{10}. To locate P_1 we note that the slope of the β characteristic joining $(0.083, 1.079)$ to $(0, t_1)$ is

$$\frac{dt}{dx}\bigg|_{\beta, -1} = \frac{5}{2(5.00) - 3(4.75)} = -1.177$$

at $(0.083, 1.079)$ and

$$\frac{dt}{dx}\bigg|_{\beta, 1} = \frac{5}{2(4.75) - 3(4.75)} = -1.053$$

at P_1. If the arc between the initial point and P_1 was a parabola, the slope of the chord would be the average of these: -1.115. The coordinate t_1 is obtained directly from the relation

$$\frac{t_1 - 1.079}{0 - 0.083} = \frac{1}{2}\left\{\frac{dt}{dx}\bigg|_{\beta, -1} + \frac{dt}{dx}\bigg|_{\beta, 1}\right\} = -1.115$$

that is, $t_1 = 1.172$. This calculation is typical of that for each boundary point.

The calculation for the location of P_2 is typical of that for an interior point. The slope of the α characteristic joining P_2 to P_1 is

$$\frac{dt}{dx}\bigg|_{\alpha, 1} = \frac{5}{3(4.75) - 2(4.75)} = 1.053$$

at P_1 and

$$\frac{dt}{dx}\bigg|_{\alpha, 2} = \frac{5}{3(4.75) - 2(4.50)} = 0.952$$

at P_2. If the arc between these two points were a parabola, the chord slope is the average of these: 1.003. Similarly,

$$\frac{dt}{dx}\bigg|_{\beta, -2} = -1.429, \qquad \frac{dt}{dx}\bigg|_{\beta, 2} = -1.25$$

with average value equal to -1.340. The point P_2 is now located analytically by solving the simultaneous equations

$$\frac{t_2 - 1.172}{x_2 - 0} = 1.003, \qquad \frac{t_2 - 1.166}{x_2 - 0.184} = -1.340$$

to obtain $x_2 = 0.103$ and $t_2 = 1.275$.

PROBLEMS

4-18 Verify that the region ACD of Fig. 4-7 is a region of constant state with $\alpha = 5, \beta = 5 - 2\epsilon, u = \epsilon, c = 1 - 0.2\epsilon$. Interpret the result physically.

4-19 Derive Eqn (4-77).

4-20 Proceeding from Eqns (4-76) through (4-77) find the coordinates of point C (Fig. 4-7), the equation for the straight line CD, and the coordinates of D.

4-21 What property characterizes the region FECD? What value does α take on the ray FE? Describe the range of α and β values in this region and specify the equation of FE.

4-22 Compute the coordinates of P_3 in Fig. 4-8.

4-23 Locate the point E in Fig. 4-8 by employing a network of parabolic arcs in which P_5 and P_7 are the only intermediate points.

Ans: $t = 2.128$ (true value 2.101)

4-24 The equations for the compression of a plastic bar are given as Eqns (1-28) of Problem 1-6. Obtain a dimensionless formulation by setting $\omega = p/2k$ and dividing all lengths by the half distance between the plates compressing the bar. Find the characteristics and the equations along the characteristics.

Ans: $\dfrac{dy}{dx}\bigg|_\alpha = \tan\psi, \quad \dfrac{dy}{dx}\bigg|_\beta = -\cot\psi;$

$d\omega + d\psi = 0$ on α characteristics and $d\omega - d\psi = 0$ on β characteristics

4-25 The physical problem for the compression of a plastic bar is shown in Fig. 4-9. Along AB, $\psi = \pi/4$ and $p = k$; along AF, $\psi = 0$; $\psi = -\pi/4$ on BD, and there

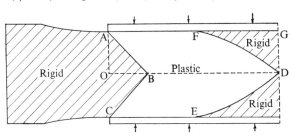

Fig. 4-9 Compression of a plastic bar

is symmetry about BD. The equations along the characteristics (Problem 4-24) are integrable. Carry out this integration and study the characteristics location qualitatively.

4-26 Organize the plastic bar computation, by the numerical method of characteristics, in the domain AODG. What are the constant state regions and simple waves, if any?

4-6   Explicit finite difference methods

For two first-order hyperbolic equations in two independent variables, the method of integrating along characteristics is usually the most convenient and most accurate process. One of the main advantages of characteristics, and a disadvantage of finite differences, is that discontinuities in the initial values may propagate along the characteristics. This situation is difficult to handle on other than the characteristic net. We discuss these difficulties later in this chapter.

If the equations are of no great complexity and are known to possess well-behaved solutions, we can employ finite difference procedures *providing* the limitations imposed by the characteristics are considered in the development. The latter point is amplified in our discussion.

Consider the pure initial value problem

$$u_{xx} = u_{tt}, \qquad u(x, 0) = f(x), \qquad u_t(x, 0) = g(x). \tag{4-81}$$

An explicit method is constructed in a manner directly analogous to those of Chapter 2. We take a rectangular net with constant intervals $h = \Delta x$, $k = \Delta t$ and write $u_{ij} = u(i\,\Delta x, j\,\Delta t)$, $-\infty < i < \infty$, $0 \le j < \infty$. Both second partial derivatives are approximated by central difference expressions given by Eqn (1-51) whose truncation error is $O(h^2)$. Thus $u_{xx} = u_{tt}$ is approximated by the explicit formula

$$U_{i,j+1} = m^2(U_{i-1,j} + U_{i+1,j}) + 2(1 - m^2)U_{i,j} - U_{i,j-1} \tag{4-82}$$

where $m = k/h = \Delta t/\Delta x$. The first initial condition specifies $U_{i,0}$, on the line $t = 0$. We can use the second condition to find values on the line $t = k$ by employing a 'false' boundary and the second-order central difference formula

$$\left.\frac{\partial u}{\partial t}\right|_{i,0} = \frac{u_{i,1} - u_{i,-1}}{2k} + O(k^2). \tag{4-83}$$

Writing $g(i\,\Delta x) = g_i$ we have the approximation

$$U_{i,1} - U_{i,-1} = 2kg_i \tag{4-84}$$

that is, when $U_{i,-1}$ appears, we replace it by its value given in Eqn (4-84), $U_{i,-1} = U_{i,1} - 2kg_i$. With $j = 0$ in Eqn (4-82), we have

$$U_{i,1} = m^2(U_{i-1,0} + U_{i+1,0}) + 2(1 - m^2)U_{i,0} - U_{i,-1}.$$

Upon replacing $U_{i,-1}$ with its value, from Eqn (4-84), and solving for $U_{i,1}$ we find

$$U_{i,1} = \tfrac{1}{2}m^2(f_{i-1} + f_{i+1}) + (1 - m^2)f_i + kg_i. \tag{4-85}$$

The computational molecule for Eqn (4-82) is shown in Fig. 4-10. Superimposed on this figure are the characteristics of the wave equation, $u_{xx} - u_{tt} = 0$, namely $t = \pm x + \text{constant}$, whose slopes are ± 1, represented by the lines AC and BC. By theorem (1) of Section 4-0, the solution is uniquely determined in the triangle ACB, provided we know the solution up to AB. If the absolute value of m (i.e. the slopes) exceeds 1, then Eqn (4-82) would provide a 'solution' in a region *not reached* by the continuous solution. We would hardly expect this result to be correct.

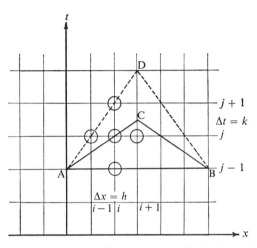

Fig. 4-10   Comparison of the finite difference characteristics AD and DB with the true characteristics AC and BC

If $|m| \leq 1$, however, it can be shown that the method will converge under the usual assumption that certain higher derivatives exist. Convergence of the solution of Eqns (4-82) and (4-83) to that of the differential problem, Eqns (4-81) as h and $k \to 0$, was first examined by Courant, Friedrichs, and Lewy [14], by Lowan [15] using operator methods, and by Collatz [16].

The general procedure is similar to that of Section 2-1. Let $z_{i,j} = u_{i,j} - U_{i,j}$ be the difference between the true solution at (i, j), $u_{i,j}$, and the finite difference solution $U_{i,j}$. By Taylor's series (Problem 4-27), we determine the truncation error for the various terms and find the difference expression

$$z_{i,j+1} = -z_{i,j-1} + m^2(z_{i-1,j} + z_{i+1,j}) + 2(1 - m^2)z_{i,j}$$
$$+ O[k^4] + O[k^2h^2]. \quad (4\text{-}86)$$

Since U agrees with u on the initial line, $z_{i,0} = 0$ for all i. If we employ Eqn (4-83) in the first time step, and note that $z_{i,0} = 0$, we find

$$z_{i,1} = O(k^3). \quad (4\text{-}87)$$

To investigate the stepwise stability of Eqn (4-86), we examine the propagating effect of a single term taking the form $\exp[(-1)^{1/2}\beta x]$, where β is any real number, say, along the line $t = 0$. The complete effect is obtainable by linear superposition of all such errors at the pivotal points. The errors are propagated according to the homogeneous form of Eqn (4-86)—that is, Eqn (4-82) with $U_{i,j}$ replaced by $z_{i,j}$. The initial condition, according to our methodology, is

$$z_{i,0} = \exp{[(-1)^{1/2}\beta i h]}. \tag{4-88}$$

Upon attempting a solution of the system by separation of variables, we try

$$z_{i,j} = \exp{[\alpha j k]} \exp{[(-1)^{1/2}\beta i h]}. \tag{4-89}$$

Setting this into Eqn (4-86), in homogeneous form, results in

$$e^{\alpha k} + e^{-\alpha k} = [2 - 4m^2 \sin^2{\tfrac{1}{2}\beta h}]$$

which can be expressed as the quadratic

$$(e^{\alpha k})^2 - 2[1 - 2m^2 \sin^2{\tfrac{1}{2}\beta h}]e^{\alpha k} + 1 = 0 \tag{4-90}$$

in $e^{\alpha k}$. If we are to avoid an increasing exponential solution as $j \to \infty$, it is necessary that $|e^{\alpha k}| \leq 1$ for all real values of β. From Eqn (4-90) we note that the *product* of the two values of $e^{\alpha k}$ is clearly 1. Thus, it follows that the magnitude of one of these must exceed 1 unless *both* magnitudes are equal to unity. That is, there exist solutions of the form of Eqn (4-89) which grow exponentially as j increases unless the discriminant of Eqn (4-90) is nonpositive—that is,

$$(1 - 2m^2 \sin^2{\tfrac{1}{2}\beta h})^2 - 1 \leq 0.$$

Thus stepwise stability, by the Fourier method, follows for all real β if

$$m^2 \leq \frac{1}{\sin^2{\tfrac{1}{2}\beta h}}$$

and this is always true if $m^2 \leq 1$—that is,

$$\Delta t/\Delta x \leq 1. \tag{4-91}$$

By employing Eqn (4-86) convergence can be verified if $|m| \leq 1$. We sketch the proof for $m = 1$ and leave the case $m < 1$ for Problem 4-28.

Note that the solution at $(i, j + 1)$ has an *interval of dependence* [see Fig. 4-1(a)] which reaches back to the initial line $j = 0$. The boundaries of this domain are the characteristic lines $x \pm t = ih \pm (j + 1)h$, and these intersect the x-axis at $x = (i - j - 1)h$ and $x = (i + j + 1)h$—that is, the interval of dependence is *finite* and of *length* $2(j + 1)h$.

Immediate application of the max norm unduly complicates the proof. Instead, Eqn (4-86) is directly employed in working backwards from the point $(i, j + 1)$ with j changed to $j - 1$. The nonvanishing terms in $z_{i,j}$ reduce to a sum of $(j + 1)$ values on the line $j = 1$, terminating at the boundary lines of the triangle of dependence. Additionally, there are $\frac{1}{2}(j + 1)j$ terms of the form Ah^4. By employing the max norm $\|z_j\| = \max_i |z_{i,j}|$ we find that

$$\|z_j\| \leq jBh^3 + \tfrac{1}{2}j(j - 1)Ah^4 \tag{4-92}$$

reducing, since $t = jh$, to

$$\|z_j\| \leq tBh^2 + \tfrac{1}{2}t^2Ah^2. \tag{4-93}$$

The error tends to zero as $h \to 0$, verifying convergence. The nature of the error, on a finite interval, is also displayed as $O(h^2)$.

The finite difference equation, corresponding to $m = 1$,

$$U_{i,j+1} = U_{i-1,j} + U_{i+1,j} - U_{i,j-1} \tag{4-94}$$

is of special interest. Not only does it correspond to maximizing the permissible time interval k for fixed h, but also it has the interesting property that any solution of the differential equation [Eqn (4-81)] satisfies the difference equation [Eqn (4-94)] exactly. To establish this we note that the general solution of Eqns (4-81) is available (Section 4-0) as $u(x, t) = f(x + t) + g(x - t)$. Thus, at a mesh point (ih, jk) we have

$$
\begin{aligned}
u(ih, jk) &= f(ih + jk) + g(ih - jk) \\
&= f[(i + j)h] + g[(i - j)h] \tag{4-95}
\end{aligned}
$$

since $h = k$ $(m = 1)$. Further, we write Eqn (4-95) as

$$u_{i,j} = f_{i+j} + g_{i-j} \tag{4-96}$$

and verify, by direct substitution, that Eqn (4-94) is satisfied. Conversely, it is possible to demonstrate that the general solution of Eqn (4-94) is

$$U_{i,j} = H_{i+j} + K_{i-j}$$

where H and K are arbitrary functions and that any solution of Eqn (4-94) agrees with a solution of Eqns (4-81) at net points.

PROBLEMS

4-27   Complete the development of Eqn (4-86).

4-28   Prove the convergence of Eqn (4-82) as h and $k \to 0$ for $m < 1$. (*Hint:* Write $k = mh$ and let $h \to 0$.)

4-29   Show that the solution of Eqn (4-94) which satisfies $U_{i,0} = f_i$ and, from Eqn (4-85), $U_{i,1} = \frac{1}{2}(f_{i-1} + f_{i+1}) + hg_i$ is

$$U_{i,j} = \tfrac{1}{2}(f_{i-j} + f_{i+j}) + h \sum_{p=-(j-1)}^{j-1} g_{i+p}$$

(*Hint:* Use induction.)

4-30 Let $u_{xx} = u_{tt}$, $0 < x < 1$, $t > 0$, be subject to the conditions $u(x, 0) = 100x^2$, $u_t(x, 0) = 200x$, $u_x(0, t) = 200t$, $u(1, t) = 100(1 + t)^2$.

(a) Take $h = k = \frac{1}{2}$ and develop the finite difference approximation.

(b) Verify both numerically and analytically that $U_{i,j}$ agrees exactly with the true solution $u_{i,j} = 100(x_i + t_j)^2$.

4-31 Consider the damped wave equation

$$u_{tt} = \lambda^2 u_{xx} - 2cu_t, \quad 0 < x < L, \quad t > 0,$$
$$u(0, t) = u(L, t) = 0, \quad u(x, 0) = f(x), \quad u_t(x, 0) = g(x)$$

where λ and c are positive constants. Show that the finite difference approximation

$$U_{i,j+1} - 2U_{i,j} + U_{i,j-1} = m^2(U_{i+1,j} - 2U_{i,j} + U_{i-1,j}) - kc(U_{i,j+1} - U_{i,j-1})$$

is stable if $0 < m \le 1$.

4-7 Overstability

A finite difference approximation may possess favorable stability properties in the sense that it will generate an approximate solution which converges to the exact solution as the net is refined, and also that undesirable error growth is not admitted. At the same time, the errors associated with the generated solution on a specific net may compare poorly with those associated with the solution generated on the same net by a less strongly stable approximation. Zajac [17] discusses an example which will clarify this concept of *overstability*. Overstability has been observed by others but little reporting of the phenomenon has occurred.

In the one-dimensional wave equation $u_{tt} = u_{xx}$, let the term u_{xx} be approximated, as before, by the divided second central difference centered at (i, j). Further, let u_{tt} be approximated by the divided second backward difference at (i, j). Then the difference approximation becomes

$$U_{i,j} - 2U_{i,j-1} + U_{i,j-2} = m^2[U_{i+1,j} - 2U_{i,j} + U_{i-1,j}] \qquad (4\text{-}97)$$

where $m = k/h$.

Since Eqn (4-97) is linear with constant coefficients, exact solutions are sought in the separated form

$$U_{i,j} = \beta^j \exp[(-1)^{1/2}\alpha i]. \qquad (4\text{-}98)$$

Equation (4-98) satisfies Eqn (4-97) if and only if α is related to β by means of the equation

$$\left(1 + 4m^2 \sin^2 \frac{\alpha}{2}\right)\beta^2 - 2\beta + 1 = 0. \qquad (4\text{-}99)$$

If we set $\cos \theta = \left[1 + 4m^2 \sin^2 \frac{\alpha}{2}\right]^{-1/2}$—that is,

$$\theta = \tan^{-1}\left(2m \sin \frac{\alpha}{2}\right) \qquad (4\text{-}100)$$

it follows that

$$\beta = \exp[\pm(-1)^{1/2}\theta] \cos \theta. \qquad (4\text{-}101)$$

Clearly $|\beta| \leq 1$ for all real values of α and m, so Eqn (4-97) is stable† for *all* values of m.

A solution of the wave equation is found to be

$$u_{i,j} = \exp\left[(-1)^{1/2}\lambda(x_i \pm t_j)\right] \tag{4-102}$$

for real λ, where $x_i = ih$, $t_j = jk$. This is more easily compared with Eqn (4-98) if the latter is rewritten, with $\alpha = h\lambda$, in the form

$$U_{i,j} = \exp\left[(-1)^{1/2}\lambda(x_i \pm \psi t_j)\right] \cos^j \theta \tag{4-103}$$

where

$$\psi = \frac{\theta}{\lambda k} = \frac{1}{h\lambda m} \tan^{-1}\left(2m \sin\frac{h\lambda}{2}\right). \tag{4-104}$$

If h is small and λ independent of h, we have by Taylor's expansion

$$\psi = 1 - \left(\frac{1 + 8m^2}{24}\right)\lambda^2 h^2 + O(h^4)$$

and

$$\cos^j \theta = \exp\left[j \ln \cos \theta\right] = \exp\left[-\frac{j}{2} \ln\left(1 + 4m^2 \sin^2\frac{\alpha}{2}\right)\right]$$

$$= \exp\left[-\frac{h}{2} m\lambda^2 t_j + O(h^3 t_j)\right]. \tag{4-105}$$

The real and imaginary parts of the exact solution, Eqn (4-102), *oscillate periodically in time without damping*. On the other hand, the corresponding terms of $U_{i,j}$ [Eqn (4-103)] *possess damped oscillations*, the amplitude being attenuated approximately as $\exp\left[-\frac{1}{2}hm\lambda^2 t_j\right]$. We also observe, from Eqn (4-103), that the period is increased approximately in the ratio $1/\psi \approx 1 + (1 + 8m^2)\lambda^2 h^2/24$.

To illustrate the phenomena more clearly, suppose that $u_{xx} = u_{tt}$ is to hold for $0 < x < 1$, with $u(0, t) = u(1, t) = 0$ and prescribed initial conditions. The exact solution is a superposition of terms of the form in Eqn (4-102), with $\lambda = n\pi$, $n = 1, 2, \ldots$, and the solution of the finite difference approximation is expressible as a superposition of the corresponding terms of the type in Eqn (4-103). The common period P of all terms in the exact solution is $P = 2$. From Eqn (4-105) the ratio of the amplitude of the rth harmonic, at time t, to its initial amplitude, is approximately

$$\exp\left[-\frac{h}{2} mn^2\pi^2 \cdot t\right] = \exp\left[-\frac{t}{2} \frac{mn^2\pi^2}{M + 1}\right] \tag{4-106}$$

† Again, the use of this term refers to the numerical instability discussed in Section 1-8. This type of stability is the *stepwise stability* as opposed to the *pointwise stability* of Section 1-8. Pointwise stability is not concerned with error growth in a given calculation on a fixed net, but with the limiting behavior of the error at a given point when an infinite sequence of calculations is developed for a sequence of progressive net refinements.

where $M + 1 = 1/h$ is the number of mesh spacings in the x direction. M is assumed large in the following argument.

For $m = 1$ and $h = 0.01$, the approximate damping ratio, Eqn (4.106), is about $\exp[-n^2t/20]$. Consequently, while no attenuation exists in the true solution, the amplitude of the fundamental mode ($n = 1$) has decreased to about 90 per cent of its initial value after one period ($P = 2$) and to about 82 per cent after two periods, and so forth. The second harmonic has corresponding values of about 67 and 45 per cent after one and two periods, respectively.

This improper attenuation renders the finite difference approximation of little use in approximating the wave equation. We make this assertion despite its stepwise stability for all values of m and despite the fact (Problem 4-32) that the approximate solution will converge to the exact solution at any fixed mesh point as the mesh is continually refined.

The phenomenon, termed *overstability* by Zajac [17], may be introduced unknowingly in other approximations. One should be especially alert when it is known that the *true solution has time damping and backward differences are used to guarantee stability for all mesh ratios m.*

PROBLEM

4-32 Verify the convergence, as h and $k \to 0$, of the approximate solution, Eqn (4-103), to the exact solution, Eqn (4-102).

4-8 Implicit methods for second-order equations

In Chapter 2 the stability advantages of implicit finite difference approximations were discussed with respect to parabolic equations. The same general observations hold for hyperbolic equations.

For the wave equation $u_{tt} = u_{xx}$, the simplest implicit system is obtained by approximating u_{tt}, as before, by a second central difference centered at i, j while u_{xx} is approximated by the *average of two* second central differences, one centered at $(i, j + 1)$ and the other at $(i, j - 1)$. Thus, one simple implicit approximation takes the form

$$U_{i,j+1} - 2U_{i,j} + U_{i,j-1} = \tfrac{1}{2}m^2\{(U_{i+1,j+1} - 2U_{i,j+1} + U_{i-1,j+1})$$
$$+ (U_{i+1,j-1} - 2U_{i,j-1} + U_{i-1,j-1})\}. \quad (4\text{-}107)$$

This approximation can also be viewed as arising from the replacement of $u_{xx}|_{i,j}$ by the expression $\tfrac{1}{2}\{u_{xx}|_{i,j+1} + u_{xx}|_{i,j-1}\}$.

The implicit nature of Eqn (4-107) is easily seen by writing out the expression to be solved on the $(j + 1)$ line in dependence on values on the two preceding lines. Thus, one finds the equation

$$-m^2U_{i+1,j+1} + 2(1 + m^2)U_{i,j+1} - m^2U_{i-1,j+1}$$
$$= 4U_{i,j} + m^2U_{i+1,j-1} - 2(1 + m^2)U_{i,j-1} + m^2U_{i-1,j-1} \quad (4\text{-}108)$$

which, as in Chapter 2, is seen to be suited to bounded space domains but not for the unbounded domains of pure initial value problems. Suppose that N mesh values are to be determined. Upon writing Eqn (4-108) for each i, $i = 1, 2, \ldots, N$, and inserting the discretized boundary conditions, the *tridiagonal nature* of the system becomes clear. Thus the Thomas algorithm of Section 2-3 may be applied to find a noniterative solution. The iterative methods of Chapter 3 could be used to obtain the solution of this system. In each row of the matrix, the diagonal coefficient is $2(1 + m^2)$ and the other two are $-m^2$. Thus the matrix is diagonally dominant and no difficulty with iterative solutions is anticipated.

When the Fourier stability method of Section 4-6 is applied—that is, when Eqn (4-89) is set into Eqn (4-107), one finds

$$e^{\alpha k} + e^{-\alpha k} = \frac{2}{1 + 2m^2 \sin^2 \tfrac{1}{2}\beta h}. \tag{4-109}$$

Since $|e^{\alpha k}| < 1$ for any positive value of m, unrestricted stepwise stability is obtained (Problem 4-33).

Equation (4-107) is a special case of a *general three-level implicit form* obtained by approximating $u_{xx}|_{i,j}$ with

$$h^{-2}[\lambda \delta_i^2 U_{i,j+1} + (1 - 2\lambda)\delta_i^2 U_{i,j} + \lambda \delta_i^2 U_{i,j-1}] \tag{4-110}$$

where λ is a 'relaxation' factor and δ^2 is the operator

$$\delta_i^2 U_{i,j} = U_{i+1,j} - 2U_{i,j} + U_{i-1,j}.$$

Note that $\lambda = 0$ gives the explicit method, $\lambda = \tfrac{1}{2}$ gives Eqn (4-107), and $\lambda = \tfrac{1}{4}$ is a method discussed by Richtmyer and Morton [18]. By application of the Fourier method it can be shown that the implicit algorithm, generated by employing Eqn (4-110), has unrestricted stability if $\lambda \geq \tfrac{1}{4}$.

Von Neumann (cf. O'Brien *et al.* [19]) introduced the difference equation

$$k^{-2}\delta_j^2 U_{i,j} = h^{-2}\delta_i^2 U_{i,j} + \omega[h^{-2}k^{-2}\delta_j^2\delta_i^2 U_{i,j}] \tag{4-111}$$

as an approximation for the wave equation. When $\omega = 0$ this is the classical explicit scheme; otherwise it is an implicit difference equation solvable at each step by using the tridiagonal algorithm. Von Neumann proved that Eqn (4-111) is unconditionally stable if $4\omega > 1$ and conditionally stable if $4\omega \leq 1$ —the stability condition in the latter case being $kh^{-1} \leq (1 - 4\omega)^{-1/2}$.

Friberg [20] and Lees [21] generalized von Neumann's result to linear hyperbolic equations with variable coefficients,

$$w_{tt} = a(x, t)w_{xx} + b(x, t)w_x + c(x, t)w_t + d(x, t)w + e(x, t). \tag{4-112}$$

In this case a term identical to the second term on the right-hand side of Eqn (4-111) is added. The stability requirements are the same.

The results of Friberg and Lees can be extended to cover the von Neumann type difference approximation to certain linear multidimensional systems (see Chapter 5 for additional information in higher dimensions). However, the linear equations that arise are no longer tridiagonal. Lees [22] develops two modifications of Eqn (4-111) for multidimensional hyperbolic systems by applying the alternating direction procedure to the standard von Neumann scheme. These modified von Neumann type difference equations are shown to be unconditionally stable if $4\omega > 1$.

PROBLEMS

4-33 Complete the Fourier stability analysis of Eqn (4-107).

4-34 Examine Eqn (4-107) for possible overstability, as discussed in Section 4-7.

4-35 Consider the nonlinear equation
$$u_{xx} = u^2 u_{tt} \quad \text{with } u(x, 0) = 1 + x^2, \quad u_t(x, 0) = 0.$$
Describe an explicit finite difference scheme for this pure initial value problem. Can the stability be analyzed by the Fourier method?

4-36 Describe an implicit method for Problem 4-35 in the bounded domain $0 < x < 1$. Suppose the boundary conditions are $u(0, t) = 1$, $u(1, t) = 0$. Is the tridiagonal algorithm applicable?

4-37 Apply the numerical method of characteristics to Problem 4-35.

4-9 Nonlinear examples

No general method exists for studying stability in the case of nonlinear equations. In practice, the best that can be accomplished is to study a 'linearization' based upon bounds for the function and its derivatives. We might do this in any case to discover the effect on the solution of small changes in the coefficients or auxiliary conditions, or perhaps to develop a useful computational algorithm. In many nonlinear cases the use of implicit methods leads to the necessity to solve sets of nonlinear algebraic equations. For such cases there is little to be gained by employing finite difference methods over that of characteristics. However, in some problems the proper use of implicit methods can lead to linear equations. We shall describe such a class, termed *time quasilinear*, later in this section.

Determination of stability bounds, for particular examples, can be accomplished by numerical experimentation. As an example, we consider the nonlinear vibrations of a string fixed at both ends. The governing equations for this system, due to Carrier [23], are
$$[T' \sin \theta]_x = \rho A u''_{tt}, \qquad [T' \cos \theta]_x = \rho A v'_{tt} \tag{4-113}$$
where
$$\theta = \tan^{-1} u'_x/(1 + v'_x) \quad \text{and} \quad T' = T_0 + EA\{[(1 + v'_x)^2 + (u'_x)^2]^{1/2} - 1\}$$

For this study we consider the simplified example obtained by neglecting the displacement v' in the x direction, and eliminate the second of Eqns (4-113). Consequently, the problem becomes

$$[T' \sin \theta]_x = \rho A u''_{tt}, \quad \theta = \tan^{-1} u'_x$$

$$T' = T_0 + EA\{(1 + u'^2_x)^{1/2} - 1\}. \tag{4-114}$$

By introducing the dimensionless variables

$$u = \frac{u'}{L}, \quad X = \frac{x}{L}, \quad T = \frac{t}{L[\rho A/T_0]^{1/2}}$$

and by setting $B = EA/T_0$ we obtain the dimensionless equation

$$\frac{1 - B + B[1 + u_x^2]^{3/2}}{[1 + u_x^2]^{3/2}} u_{XX} = u_{TT}. \tag{4-115}$$

Equation (4-115) is subject to the auxiliary conditions

$$u(0, T) = u(1, T) = 0, \quad u_T(X, 0) = 0, \quad u(X, 0) = 4X(1 - X).$$

When we approximate Eqn (4-115) by the same second-order central differences employed to obtain Eqn (4-82) and a first-order forward difference is used for u_x, the calculated stability threshold was approximately $k/h = 0.55$.

The use of implicit methods to approximate nonlinear equations need not always generate nonlinear algebraic equations. One such general class was discussed by Ames *et al.* [6] in their studies of the nonlinear transverse vibrations of a traveling threadline. Three 'simple' nonlinear models were examined and compared with experimental results. These were

Zaiser model:

$$u_{tt} + 2V_s(1 + u_x^2)^{-1/2}u_{xt} + (V_s^2 - C^2)(1 + u_x^2)^{-1}u_{xx} = 0. \tag{4-116}$$

Mote model:

$$u_{tt} + 2V_s(1 + \delta u_x^2)^{-3/2}u_{xt} + [V_s^2(1 - \delta)(1 - 3\delta u_x^2)(1 + \delta u_x^2)^{-3}$$
$$- 3\delta V_s(1 + \delta u_x^2)^{-5/2}u_x u_t - k^2 C^2 - C^2(1 - k^2)(1 + u_x^2)^{-3/2}]u_{xx} = 0. \tag{4-117}$$

and Hard Spring model:

$$u_{tt} + 2Vu_{xt} + [V^2 - C^2(1 + \nu u_x^2)]u_{xx} = 0. \tag{4-118}$$

All of the aforementioned equations are special cases of the general nonlinear† dimensionless equation

$$u_{tt} + f(x, t, u, u_x, u_{xx})u_{xt} + g(x, t, u, u_x, u_{xx})u_t$$
$$= P(x, t, u, u_x, u_{xx}) \tag{4-119}$$

† This equation is *not* quasilinear in the sense of our previous definition.

which we shall call *time quasilinear*. In the region $0 < x < 1, t > 0$, we seek a solution of Eqn (4-119) subject to the auxiliary conditions $u(0, t) = F(t)$, $u(1, t) = G(t), u(x, 0) = H(x), u_t(x, 0) = J(x)$.

A discretization of Eqn (4-119) is introduced by means of the second-order finite difference approximates, Eqn (1-49), for first derivatives and Eqn (1-51) for second derivatives, both centered at (i, j), and u_{xt} is approximated by

$$(4hk)^{-1}[U_{i+1,j+1} - U_{i+1,j-1} - U_{i-1,j+1} + U_{i-1,j-1}]. \qquad (4\text{-}120)$$

The result of this approximation leads to the algebraic equation

$$a_i U_{i-1,j+1} + b_i U_{i,j+1} + c_i U_{i+1,j+1} = d_i \qquad (4\text{-}121)$$

where

$$a_i = -(4hk)^{-1}f_{i,j}, \qquad b_i = k^{-2} + (2k)^{-1}g_{i,j}, \qquad c_i = -a_i$$

$$d_i = P_{i,j} + (4hk)^{-1}(U_{i+1,j-1} - U_{i-1,j-1})f_{i,j} \qquad (4\text{-}122)$$

$$+ (2k)^{-1}U_{i,j-1}g_{i,j} + k^{-2}(2U_{i,j} - U_{i,j-1})$$

and the notation $f_{i,j}$ means

$$f_{i,j} = f[ih, jk, U_{i,j}, U_x|_{i,j}, U_{xx}|_{i,j}].$$

The nine-point computational molecule of Eqn (4-121) generates an implicit *linear* tridiagonal system at each time step which is explicitly solvable, for the values at the mesh points in the $(j + 1)$ time row, by the Thomas algorithm, thereby eliminating matrix operations. The above algorithm was employed to solve Eqns (4-116), (4-117), and (4-118) subject to the (physical) auxiliary conditions

$$u(0, t) = 0, \qquad u(1, t) = \frac{A_0}{L} \sin \frac{\omega \pi}{\omega_1^{(s)}} t, \qquad u(x, 0) = u_t(x, 0) = 0$$

where A_0/L and $\omega/\omega_1^{(s)}$ were experimental values. The actual computations are shown in Fig. 4-11 compared to the linear computations and experiments. All nonlinear models displayed a 'jump'†—the Hard Spring model agrees very well with the experimental results.

4-10   Simultaneous first-order equations—explicit methods

Finite difference methods can also be used for single or simultaneous first-order equations, provided convergence and stability requirements are considered. Convergence is usually guaranteed if the interval ratio is chosen so that the region of finite difference determination lies completely within that of

---

† For a discussion of 'jump' or 'overhang' phenomena in mechanics, see Bolotin [24].

the differential equations. To begin this discussion let us consider the wave equation $u_{xx} = u_{tt}$. Upon introducing the auxiliary variables

$$v = u_t, \qquad w = u_x \qquad \qquad (4\text{-}123)$$

we have the simple simultaneous equations

$$v_t = w_x, \qquad w_t = v_x. \qquad \qquad (4\text{-}124)$$

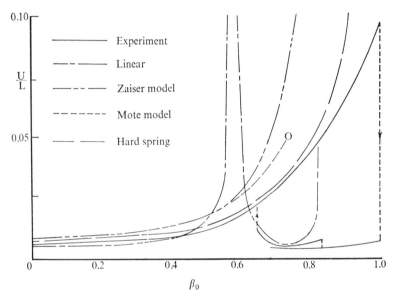

Fig. 4-11   Numerical solutions of 'time quasilinear systems' compared to experimental data

Suppose that the initial values of v and w are specified on the interval $0 < x < 1$ of the initial line $t = 0$. We seek to find v and w at other points in the triangular-shaped *domain* of *determinancy*, shown in Fig. 4-12, bounded by the characteristics, $x - t = 0$ and $x + t = 1$.

If it is desired to use the simplest possible explicit finite difference approximation, we simulate the t derivative by the first forward difference in the t direction. In the x direction we *apparently* have a choice between forward, backward, or central differences, but to remain within the domain of determinancy, a backward difference must not be employed near the line $x - t = 0$, nor a forward difference near $x + t = 1$. On the other hand, central differences *appear* to be satisfactory everywhere inside the domain.† To continue the solution, as t increases, boundary values of v or w need to be specified on the lines $x = 0$ and $x = 1$, for $t > 0$. Once this extension of the domain of

† The situation is more complicated if the characteristics are curved, perhaps depending upon the solution. However, the concepts are the same.

determinancy is accomplished, the solution is uniquely determined (see Section 4-0).

If the function u is sufficiently smooth, the above analysis suggests the explicit formulae

$$(V_{i,j+1} - V_{i,j})/k = (W_{i+1,j} - W_{i-1,j})/2h$$
$$(W_{i,j+1} - W_{i,j})/k = (V_{i+1,j} - V_{i-1,j})/2h. \qquad (4\text{-}125)$$

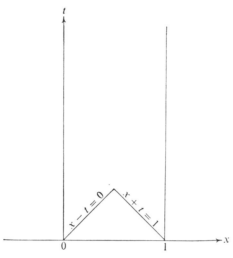

Fig. 4-12 Domain of determinancy of initial conditions

Alternatively, once the solution has been computed on one line above the initial line, we could switch to central differences in the t direction, replacing Eqn (4-125) by

$$(V_{i,j+1} - V_{i,j-1})/2k = (W_{i+1,j} - W_{i-1,j})/2h$$
$$(W_{i,j+1} - W_{i,j-1})/2k = (V_{i+1,j} - V_{i-1,j})/2h. \qquad (4\text{-}126)$$

To avoid difference quotients over the double interval $2h$, the use of *midpoints* of the interval is recommended for one of the functions. Thus we write $W_{i+1/2,j}$ for example. This is purely a matter of notation. Sometimes authors also treat the time intervals in the same way, resulting in a more symmetric appearance and an obviously correct centering. Herein we use this notation only in the space direction. Thus, for example, the scheme of Courant *et al.* [14]

$$(V_{i,j+1} - V_{i,j})/k = (W_{i+1/2,j} - W_{i-1/2,j})/h$$
$$(W_{i-1/2,j+1} - W_{i-1/2,j})/k = (V_{i,j+1} - V_{i-1,j+1})/h \qquad (4\text{-}127)$$

is equivalent to the usual scheme, Eqn (4-82), if one identifies

$$V_{i,j} = (U_{i,j} - U_{i,j-1})/k \qquad (4\text{-}128)$$

and

$$W_{i-1/2,j} = (U_{i,j} - U_{i-1,j})/h \qquad (4\text{-}129)$$

Equation (4-128) is recognized as a backward divided difference for $\partial u/\partial t$ at a mesh point, while Eqn (4-129) is a central divided difference approximation for $\partial u/\partial x$ at a midpoint.

Borrowing from the second-order process used to integrate along the characteristics, we can modify Eqns (4-125) to produce the new explicit formula

$$\frac{1}{k}[V_{i,j+1} - \tfrac{1}{2}(V_{i+1,j} + V_{i-1,j})] = \frac{1}{2h}[W_{i+1,j} - W_{i-1,j}]$$

$$\frac{1}{k}[W_{i,j+1} - \tfrac{1}{2}(W_{i+1,j} + W_{i-1,j})] = \frac{1}{2h}[V_{i+1,j} - V_{i-1,j}].$$

(4-130)

Comparing this with Eqns (4-125), we note that $V_{i,j}$ is replaced by the mean of $V_{i+1,j}$ and $V_{i-1,j}$ with a similar replacement for W in the second equation.

Other methods have been employed to produce stability for larger values of m. In particular, the arithmetic is not more complicated for the approximations

$$\frac{1}{k}(V_{i,j+1} - V_{i,j}) = \frac{1}{2h}[W_{i+1,j} - W_{i-1,j}]$$

$$\frac{1}{k}(W_{i,j+1} - W_{i,j}) = \frac{1}{2h}[V_{i+1,j+1} - V_{i-1,j+1}].$$

(4-131)

which are also explicit.

Each set of finite difference formulae must be examined for stability. Here we sketch the results obtained by the Fourier method. The propagating effect of a line of errors can be examined by considering the effect of a single term and then superposing the results. The errors $\epsilon_{i,j}$ in $W_{i,j}$ and $\eta_{i,j}$ in $V_{i,j}$ are propagated according to the homogeneous forms of the particular equation. If we substitute

$$\epsilon_{i,j} = A \exp [\alpha jk] \exp [(-1)^{1/2}\beta ih]$$

$$\eta_{i,j} = B \exp [\alpha jk] \exp [(-1)^{1/2}\beta ih]$$

(4-132)

into the approximations given above, we find the following results ($m = k/h > 0$):

Equation	m, α, β relation	Stability condition
(4-125)	$(e^{\alpha k} - 1)^2 + m^2 \sin^2 \beta h = 0$	Always unstable
(4-126)	$(e^{\alpha k} - e^{-\alpha k})^2 + 4m^2 \sin^2 \beta h = 0$	$m \le 1$
(4-127)	Problem 4-38	$m \le 1$
(4-130)	$(e^{\alpha k} - \cos \beta h)^2 + m^2 \sin^2 \beta h = 0$	Problem 4-40
(4-131)	$(e^{\alpha k} - 1)^2 + m^2 e^{\alpha k} \sin^2 \beta h = 0$	$m \le 2$

These results indicate that care must be exercised in adopting finite difference formula for first-order equations. Seemingly obvious schemes, of which

Eqns (4-125) are an example, lead to impractical methods. The situation is even more complicated in nonlinear problems. Despite the generality of the convergence theory available in Richtmyer and Morton [18], for example, all is not straightforward with nonlinear equations, *even if their solutions are sufficiently smooth.*† An example, due to Richtmyer [25], demonstrates the problems that may arise in quite simple cases.

The one-dimensional momentum equation from gas dynamics, usually seen in the form $u_t + uu_x = 0$, is used as a nonlinear test example for many finite difference simulations. In some ways it is more convenient to write it in conservation form (see Section 4-13)

$$\frac{\partial u}{\partial t} + \frac{\partial}{\partial x}\left(\frac{1}{2}u^2\right) = 0. \tag{4-133}$$

On the domain $0 \le x \le 1$, $t \ge 0$, with initial and boundary conditions

$$u(x, 0) = x, \quad 0 \le x \le 1$$

$$u(0, t) = 0, \quad t \ge 0$$

Eqn (4-133) has the smooth exact solution

$$u(x, t) = x/(1 + t). \tag{4-134}$$

We attempt to solve this problem using the 'leapfrog' finite difference approximation

$$U_{i,j+1} - U_{i,j-1} + \frac{m}{2}[(U_{i+1,j})^2 - (U_{i-1,j})^2] = 0 \tag{4-135}$$

with $m = k/h$. For the linearized equation (Problem 4-42) the stability condition is

$$m \max_{x,t} (|u|) < 1. \tag{4-136}$$

Even if Eqn (4-136) is satisfied, Eqn (4-135) has solutions which 'blow up' like $\exp(\text{const.}/k)$ as $k \to 0$ for fixed t. To show this, consider a solution of the form

$$U_{i,j} = C^j \cos \tfrac{1}{2}\pi i + S^j \sin \tfrac{1}{2}\pi i + R^j \cos \pi i + V. \tag{4-137}$$

Upon substitution into Eqn (4-135), the recurrence relations

$$C^{j+1} - C^{j-1} = 2mS^j(R^j - V)$$

$$S^{j+1} - S^{j-1} = 2mC^j(R^j + V) \tag{4-138}$$

$$R^{j+1} - R^{j-1} = 0$$

are obtained.

† This shows the measure of the limitation of the presently available ideas of stability and convergence when applied to practical computations.

The third equation provides the information that R^j takes only two constant values, say A and B—that is, $R^{2n} = A$, $R^{2n+1} = B$. To obtain an expression for C^j we combine the first two relations of Eqns (4-138) (Problem 4-43) and develop

$$C^{j+2} - 2C^j + C^{j-2} = 4m^2(A + V)(B - V)C^j. \qquad (4\text{-}139)$$

Upon multiplying Eqn (4-139) by C^{-j+2} we obtain a quadratic in (C^2) and observe that all the solutions are bounded if $4m^2(A + V)(B - V)$ lies between -4 and 0. Otherwise, there is a solution C^j such that $|C^j|$ grows exponentially with j; that is, with t/k.

When can this exponential solution occur? Using Eqn (4-137) the stability condition, Eqn (4-136), becomes

$$m\{|V| + \max(|A|, |B|)\} < 1. \qquad (4\text{-}140)$$

When this holds, the coefficient of C^j, on the right-hand side of Eqn (4-139), *can never be less than* -4. If $|A| < |V|$ and $|B| < |V|$, the coefficient can never be greater than zero (Problem 4-44). Suppose the constant V in Eqn (4-137) represents a smooth solution upon which the other terms are superposed as a perturbation. If the perturbation amplitude is less than a threshold set by $|A| < |V|$, $|B| < |V|$, it will not grow, but if it is initially above this, we expect an exponential blow up.

Richtmyer and Morton [26] and Stetter (see Richtmyer and Morton [18]) ran test calculations on this problem. Use of the leapfrog simulation for the nonlinear problem yielded solutions in several cases which 'blew up.' For the corresponding linearized problem, the method either displayed stability or a slight error growth. The leapfrog technique is only *marginally stable*, even in the linearized case, although it probably provides a convergent approximation in both cases for sufficiently small k—that is, when the truncation error of the difference scheme generates sufficiently small perturbations.

PROBLEMS

4-38 Verify that Eqns (4-127) are equivalent to the simple explicit scheme, Eqn (4-82), for the wave equation. Show that Eqns (4-127) are stepwise stable if $k/h \leq 1$. Employ the Fourier method.

4-39 Examine the stability of the finite difference approximation, Eqn (4-125), by the Fourier method.

4-40 Find the stability condition for Eqns (4-130). *Ans: $m \leq 1$*

4-41 Consider the overstability question for Eqns (4-131).

4-42 Linearize $u_t + uu_x = 0$ by considering $u_t + \alpha u_x = 0$, where $\alpha = \max_{x,t} (|u|)$. Verify the stability condition, Eqn (4-136), if the analogous leapfrog method is employed for this linearized equation.

4-43 Combine the first two relations of Eqns (4-138) to develop Eqn (4-139). Find the equation for S^j.

4-44 Verify the two observations concerning the size of the coefficient $4m^2(A + V)(B - V)$ in Eqn (4-139).

4-11 An implicit method for first-order equations

The implicit scheme

$$V_{i,j+1} - V_{i,j} = \frac{m}{2} [W_{i+1/2,j} - W_{i-1/2,j} + W_{i+1/2,j+1} - W_{i-1/2,j+1}]$$

$$(4\text{-}141)$$

$$W_{i-1/2,j+1} - W_{i-1/2,j} = \frac{m}{2} [V_{i,j+1} - V_{i-1,j+1} + V_{i,j} - V_{i-1,j}]$$

is equivalent to the implicit equation [Eqn (4-108)] previously discussed for the wave equation. We therefore expect stability for all values of $m = k/h$ and this is easily verified (Problem 4-45). The solution of this implicit system is obtainable by means of the Thomas algorithm of Section 2-3. Thus Eqns (4-141) provide a practical system for solving the two simultaneous first-order equations $v_t = w_x, v_x = w_t$.

PROBLEM

4-45 Establish the unconditional stability of Eqns (4-141).

4-12 Hybrid methods for first-order equations

The natural method of characteristics has several disadvantages. One of these becomes clear in the event that spatial distributions of the dependent variables are required at a *fixed time*. Two-dimensional interpolation in the characteristic net is then required and this can be complicated, depending upon the form of the equations. A procedure attributable to Hartree [27] avoids this difficulty by defining the mesh points in advance in space and time, and by interpolating as the computation advances. Consequently, the interpolation is one dimensional.

Since this method is *hybrid* and is related to a similar but simpler scheme of Courant *et al.* [28] we consider it here.† Both methods require the reduction of the system, Eqns (4-6), to the characteristics and the equations along the characteristics [Eqns (1-12) and (4-62)].

Suppose that a fixed rectangular grid is imposed on the integration domain

---

† The Hartree method can also be applied to second-order systems with only minor changes (Problem 4-46).

with $\Delta x = h$ and $\Delta y = k$ as in Fig. 4-13. If the governing equations are Eqns (4-6), then those for the characteristics may be written

$$\frac{dx}{dy} = F_\alpha, \qquad \frac{dx}{dy} = F_\beta \qquad (4\text{-}142)$$

and the relations along the characteristics are

$$dv + G_\alpha\,du + H_\alpha\,dy = 0$$
$$dv + G_\beta\,du + H_\beta\,dy = 0. \qquad (4\text{-}143)$$

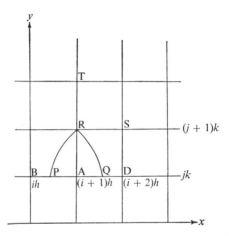

Fig. 4-13   An illustration of Hartree's hybrid method

With reference to Fig. 4-13, suppose that the solution is known at the mesh points on the line $y = jk$ (this could be the initial line) and R, S, ... are equally spaced along the next line, $y = (j + 1)k$. Draw the α and β characteristics, through R, back to their intersection with the first line $y = jk$. These two points of intersection are unknown. The equations relating the solution values at R, P, and Q are still given by formulae similar to Eqns (4-58) and (4-59), provided some obvious changes in notation are made. However, we *know* the values of the coordinates x_R, y_R, y_A and y_B and wish to calculate u_R, v_R, x_P, and x_Q. The four equations [(4-58) and (4-59)] suffice for the determination of these unknowns, except that interpolation for the values of u and v at P and Q are necessary at each step.

Hartree's computational form is

$$x_R - x_P = \tfrac{1}{2}[F_\alpha(R) + F_\alpha(P)]k$$
$$x_R - x_Q = \tfrac{1}{2}[F_\beta(R) + F_\beta(Q)]k$$
$$v_R - v_P + \tfrac{1}{2}[G_\alpha(R) + G_\alpha(P)](u_R - u_P) + \tfrac{1}{2}[H_\alpha(R) + H_\alpha(P)]k = 0 \qquad (4\text{-}144)$$
$$v_R - v_Q + \tfrac{1}{2}[G_\beta(R) + G_\beta(Q)](u_R - u_Q) + \tfrac{1}{2}[H_\beta(R) + H_\beta(Q)]k = 0$$

with a second-order truncation error. This system must be solved by iteration using interpolated values at P and Q.

Courant *et al.* [28] suggested two schemes which, on the rectangular grid of Fig. 4-13, are less accurate, first-order systems. The first of these,

$$
\begin{aligned}
x_R - x_P &= F_\alpha(A)k \\
x_R - x_Q &= F_\beta(A)k
\end{aligned}
\tag{4-145}
$$

$$
\begin{aligned}
v_R - v_P + G_\alpha(A)(u_R - u_P) + H_\alpha(A)k &= 0 \\
v_R - v_Q + G_\beta(A)(u_R - u_Q) + H_\beta(A)k &= 0
\end{aligned}
\tag{4-146}
$$

is essentially a simple Euler method. In this system x_P and x_Q are obtained immediately from Eqns (4-145) (note the evaluation of F_α and F_β at A). After interpolation, to get function values of u and v at P and Q (using, say, linear interpolation with the three adjacent mesh points B, A, D), we may calculate u_R and v_R from the simultaneous Eqns (4-146), thus requiring no iteration.

The second method of Courant *et al.*,† of comparable accuracy to the first, involves rewriting Eqns (4-143) in the canonical form

$$
\begin{aligned}
\left\{ \frac{\partial v}{\partial y} + F_\alpha \frac{\partial v}{\partial x} \right\} + G_\alpha \left\{ \frac{\partial u}{\partial y} + F_\alpha \frac{\partial u}{\partial x} \right\} + H_\alpha &= 0 \\
\left\{ \frac{\partial v}{\partial y} + F_\beta \frac{\partial v}{\partial x} \right\} + G_\beta \left\{ \frac{\partial u}{\partial y} + F_\beta \frac{\partial u}{\partial x} \right\} + H_\beta &= 0
\end{aligned}
\tag{4-147}
$$

and replacing the derivatives by differences. Thus we have

$$
\left\{ \frac{v_R - v_A}{k} + F_\alpha(A) \frac{v_A - v_B}{h} \right\} + G_\alpha(A) \left\{ \frac{u_R - u_A}{k} + F_\alpha(A) \frac{u_A - u_B}{h} \right\} \\
+ H_\alpha(A) = 0 \quad (4\text{-}148)
$$

and

$$
\left\{ \frac{v_R - v_A}{k} + F_\beta(A) \frac{v_D - v_A}{h} \right\} + G_\beta(A) \left\{ \frac{u_R - u_A}{k} + F_\beta(A) \frac{u_D - u_A}{h} \right\} \\
+ H_\beta(A) = 0. \quad (4\text{-}149)
$$

These are so constructed that in Eqn (4-148), corresponding to the forward facing (or α) characteristic, the space derivative is replaced by a *backward* difference. In the backward facing (or β) characteristic, the space derivative is replaced by a *forward* difference. Having obtained x_P and x_Q from Eqns (4-145) we may now immediately obtain v_R and u_R from Eqns (4-148) and (4-149), without iteration.

---

† This is known as the method of 'forward and backward space differences' when applied to the transport equation (Richtmyer and Morton [18], p. 238).

Both of the Courant–Isaacson–Rees *explicit* methods are first order. The established stability condition requires that both the relations

$$|F_\alpha| \frac{k}{h} < 1 \quad \text{and} \quad |F_\beta| \frac{k}{h} < 1 \tag{4-150}$$

be satisfied.

PROBLEMS

4-46 Describe the application of Hartree's method to the second-order equation, Eqn (4-52).

4-47 Employ the second method of Courant–Isaacson–Rees on the system of Eqns (4-38).

4-48 Discuss the advantages and disadvantages of the three methods. Note that in the second method of Courtant *et al.* the characteristic directions must be determined but the special properties of characteristics are not then employed!

4-13   Gas dynamics in one-space variable

The differential equations governing the flow of gases† are based on the conservation of mass (continuity equation), momentum (momentum equations), and energy (energy equation). These, together with appropriate auxiliary conditions, determine the (mathematical) state of the gas. Gas flows are often characterized by *internal discontinuities*; that is, *shocks* may appear in the gas. Interaction of shocks may give rise to the development of discontinuities separating regions of the same fluid in different thermodynamic states. Special internal boundary conditions (jump conditions) are therefore necessary. The usual set of conditions is the Rankine–Hugoniot relations (see, for example, Courant and Friedrichs [3]), but their application is complicated when the surfaces, on which they are to be applied, are in motion. This motion is usually not known in advance but is determined by the differential equations and the jump conditions themselves. The resulting numerical work is highly implicit. In the next few sections we discuss several finite difference methods for calculating the *smooth portion* of the flow. Techniques for dealing with shocks will be discussed in Chapter 5.

The equations for the flow of a one-dimensional inviscid fluid (gas) may be expressed in several ways, the two fundamental methods being the Eulerian and the Lagrangian. These are equivalent, but each has its own advantages and disadvantages. In the *Eulerian form*, the independent space variables are referred to a *fixed* spatial coordinate system. The fluid is visualized as moving

† We use the term 'gas dynamics' when the effect of viscosity is neglected and 'fluid mechanics' for the general subject.

through this fixed reference frame and is characterized by a time-dependent velocity field which is to be determined by solving an initial value problem. Alternatively, if the independent space variables are referred to a reference frame fixed in the fluid and undergoing all the distortion and motion of the fluid, then the fluid particles are *permanently* identified by these variables, called *Lagrangian* variables. Particle positions in space are among the dependent variables we wish to determine. *Of course, the two forms are equivalent but this may not be true of the two solutions obtained from the distinct finite difference approximations.*

The Lagrangian form is generally conceded to provide more information since it describes where each particle of fluid came from originally. Additionally, until fairly recently (1960), the only known stable difference approximations for the Eulerian forms were less accurate than those for the Lagrangian forms. Consequently, the latter were preferred.

Let the properties of the gas be described by the density ρ, pressure p, internal energy per unit mass e, and fluid velocity u. If viscous forces, body forces (e.g. gravity and other external fields), heat conduction, and energy sources are absent, the one-dimensional Euler form is

$$\rho_t + u\rho_r + \rho r^{-\alpha}(r^\alpha u)_r = 0, \qquad u_t + uu_r + \rho^{-1}p_r = 0$$

$$e_t + ue_r + p\rho^{-1}r^{-\alpha}(r^\alpha u)_r = 0, \qquad p = p(e, \rho) \tag{4-151}$$

where r denotes the spatial coordinate and α is a constant depending upon the problem geometry ($\alpha = 0$ for slab symmetry, $r = x$; $\alpha = 1$ for cylindrical symmetry, $r = (x^2 + y^2)^{1/2}$; $\alpha = 2$ for spherical symmetry, $r = (x^2 + y^2 + z^2)^{1/2}$).

The Lagrange form of Eqns (4-151), with x the position of a particle at time zero ($x = r\,(t = 0)$), ρ_0 the initial density, and $V = 1/\rho$, is

$$\rho_0 x^\alpha\, dx = \rho r^\alpha\, dr \tag{4-152}$$

$$V_t - \rho_0^{-1}x^{-\alpha}(r^\alpha u)_x = 0, \qquad u_t + \rho_0^{-1}(r/x)^\alpha p_x = 0 \tag{4-153}$$

$$e_t + pV_t = 0, \qquad r_t = u, \qquad p = p(e, V). \tag{4-154}$$

One can simplify this system by introducing the alternate Lagrange coordinate y as a mass instead of a length, so that

$$dy = \rho_0 x^\alpha\, dx = V^{-1}r^\alpha\, dr. \tag{4-152a}$$

Hence, Eqns (4-153) take the simpler forms

$$V_t - (r^\alpha u)_y = 0, \qquad u_t + r^\alpha p_y = 0. \tag{4-155}$$

One of the equations of the Lagrange form is redundant since there are *six* [(Eqns (4-152)–(4-154)] to solve for the *five* unknowns, V, u, e, p, and r. Most calculations have been accomplished by omitting the first of Eqns (4-153), obtaining V from Eqn (4-152) as $V = d(r^{\alpha+1})/\rho_0 d(x^{\alpha+1})$ and r from

$r_t = u$. When this equation is used for V, automatic mass conservation is ensured. If the first of Eqns (4-153) is employed for V, small variations in mass tend to occur.

Both systems have approximately the same complexity. The major disadvantage of the Eulerian system arises when interfaces (shocks) occur separating fluids of different density. The Lagrangian system does not have the 'spatial' coordinate mesh fixed in advance and may *require* refinement of the mesh as the computation advances. This possibility of 'regridding' arises since the Lagrange form is constructed so that the mass between two successive mesh points is (approximately) conserved.

4-14 Eulerian difference equations

The choice and construction of finite difference approximations depends on such factors as accuracy, stability, and the proper incorporation of important physical principles—for example, conservation laws. We shall describe several of the methods and their history in this section.

(a) Method of Courant *et al.* [28]

This *first-order* procedure is the obvious extension of the method given by Eqns (4-147)–(4-149). Time differences are *forward* and each equation, after reduction to canonical or characteristic form, is approximated by a forward or backward space difference according to the sign of the slope of the corresponding characteristic (positive → backward, negative → forward). Stability and convergence have been established for a linearized version leading to the stability condition

$$(|u| + c)m \le 1, \qquad m = \frac{k}{h} = \frac{\Delta t}{\Delta r}. \qquad (4\text{-}156)$$

Here $c(r, t)$ is the local isentropic sound speed. For isentropic flow, $de + d(1/\rho) = 0$ or, rewriting $p = p(e, \rho)$ as $e = F(p, \rho)$, we have

$$\frac{\partial F}{\partial p} dp + \frac{\partial F}{\partial \rho} d\rho - \frac{p}{\rho^2} d\rho = 0.$$

Since the sound speed $c^2 = dp/d\rho$ along an adiabat, there follows

$$c^2 = \frac{p/\rho^2 - \partial F/\partial \rho}{\partial F/\partial p}.$$

Ansorge [29] has constructed higher order methods of this type for quasilinear hyperbolic systems by analogy with the Adams predictor–corrector methods for ordinary differential equations.

PROBLEM

4-49 With $\alpha = 0$, write Eqns (4-151) in first-order finite difference form using the Courant–Isaacson–Rees technique. (Note that the equations must first be converted to the canonical form of Eqns (4-147).)

(b) Lelevier's scheme

The most commonly used schemes for both systems have been designed after the method [Eqn (4-82)] for the wave equation. As noted in Section 4-10, direct transfer of this idea can lead to an unstable procedure. We must exercise care in the manner of discretizing the terms involving $u(\ )_r$ as recognized by Lelevier (see Richtmyer and Morton [18, p. 292]). Thus, the approximation

$$u_{i,j+1}(2\Delta r)^{-1}(p_{i+1,j} - p_{i-1,j}) \tag{4-157}$$

for $u\, \partial p/\partial r$ is unsatisfactory, leading to an *always unstable* situation [compare Eqns (4-125)].

Lelevier corrected this difficulty by using a forward difference for terms in $u(\ )_r$ when $u < 0$ and a backward difference when $u > 0$. The resulting scheme is stable if

$$(|u| + c)\frac{\Delta t}{\Delta r} < 1 \tag{4-158}$$

a result borne out by calculations.

The Lelevier finite difference equations† for slab symmetry [Eqns (4-151), with $\alpha = 0$] are

$$p_{i+1/2,j+1} - p_{i+1/2,j} + \left(\frac{\Delta t}{\Delta r}\right)u_{i,j+1}\,\underline{(p_{i+1/2,j} - p_{i-1/2,j})}$$

$$= -p_{i+1/2,j}\left(\frac{\Delta t}{\Delta r}\right)(u_{i+1,j+1} - u_{i,j+1}) \quad \text{if } u_{i,j+1} \geq 0$$

$$p_{i,j}\left[u_{i,j+1} - u_{i,j} + \left(\frac{\Delta t}{\Delta r}\right)u_{i,j}\,\underline{(u_{i,j} - u_{i-1,j})}\right] \tag{4-159}$$

$$= -\frac{\Delta t}{\Delta r}(p_{i+1/2,j} - p_{i-1/2,j}) \quad \text{if } u_{i,j} \geq 0$$

$$p_{i,j}\left[e_{i+1/2,j+1} - e_{i+1/2,j} + \left(\frac{\Delta t}{\Delta r}\right)u_{i,j+1}\,\underline{(e_{i+1/2,j} - e_{i-1/2,j})}\right]$$

$$= -p_{i+1/2,j}\left(\frac{\Delta t}{\Delta r}\right)(u_{i+1,j+1} - u_{i,j+1}) \quad \text{if } u_{i,j+1} \geq 0.$$

If the velocity ($u_{i,j}$ or $u_{i,j+1}$) is negative, a forward space difference is used in the three underlined places on the left-hand side of Eqns (4-159). Coupled with Eqns (4-159) is the energy relation

$$e_{i+1/2,j+1} = F(p_{i+1/2,j+1}, \rho_{i+1/2,j+1}). \tag{4-160}$$

The scheme is explicit and has first-order accuracy, which is not usually adequate for most applications.

---

† In the remainder of this book we shall employ the same symbols in the finite difference approximations as are used in the original differential equations.

(c) Conservation form and Lax–Wendroff schemes

Since the equations of gas dynamics (and fluid mechanics in general) are based upon conservation laws, researchers in fluid mechanics have often found it convenient to use a form of the equations, called the *divergence form*, which clearly displays the conserved quantities (mass, momentum, and energy). It therefore seems reasonable to try to preserve these conservation properties in the finite difference approximations.† Lax [30, 31] first considered finite difference approximations based on the conservation-law form.

A system of equations

$$\mathbf{w}_t + [\mathbf{f}(\mathbf{w})]_r = 0 \tag{4-161}$$

—where $\mathbf{w}$ is a vector function (of r and t) with n components and $\mathbf{f}$ is a non-linear vector function, with n components, of the vector $\mathbf{w}$—is called a system of *conservation-law form*.

In the Eulerian system the conserved quantities are mass, momentum, and energy per unit volume, represented by

$$\rho, \mathbf{m} = \rho\mathbf{u} \quad \text{and} \quad E = \rho(e + \tfrac{1}{2}\mathbf{u}^2) \tag{4-162}$$

respectively. Since the general three-dimensional Eulerian equations are expressible in the form

$$\rho_t = -\nabla\cdot(\rho\mathbf{u})‡$$
$$(\rho\mathbf{u})_t = -\mathbf{u}\nabla\cdot(\rho\mathbf{u}) - \rho(\mathbf{u}\cdot\nabla)\mathbf{u} - \nabla p \tag{4-163}$$
$$(\rho E)_t = -\nabla\cdot(\rho E\mathbf{u}) - \nabla\cdot(p\mathbf{u})$$

it follows (Problem 4-50) that Eqns (4-151) can only be put in the conservation-law form of Eqn (4-161) if $\alpha = 0$ (slab symmetry), and in that case they become

$$\frac{\partial}{\partial t}\mathbf{U} + \frac{\partial}{\partial r}\mathbf{F}(\mathbf{U}) = 0 \tag{4-164}$$

where $\mathbf{U}$ and $\mathbf{F}(\mathbf{U})$ are the vectors

$$\mathbf{U} = \begin{bmatrix} \rho \\ m \\ E \end{bmatrix}, \quad \mathbf{F}(\mathbf{U}) = \begin{bmatrix} m \\ (m^2/\rho) + p \\ (E + p)m/\rho \end{bmatrix}. \tag{4-165}$$

The pressure is calculated from the 'equation of state'

$$p = p(e, \rho) = p\left(\frac{E}{\rho} - \frac{m^2}{2\rho^2}, \rho\right). \tag{4-166}$$

The original suggestion of Lax [30] was the 'staggered' scheme previously

---

† The argument as to the necessity of this form still continues (1969) and some insist, for the *Navier–Stokes equations*, that nothing is gained (see, for example, Chorin [32]).
‡ The *bold face* denotes the appropriate vector.

given in Eqns (4-130) for the linear case. When applied to Eqn (4-161) we have

$$k^{-1}\{w_{i,j+1} - \tfrac{1}{2}(w_{i+1,j} + w_{i-1,j})\} + (2h)^{-1}\{f_{i+1,j} - f_{i-1,j}\} = 0 \qquad (4\text{-}167)$$

where $f_{i,j}$ is an abbreviation for $f(U_{i,j})$. The staggering enables central space differences and forward time differences to be used without developing instability. These staggered schemes are usually stable if $a\,\Delta t/\Delta r < 1$, where a is the local speed of sound. In essence, the replacement of $w_{i,j}$ by the means of neighboring values, in Eqn (4-167), has the effect of introducing a dissipative term. We shall meet this idea again in shock calculations.

Longley [33] and Lax [34] confirm that conservative approximations represent the physical facts more appropriately than nonconservative schemes. Payne [35] attempted the Lax scheme on the equations with cylindrical symmetry ($\alpha = 1$), which cannot be put into conservation form. The additional term, $r(\partial p/\partial r)$, when approximated by $r_{i,j}(p_{i+1,j} - p_{i-1,j})/2\Delta r$ to retain the staggered effect, generates a scheme unstable at the axis $r = 0$. Payne's stable representation is not in staggered form, thus showing that it is difficult to extend successful schemes in plane geometry into other geometries. Roberts [36] used the Lax scheme in spherical symmetry, but none of the equations took the conservation-law form.

Probably the most successful method, which has also generated much additional research, is that of Lax and Wendroff [37, 38]. Their scheme can also be employed for the Lagrangian equations in slab symmetry. With reference to Eqn (4-164) the Lax–Wendroff equations start from the Taylor's series in t, ($k = \Delta t$)

$$U_{i,j+1} = U_{i,j} + k\left(\frac{\partial U}{\partial t}\right)_{i,j} + \frac{k^2}{2}\left(\frac{\partial^2 U}{\partial t^2}\right)_{i,j} + \cdots \qquad (4\text{-}168)$$

The indicated t derivatives are replaced by r derivatives from Eqn (4-164). Clearly

$$\frac{\partial U}{\partial t} = -\frac{\partial F}{\partial r}$$

and $\quad \dfrac{\partial^2 U}{\partial t^2} = -\dfrac{\partial}{\partial t}\dfrac{\partial F}{\partial r} = -\dfrac{\partial}{\partial r}\dfrac{\partial F}{\partial t} = -\dfrac{\partial}{\partial r}\left(A\dfrac{\partial U}{\partial t}\right) = \dfrac{\partial}{\partial r}\left(A\dfrac{\partial F}{\partial r}\right)$

where the matrix $A = A(U)$ is the Jacobian of $F(U)$ with respect to U; that is, $A = (A_{ij})$ where $A_{ij} = \partial F_i/\partial U_j$. The r derivatives are then approximated by differences to give the *second-order accurate* Lax–Wendroff scheme

$$U_{i,j+1} = U_{i,j} - \frac{1}{2}\frac{k}{h}\,(F_{i+1,j} - F_{i-1,j})$$

$$+ \frac{1}{2}\left(\frac{k}{h}\right)^2\{A_{i+1/2,j}(F_{i+1,j} - F_{i,j}) - A_{i-1/2,j}(F_{i,j} - F_{i-1,j})\} \qquad (4\text{-}169)$$

where $h = \Delta r$, and $A_{i+1/2,j} = A[\tfrac{1}{2}(U_{i+1/2,j} + U_{i,j})]$.

The formal Lax–Wendroff scheme, Eqn (4-169), is complicated by the appearance of the Jacobian matrix A. A two-step modification, which has second-order accuracy and reduces to the form of Eqn (4-169) when A is constant (see Problem 4-52), is simpler to use since A does not appear. First, predicted values are calculated at the *centers* of rectangular meshes by means of the first-order accurate formula

$$\mathbf{U}_{i+1/2, j+1/2} = \frac{1}{2}(\mathbf{U}_{i+1, j} + \mathbf{U}_{i, j}) - \frac{k}{2h}(\mathbf{F}_{i+1, j} - \mathbf{F}_{i, j}) \qquad (4\text{-}170)$$

and the final corrected values are found from

$$\mathbf{U}_{i, j+1} = \mathbf{U}_{i, j} - \frac{k}{h}(\mathbf{F}_{i+1/2, j+1/2} - \mathbf{F}_{i-1/2, j+1/2}) \qquad (4\text{-}171)$$

which is second-order accurate, since the difference on the right-hand side is central.

The stability condition for the Lax–Wendroff scheme is established without difficulty in the constant coefficient case, $\mathbf{F} = A\mathbf{U}$. In the nonlinear conservation-law form the matrix A, obtained from Eqns (4-165), is cumbersome but its eigenvalues can be found indirectly by the following artifice. The original Eulerian equations with slab symmetry are [Eqns (4-151) with $\alpha = 0$]:

$$\rho_t + u\rho_r + \rho u_r = 0, \qquad u_t + uu_r + \rho^{-1}p_r = 0$$
$$e_t + ue_r + p\rho^{-1}u_r = 0, \qquad p = p(e, \rho) \qquad (4\text{-}172)$$

and these are also of the form $\partial \mathbf{U}/\partial t + A_1(\partial \mathbf{U}/\partial r) = 0$, but with a different matrix A_1. The eigenvalues of the conservation-law matrix and the matrix A_1 are the same, since the reciprocals of these eigenvalues are the slopes of the characteristics in the (x, t) plane.

These eigenvalues can be found by solving the equation of state for $e = e(p, \rho)$ and substituting this into the equation for e in Eqns (4-172). The ρ derivatives are eliminated from this equation using the first (continuity) equation of (4-172), so that

$$p_t + up_r + \rho c^2 u_r = 0 \qquad (4\text{-}173)$$

where $c^2 = dp/d\rho$. Using the first two equations of (4-172), together with Eqn (4-173), the system takes the required form

$$\begin{bmatrix} \rho_t \\ u_t \\ p_t \end{bmatrix} + \begin{bmatrix} u & \rho & 0 \\ 0 & u & \rho^{-1} \\ 0 & \rho c^2 & u \end{bmatrix} \begin{bmatrix} \rho_r \\ u_r \\ p_r \end{bmatrix} = 0. \qquad (4\text{-}174)$$

The eigenvalues of A_1 are $u, u + c, u - c$, so the stability condition of the Lax–Wendroff scheme is

$$(|u| + c)k/h < 1 \qquad (4\text{-}175)$$

as for the other Eulerian difference methods.

Modifications and extensions of the original Lax–Wendroff ideas have been developed by a number of authors. Strang [39] also considered the Lax–Wendroff technique, and the application of Runge–Kutta type methods to integrate Eqn (4-164). Richtmyer [40], Gary [41], and Gourlay and Morris [42] have introduced implicit methods for Eqn (4-164) and for the differentiated form

$$\frac{\partial \mathbf{U}}{\partial t} + A(\mathbf{U}) \frac{\partial \mathbf{U}}{\partial r} = 0 \qquad (4\text{-}176)$$

where A is the Jacobian matrix. The paper of Gourlay and Morris also contains some new explicit predictor–corrector methods and extensions to higher dimension.

PROBLEMS

4-50 Show that *only* in the one-dimensional case of slab symmetry ($\alpha = 0$) can Eqns (4-151) be put in conservation-law form.

4-51 Develop the conservation-law form of the Lagrangian equations in the case of slab symmetry ($\alpha = 0$).

Ans: $\quad V_t = V_0 u_x, \qquad u_t = -V_0 p_x, \qquad \epsilon_t = -V_0(pu)_x$

where $\epsilon = e + \frac{1}{2}u^2$ is the *total* energy per unit mass.

4-52 If A is a constant matrix—that is, $\mathbf{F} = A\mathbf{U}$—what form does the Lax–Wendroff scheme [Eqn (4-169)] become? What order accuracy is immediately observable?

Ans: $\quad \mathbf{U}_{i,j+1} = \mathbf{U}_{i,j} - \frac{1}{2} A \frac{k}{h} (\mathbf{U}_{i+1,j} - \mathbf{U}_{i-1,j})$

$$+ \frac{1}{2}\left(A\frac{k}{h}\right)^2 (\mathbf{U}_{i+1,j} - 2\mathbf{U}_{i,j} + \mathbf{U}_{i-1,j})$$

4-53 Verify that Eqns (4-170) and (4-171) yield the result of Problem 4-52 if $\mathbf{F} = A\mathbf{U}$, A constant.

4-15 Lagrangian difference equations

Of course, the Lagrangian formulation is also hyperbolic and the difference equations might be taken as those of Courant *et al.* [28], but it is of first-order accuracy. In practice the scheme for the wave equation [Eqns (4-127)] or the Lax–Wendroff technique has provided the foundation for the finite difference simulation of the Lagrangian system. One of the usual means for simulating (see Eqns (4-152)–(4-154)) the Lagrangian system

$$r_t = u, \qquad e_t + pV_t = 0, \qquad u_t + V_0(r/x)^\alpha p_x = 0$$

$$V = V_0(r/x)^\alpha \frac{\partial r}{\partial x}, \qquad p = f(e, V) \qquad (4\text{-}177)$$

is

$$V_{i+1/2,j+1} = V_0 \frac{(r_{i+1,j+1})^{\alpha+1} - (r_{i,j+1})^{\alpha+1}}{x_{i+1}^{\alpha+1} - x_i^{\alpha+1}}$$

$$\frac{u_{i,j+1/2} - u_{i,j-1/2}}{\Delta t} = -V_0 \left(\frac{r_{i,j}}{x_i}\right)^\alpha \frac{p_{i+1/2,j} - p_{i-1/2,j}}{\Delta x}$$

$$\frac{e_{i+1/2,j+1} - e_{i+1/2,j}}{\Delta t} = -p_{i+1/2,j} \frac{V_{i+1/2,j+1} - V_{i+1/2,j}}{\Delta t} \qquad (4\text{-}178)$$

$$\frac{r_{i,j+1} - r_{i,j}}{\Delta t} = u_{i,j+1/2}$$

$$p_{i+1/2,j+1} = f(e_{i+1/2,j+1}, V_{i+1/2,j+1})$$

Here, x_i denotes the ith net point of the Lagrangian net. For the system to be correct to $O[(\Delta t)^2] + O[(\Delta x)^2]$, the third equation must be centered by *replacing* $p_{i+1/2,j}$ by some better approximation, such as $\frac{1}{2}[p_{i+1/2,j+1} + p_{i+1/2,j}]$. The consequence of this is the appearance of two unknowns $p_{i+1/2,j+1}$ and $e_{i+1/2,j+1}$ in the third equation. An iterative procedure is then required to solve the third equation simultaneously with the fifth.

If the dependent variables of Eqns (4-178) are determined in the proper order, the system is effectively explicit. If there are boundaries† at $i = 0$ and $i = I$ and all quantities are known at $t = j\,\Delta t$, one order might be:

   (i) Calculate $u_{i,j+1/2}$, $i = 1, 2, \ldots, I - 1$, from the second of Eqns (4-178);
  (ii) Calculate $u_{0,j+1/2}$, $u_{I,j+1/2}$ from boundary conditions;
 (iii) Calculate $r_{i,j+1}$, $i = 0, 1, \ldots, I$, from the fourth of Eqns (4-178);
 (iv) Calculate $V_{i+1/2,j+1}$, $i = 0, 1, \ldots, I - 1$, from the first of Eqns (4-178);
  (v) Calculate $e_{i+1/2,j+1}$, $i = 0, 1, \ldots, I - 1$, from the third of Eqns (4-178);
 (vi) Calculate $p_{i+1/2,j+1}$, $i = 0, 1, \ldots, I - 1$, from the fifth of Eqns (4-178).

Stability of this system has been analyzed by small perturbations about a constant state (Richtmyer and Morton [18]) leading to the condition

$$\frac{V_0}{V} \left(\frac{r}{x}\right)^\alpha c \frac{\Delta t}{\Delta x} \le 1. \qquad (4\text{-}179)$$

---

† Typical boundary conditions at $i = 0$ are: *rigid wall*, $u_{0,j} = 0$ all j; *free surface*, $p_{1/2,j} + p_{-1/2,j} = 0$ all j—that is, the interpolated value of p vanishes at $i = 0$. Boundary conditions can also be applied at $i = I - \frac{1}{2}$ as follows: *rigid wall*, $u_{I,j} + u_{I-1,j} = 0$; *free surface*, $p_{I-1/2,j} = 0$.

A more natural criterion is obtained if we use

$$\Delta r = \frac{V}{V_0} \left(\frac{x}{r}\right)^\alpha \Delta x$$

where Δr is the difference of the values of $r(x, t)$ for adjacent net points r_i and r_{i+1}. This is the actual distance between particles labeled r_i and r_{i+1}. Therefore Eqn (4-179) becomes

$$\frac{c\,\Delta t}{r_{i+1,j} - r_{i,j}} \le 1 \quad \text{for all } i, j \tag{4-180}$$

a result that can be automatically tested as the computation progresses. The sound speed c also depends upon i and j and is calculated from the equation of state.

In most gas dynamics problems implicit methods are seldom justified on the basis of increased stability. For most problems significant changes occur in time $\Delta t = \Delta x/c$, so the use of longer time intervals is not a motivation. However, in astrophysics and meteorology, they have been found useful. Their application is discussed by Gary [41], Turner and Wendroff [43], and Gourlay and Morris [42].

PROBLEMS

4-54   The conservation law of the Lagrangian equations in slab symmetry was developed in Problem 4-51, in the variables V, u, and ϵ. Rewrite the third equation in an alternative form. Taking V, u, and p as the components of $\mathbf{U}$, put the system in the form $\partial \mathbf{U}/\partial t + A(\partial \mathbf{U}/\partial x) = 0$. From the resulting matrix A, find the eigenvalues and infer the stability condition for the Lax–Wendroff method.

$$Ans: \quad A = \begin{bmatrix} 0 & -V_0 & 0 \\ 0 & 0 & V_0 \\ 0 & V_0(cV^{-1})^2 & 0 \end{bmatrix}; \quad \frac{V_0}{V} c \frac{\Delta t}{\Delta x} < 1$$

4-55   Find the canonical form of the Lagrangian equations in slab symmetry.

4-16   Hopscotch methods for conservation laws

This and the next section supplement the material of Section 4-13(c) on methods for conservation laws

$$u_t + [f(u)]_x = 0 \tag{4-181}$$

$$u(x, 0) = g(x), \quad a \le x \le b \tag{4-182}$$

$$u(0, t) = h(t)\dagger \tag{4-183}$$

† The exact form of the boundary conditions does not concern us here since their influence will not be examined. The analysis, including that of stability, will be done as though an initial value problem is being studied.

where u and f are n vectors. Here we consider both first-order and second-order methods which are similar in nature to the hopscotch schemes introduced by Gourlay [44] for the solution of parabolic equations. A good survey paper for hyperbolic systems is due to Gourlay and Morris [45] with further discussion in Gourlay and McGuire [46].

Let $\Delta x = h$, $\Delta t = k$, and $u_{i,j} = u(ih, jk)$ be the grid parameters with the grid ratio $m = k/h$ held constant. The usual difference operators will be used supplemented by

$$H_x u_{i,j} = u_{i+1,j} - u_{i-1,j}. \tag{4-184}$$

The classic first-order method due to Lax [30] is

$$u_{i,j+1} = u_{i,j} - (m/2)H_x f_{i,j} + \sigma \delta_x^2 u_{i,j} \tag{4-185}$$

where σ, the coefficient of the pseudoviscous term, is chosen to obtain the best possible shock resolution (see Section 5-2(b) for discussion). The stability requirement obtained by Fourier analysis of the linearized scheme is $m|\lambda| \leq \sqrt{(2\sigma)}$, $0 \leq \sigma \leq \frac{1}{2}$ where $|\lambda|$ is the maximum modulus eigenvalue of the Jacobian matrix A of partial derivatives of f with respect to u.

For good shock resolution, a small value of σ is needed. Thus a small value of m has to be chosen, thus increasing the number of steps required in the computation. In order to calculate with a scheme possessing a value of $m|\lambda|$ near the theoretical Courant–Friedrichs–Lewy upper limit of 1, yet having considerable freedom in the value of the pseudoviscosity parameter σ a 'hopscotch' method can be employed. The resulting *explicit hopscotch Lax* scheme

$$u_{i,j+1} + \theta_{i,j+1}[(m/2)H_x f_{i,j+1} - \sigma \delta_x^2 u_{i,j+1}]$$
$$= u_{i,j} - \theta_{i,j}[(m/2)H_x f_{i,j} - \sigma \delta_x^2 u_{i,j}] \tag{4-186}$$

where

$$\theta_{i,j} = \begin{cases} 1 & \text{if } i+j \text{ is odd} \\ 0 & \text{if } i+j \text{ is even} \end{cases}$$

is stable up to the C-F-L limit for any positive value of σ, thus permitting an arbitrary variation in the amount of viscosity used in the computation.

For $i + j$ odd, (4-186) is (4-185); while for $i + j$ even, (4-186) becomes the implicit scheme

$$u_{i,j+1} + [(m/2)H_x f_{i,j+1} - \sigma \delta_x^2 u_{i,j+1}] = u_{i,j} \tag{4-187}$$

which appears to require the solution of a nonlinear tridiagonal system. That this is not so follows from the computational strategy for fixed j: Apply (4-186) for those values of i where $i + j$ is odd, thereby calculating alternate points explicitly; now apply (4-186) for those i with $i + j$ even, making use of those values previously calculated explicitly. The entire algorithm is thus an explicit computation!

A 'fast' version† of the algorithm is obtained by applying (4-186) over a time interval $2k$—that is, two applications give

$$u_{i,j+1} + \theta_{i,j+1}[(m/2)H_x f_{i,j+1} - \sigma \delta_x^2 u_{i,j+1}]$$
$$= u_{i,j} - \theta_{i,j}[(m/2)H_x f_{i,j} - \sigma \delta_x^2 u_{i,j}] \qquad (4\text{-}188)$$

$$u_{i,j+2} + \theta_{i,j+2}[(m/2)H_x f_{i,j+2} - \sigma \delta_x^2 u_{i,j+2}]$$
$$= u_{i,j+1} - \theta_{i,j+1}[(m/2)H_x f_{i,j+1} - \sigma \delta_x^2 u_{i,j+1}]. \qquad (4\text{-}189)$$

The right-hand side of (4-189) can be rewritten as

$$2u_{i,j+1} - \{u_{i,j+1} + \theta_{i,j+1}[(m/2)H_x f_{i,j+1} - \sigma \delta_x^2 u_{i,j+1}]\}.$$

Then (4-189), using (4-188), becomes

$$u_{i,j+2} + \theta_{i,j+2}[(m/2)H_x f_{i,j+2} - \sigma \delta_x^2 u_{i,j+2}]$$
$$= 2u_{i,j+1} - \{u_{i,j} - \theta_{i,j}[(m/2)H_x f_{i,j} - \sigma \delta_x^2 u_{i,j}]\}. \qquad (4\text{-}190)$$

For those points where $\theta_{i,j+2} = \theta_{i,j} = 0$, that is, when $i + j$ is even, (4-190) reduces to $u_{i,j+2} = 2u_{i,j+1} - u_{i,j}$. Since this does not apply at all grid points $i + j$, with j fixed, the inherent instability of the recursion will not appear. The calculations at those points where $\theta_{i,j+2} = \theta_{i,j} = 1$ uses the preceding values and (4-190).

Gourlay [47] has observed that (4-188) and (4-189) are equivalent to a three-level scheme on two interlacing grids

$$(1 + 2\sigma)u_{i,j+2} = (1 - 2\sigma)u_{i,j} - mH_x f_{i,j+1} + 4\sigma\mu_x u_{i,j+1}. \qquad (4\text{-}191)$$

Thus the stability of (4-188) and (4-189) away from the boundaries may be studied by means of (4-191) by Fourier analysis. The stability condition requires that the roots ρ of the quadratic equation

$$(1 + 2\sigma)\rho^2 + \{2im\lambda \sin \theta - 4\sigma \cos \theta\}\rho - (1 - 2\sigma) = 0$$

have modulus less than or equal to one for all θ, which is true for $m|\lambda| \leq 1$, $\sigma \geq 0$.

Just as the Lax method is used as a basis for the hopscotch–Lax method, the two-step version, (4-170) and (4-171), of Lax–Wendroff forms the foundation of the hopscotch–Lax–Wendroff scheme. To obtain the scheme consider

$$\left.\begin{array}{l} \bar{u}_{i \pm 1/2, j+1} = \mu_x u_{i \pm 1/2, j} - (m/2)\,\delta_x f_{i \pm 1/2, j} \\ u_{i,j+1} = u_{i,j} - m\,\delta_x \bar{f}_{i,j+1} \end{array}\right\} \quad j + i \text{ odd} \qquad (4\text{-}192)$$

$$\left.\begin{array}{l} \bar{u}_{i \pm 1/2, j+1} = \mu_x u_{i \pm 1/2, j+1} - (m/2)\,\delta_x f_{i \pm 1/2, j} \\ u_{i,j+1} = u_{i,j} - m\,\delta_x \bar{f}_{i,j+1} \end{array}\right\} \quad j + i \text{ even} \qquad (4\text{-}193)$$

† This is very similar to the fast version of the hopscotch method for parabolic equations (see Gourlay [44]).

where $\bar{f}_{i,j+1} = f(\bar{u}_{i,j+1})$, with $\bar{u}_{i,j+1}$ an intermediate solution. In terms of the $\theta_{i,j}$ used before, these equations can be written

$$\bar{u}_{i\pm1/2,j+1} = \theta_{i,j+1}[\mu_x u_{i\pm1/2,j+1} - (m/2)\,\delta_x f_{i\pm1/2,j+1}]$$
$$+ \theta_{i,j}[\mu_x u_{i\pm1/2,j} - (m/2)\,\delta_x f_{i\pm1/2,j}] \qquad (4\text{-}194)$$

$$u_{i,j+1} = u_{i,j} - (m/2)\,\delta_x \bar{f}_{i,j+1} \qquad (4\text{-}195)$$

where it is understood that (4-194) and (4-195) are solved first for $\theta_{i,j+1} = 0$ and then with $\theta_{i,j} = 0$. This order is dictated by (4-192) and (4-193). The method is truly implicit for half the points $u_{i,j+1}$—that is, for $\theta_{i,j+1} = 1$— and is second-order accurate.

But to solve (4-194) and (4-195) an additional technique must be introduced. Half the points must be calculated iteratively as follows. Write (4-194) and (4-195) in the form

$$\bar{u}^{(k)}_{i\pm1/2,j+1} = \theta_{i,j+1}[\mu_x u^{(k)}_{i\pm1/2,j+1} - (m/2)\,\delta_x f^{(k)}_{i\pm1/2,j+1}]$$
$$+ \theta_{i,j}[\mu_x u_{i\pm1/2,j} - (m/2)\,\delta_x f_{i\pm1/2,j}] \qquad (4\text{-}196)$$

$$u^{(k+1)}_{i,j+1} = u_{i,j} - (m/2)\,\delta_x \bar{f}^{(k)}_{i,j+1}$$

where the iteration superscript k applies only to those points for which $\theta_{i,j+1} = 1$. The case $k = 1$ is solved by using an average of the $u_{i,j+1}$ calculated from (4-194) and (4-195) with $\theta_{i,j+1} = 0$, namely

$$u_{i,j+1} = \tfrac{1}{2}(u_{i+1,j+1} + u_{i-1,j+1}).$$

In practice it has been observed that k need only take values up to two or three.

The stability condition may be obtained by examining that for hopscotch–Lax, with $\sigma = m^2\lambda^2/2$ in the quadratic equation governing its stability. Thus the scheme is stable for $m|\lambda| \le 1$.

A generalized hopscotch–Lax–Wendroff method is possible but has a region of stability of considerably smaller size than the aforementioned method. In practice the scheme is disappointing.

4-17 Explicit–implicit schemes for conservation laws

An important feature of explicit methods for hyperbolic equations is that they must satisfy the C-F-L convergence condition, which restricts the mesh ratio $m = k/h = \Delta t/\Delta x$. An additional property for nonlinear systems is that their linearized versions—e.g. $u_t + Au_x = 0$, where A is a constant matrix, for (4-181)—must be dissipative in the sense of Kreiss [18]. Here, a scheme is dissipative of order $2r$, with r a positive integer, if there exists $\delta > 0$ such that

$$|l(\alpha)| \le 1 - \delta|\alpha|^{2r} \quad \text{for every } |\alpha| \le \pi$$

where l is an eigenvalue of the amplification matrix and α is the Fourier variable of the stability analysis.

To relieve these stability restrictions one is naturally led to consideration of implicit methods and their usually larger ranges of stability. But the advantages of an increased stability range are unfortunately offset by two disadvantages. First, the implicit methods for nonlinear problems require either that a system of nonlinear equations be solved or an iterative technique be applied at each time step. Second, with the exception of a method due to McGuire and Morris [48], the implicit methods are nondissipative and are of questionable value for nonlinear hyperbolic systems where discontinuities may evolve.

Success of the hopscotch methods (Section 4-16) in combining explicit and implicit procedures for both parabolic and hyperbolic systems suggested a general classification of *explicit–implicit schemes* to McGuire and Morris [49]. The two methods are combined in such a way that the new scheme preserves the dissipation and ease of solution associated with explicit methods while retaining an increased stability inherited from the implicit methods.

Consider the class of second-order accurate explicit schemes (McGuire and Morris [50])

$$\bar{u}_{i,j+a} = (u_{i+1/2,j} + u_{i-1/2,j})/2 - am(f_{i+1/2,j} - f_{i-1/2,j}) \qquad (4\text{-}197)$$

$$u_{i,j+1} = u_{i,j} - (m/2)\left[\left(1 - \frac{1}{2a}\right)(f_{i+1,j} - f_{i-1,j})\right.$$

$$\left. + \frac{1}{a}(\bar{f}_{i+1/2,j+a} - \bar{f}_{i-1/2,j+a})\right] \qquad (4\text{-}198)$$

where $a \neq 0$ is stable, in the linearized sense, if† $m|\lambda| \leq 1$ and dissipative, in the linearized sense, if $0 < m|\Gamma| < 1$ for all eigenvalues Γ of A. The class of implicit methods to be combined with (4-197) and (4-198) are extensions of those equations introduced by McGuire and Morris [49]. Consider (4-197) with

$$u_{i,j+1} = u_{i,j} - (m/2)[(\tfrac{1}{2} + d(a - 1))(f_{i+1,j} - f_{i-1,j})$$

$$+ (\tfrac{1}{2} - ad)(f_{i+1,j+1} - f_{i-1,j+1}) + 2d(\bar{f}_{i+1/2,j+a} - \bar{f}_{i-1/2,j+a})$$

$$(4\text{-}199)$$

which is known to be second-order accurate and stable provided

$$ad > 0, \qquad m|\lambda| \leq 1/(2ad)^{1/2}. \qquad (4\text{-}200)$$

† Here, as previously, $|\lambda|$ is the maximum modulus eigenvalue of A.

The case $ad = 0$ is the Crank–Nicolson scheme for hyperbolic systems and hence is unconditionally stable. The class (4-199) is also dissipative of order 4, provided

$$ad > 0, \qquad 0 < m|\Gamma| < 1/(2ad)^{1/2}. \qquad (4\text{-}201)$$

for all eigenvalues Γ of A. The class of explicit methods (4-197) and (4-198) occurs when $ad = \frac{1}{2}$.

An explicit–implicit class of methods is obtained by combining the aforementioned methods as follows:

Employ (4-197), (4-198) *at mesh points with* $j + i$ *odd and then use* (4-197) *and* (4-199) *at the other mesh points.* (4-202)

Now the explicit method uses the points $(i, j + 1)$, $(i - 1, j)$, (i, j), $(i + 1, j)$, while the implicit method uses these plus $(i - 1, j + 1)$ and $(i + 1, j + 1)$. The combined method uses points as shown in Figure 4-14, and it is clearly

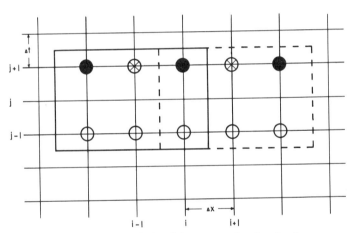

Fig. 4-14 Explicit–implicit computational molecule

computationally explicit. Application of (4-197), (4-198) at odd points of time level j (mesh points with $j + i$ odd) then makes the application of implicit method (4-197), (4-199) at the time level j an explicit process. The combined method is *not* of hopscotch type since (4-197), (4-199) is not the implicit version of (4-197), (4-198). Since both the explicit and implicit schemes have local truncation errors $O(k^3)$, so does the combined scheme and hence it is second-order accurate.

It is not an easy task to establish the stability of the combined scheme. However, this is done in [49] with the result that stability is ensured, in the linearized sense, for $m|\lambda| \le 1$ if and only if $|ad| \le \frac{1}{2}$.

Suggestions for explicit–implicit schemes for the system of conservation laws

$$u_t + [f(u)]_x + [g(u)]_y = 0$$

in two space variables are given in [49].

Transients in air–water have been calculated using an explicit–implicit method by Padmanabhan *et al.* [51]. Optimum values of a and d are determined and the results are compared with those using the Lax–Wendroff method. The pressure rise was more accurately matched with the explicit–implicit method and the computation was shorter, but there was more 'noise' in the pressure pulse.

PROBLEMS

4-56 Give a hopscotch algorithm for an initial value problem of the diffusion equation $u_t = u_{xx}$, $u(x, 0) = f(x)$. Analyze the stability of the algorithm.

4-57 Give a hopscotch algorithm for an initial value problem of the wave equation $u_{tt} = u_{xx}$, $u(x, 0) = f(x)$, $u_t(x, 0) = g(x)$.

4-58 Write out the fast version of hopscotch (4-188) and (4-189) if $f(u) = u$.

4-59 Describe the explicit–implicit algorithm if $f(u) = u$. Compare with the corresponding hopscotch version.

4-60 Give a possible explicit–implicit algorithm for $u_{tt} = u_x u_{xx}$.

REFERENCES

1. Bernstein, D. L. *Existence Theorems in Partial Differential Equations.* Princeton University Press, Princeton, N.J., 1950.
2. Garabedian, P. R. *Partial Differential Equations.* Wiley, New York, 1964.
3. Courant, R. and Friedrichs, K. O. *Supersonic Flow and Shock Waves.* Wiley (Interscience), New York, 1948.
4. Courant, R. and Hilbert, D. *Methods of Mathematical Physics*, vol. 2. Wiley (Interscience), New York, 1962.
5. Ames, W. F. *Nonlinear Partial Differential Equations in Engineering.* Academic Press, New York, 1965.
6. Ames, W. F., Lee, S. Y., and Zaiser, J. N. *Int. J. Nonlinear Mech.*, **3**, 449, 1968.
7. Swope, R. D. and Ames, W. F. *J. Franklin Inst.*, **275**, 36, 1963.
8. Ames, W. F. and Vicario, A. A., Jr. *Proc. 11th Midwest Applied Mech. Conf., Ames, Iowa*, 1969.
9. Jeffrey, A. and Taniuti, T. *Nonlinear Wave Propagation with Applications to Physics and Magnetohydrodynamics.* Academic Press, New York, 1964.
10. Massau, J. *Memoire sur l'Integration Graphique des Equations aux Derivées Partielles.* F. Meyer-Van Loo, Ghent, Belgium, 1899.

11. Crandall, S. H. *Engineering Analysis—A Survey of Numerical Procedures.* McGraw-Hill, New York and London, 1956.
12. Sauer, R. *Anfangswertprobleme bei Partiellin Differentialgleichungen.* Springer-Verlag, Berlin, 1952.
13. Thomas, L. H. *Communs pure appl. Math.,* **7**, 195, 1954.
14. Courant, R., Friedrichs, K., and Lewy, H. *Math. Ann.,* **100**, 32, 1928.
15. Lowan, A. N. The operator approach to problems of stability and convergence of solutions of difference equations and the convergence of various iteration processes. *Scripta Mathematica.* Washington Office of Technical Services, 1957.
16. Collatz, L. *The Numerical Treatment of Differential Equations,* 3rd edit. Springer-Verlag, Berlin, 1960.
17. Zajac, E. E. *J. Math. Phys.,* **43**, 51, 1964.
18. Richtmyer, R. D., and Morton, K. W. *Difference Methods for Initial Value Problems,* 2nd edit. Wiley (Interscience), New York, 1967.
19. O'Brien, G. G., Hyman, M. A., and Kaplan, S. *J. Math. Phys.,* **29**, 223, 1951.
20. Friberg, J. *Nord Tidskr. Inform. Behandl.,* **1**, 69, 1961.
21. Lees, M. *Pacif. J. Math.,* **10**, 213, 1960.
22. Lees, M. *J. Soc. ind. appl. Math.,* **10**, 610, 1962.
23. Carrier, G. F. *Q. appl. Math.,* **3**, 157, 1945.
24. Bolotin, V. V. *The Dynamic Stability of Elastic Systems.* Holden Day Inc., San Francisco, 1964.
25. Richtmyer, R. D. The stability criterion of Godunov and Ryabenkii for difference schemes, *Rept No. NYO-1480-4.* Courant Institute of Mathematical Sciences, New York University, New York, 1964. See also Richtmyer and Morton [18].
26. Richtmyer, R. D., and Morton, K. W. Stability studies for difference equations: (I) Nonlinear instability; (II) Coupled sound and heat flow, *Rept No. NYO 1480-5.* Courant Institute of Mathematical Sciences, New York University, New York, 1964.
27. Hartree, D. R. *Numerical Analysis,* 2nd edit. Oxford University Press, London and New York, 1958.
28. Courant, R., Isaacson, E., and Rees, M. *Communs pure appl. Math.,* **5**, 243, 1952.
29. Ansorge, R. *Num. Math.,* **5**, 443, 1963.
30. Lax, P. D. *Communs pure appl. Math.,* **7**, 159, 1954.
31. Lax, P. D. Nonlinear hyperbolic systems of conservation laws, in *Nonlinear Problems* (R. E. Langer, ed.), p. 3. University of Wisconsin Press, Madison, Wisconsin, 1963.
32. Chorin, A. J. On the convergence of discrete approximations to the Navier-Stokes equations, *AEC Res. Dev. Rept No. NYO 1480–106.* Courant Institute of Mathematical Sciences, New York University, New York, 1968.
33. Longley, H. J. Methods of differencing in Eulerian hydrodynamics, *Rept No. LA-2379.* Los Alamos Scientific Laboratory, New Mexico, 1960.
34. Lax, P. D. *Communs pure appl. Math.,* **10**, 537, 1957.
35. Payne, R. B. *J. Fluid Mech.,* **2**, 185, 1957.
36. Roberts, L. *J. Math. Phys.,* **36**, 329, 1958.
37. Lax, P. D. and Wendroff, B. *Communs pure appl. Math.,* **13**, 217, 1960.
38. Lax, P. D. and Wendroff, B. *Communs pure appl. Math.,* **17**, 381, 1964.
39. Strang, W. G. *Num. Math.,* **6**, 37, 1964.
40. Richtmyer, R. D. A survey of difference methods for nonsteady fluid dynamics, *NCAR Tech. Note 63-2,* 1963.

41. Gary, J. *Maths Comput.*, **18**, 1, 1964.
42. Gourlay, A. R. and Morris, J. Ll. *Maths. Comput.*, **22**, 28, 1968.
43. Turner, J. and Wendroff, B. An unconditionally stable implicit difference scheme for the hydrodynamical equations, *Rept No. LA-3007*. Los Alamos Scientific Laboratory, Los Alamos, New Mexico, 1964.
44. Gourlay, A. R. *J. Inst. Maths. Appls.*, **6**, 375, 1970.
45. Gourlay, A. R. and Morris, J. L. *IBM J. Res. Develop.*, **16**, 349, 1972.
46. Gourlay, A. R. and McGuire, G. R. *J. Inst. Maths. Appl.*, **7**, 216, 1971.
47. Gourlay, A. R. *Proc. R. Soc.* **A323**, 219, 1971.
48. McGuire, G. R. and Morris, J. L. *J. Comp. Phys.*, **14**, 126, 1974.
49. McGuire, G. R. and Morris, J. L. *Maths. Comp.*, **29**, 407, 1975.
50. McGuire, G. R. and Morris, J. L. *J. Comp. Phys.*, **11**, 531, 1973.
51. Padmanabhan, M., Martin, S. C., and Ames, W. F. Transients in air–water mixtures, in preparation.

5
Special topics

5-0 Introduction

The ENIAC† of Eckert–Mauchly not only initiated a new age in man's technological abilities but it also stimulated extensive interest in numerical analysis. The subsequent second and higher generations of computers have been a steady improvement in memory capacity, speed, and flexibility. Consequently, very ambitious numerical work is now possible which, twenty years ago, was unthinkable. This, in turn, has greatly accelerated the tempo of research in numerical analysis aimed at extending the range, accuracy, and novelty of the available methods. Of course, the present volume cannot be complete, but it can be a guide and stimulus for further study.

This chapter is roughly divided into three parts. Sections 5-1 to 5-3 discuss some of the present methods for the treatment of singularities, shocks, and eigenvalue problems; Sections 5-4 to 5-6 describe techniques in higher dimensions; and Sections 5-7 to 5-9 are devoted to some special topics of great interest to engineers and physical scientists (fluid mechanics, nonlinear vibrations and coupled phenomena). A large number of references, with brief descriptions of the research described, will be included. We hope that this chapter will also act as a bridge to more advanced works and to additional physical problems.

5-1 Singularities

Estimates of the errors between the solution of the differential equation and that of finite difference approximation depend on the boundedness of partial derivatives of some order. These estimates *cannot be valid* at or near a singular point on the boundary or in the interior of the integration domain.

In rectangular coordinates the most common form of *boundary singularity* occurs at the corners of the integration domain where the initial line meets the vertical boundary. To illustrate this situation in a parabolic problem, consider the integration domain shown in Fig. 5-1, with initial line $t = 0$, and boundaries at $x = 0$ and $x = 1$. For the diffusion equation

$$u_t = u_{xx} \tag{5-1}$$

the simplest initial and boundary conditions are

$$u(x, 0) = f(x), \qquad u(0, t) = g(t), \qquad u(1, t) = h(t). \tag{5-2}$$

† The first stored program electronic digital computer was developed by J. Mauchly and R. Eckert for the Aberdeen Proving Grounds of the U.S. Army in the later years of World War II (1939–1945).

We usually take these to mean

$$\lim_{t \to 0} u(x, t) = f(x), \quad 0 < x < 1$$

$$\lim_{x \to 0} u(x, t) = g(t), \quad t > 0, \quad x > 0 \tag{5-3}$$

$$\lim_{x \to 1} u(x, t) = h(t), \quad t > 0, \quad x < 1$$

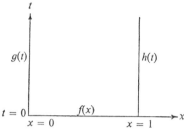

Fig. 5-1 A boundary singularity in a parabolic problem

The 'corners' at $(0, 0)$ and $(1, 0)$ will often involve some form of discontinuity. One of the most 'violent' of these is a difference in function value—that is,

$$\lim_{x \to 0} f(x) \neq \lim_{t \to 0} g(t)$$

or $\tag{5-4}$

$$\lim_{t \to 0} h(t) \neq \lim_{x \to 1} f(x).$$

Weaker types of discontinuities can also occur and cause computational problems. As an example, suppose $f(x) = 1$ and $g(t) = e^{-t}$ in Eqns (5-2). It is clear that $f(0) = 1 = g(0)$, thus there is no difference in function value at $(0, 0)$. However, a more subtle singularity occurs since $\partial g/\partial t \neq \partial^2 f/\partial x^2$ at $x = 0, t = 0$.

Finite difference methods have questionable value in the neighbourhood of singular points since a 'region of infection' lies adjacent to such points. The effect of such discontinuities does not penetrate deeply into the field of integration *provided the methods we use are stable*, in which case the errors introduced by the discontinuity decay. There are serious difficulties in obtaining accurate solutions near the points of discontinuity, and we must often abandon finite difference processes to obtain this accuracy.

When the discontinuity is of the *function difference* form, Eqns (5-4), the sensitivity of the finite difference approximation to this discontinuity can be reduced by using the value $\frac{1}{2}(f(0) + g(0))$ at $(0, 0)$. Crandall [1] explored the accuracy obtained by approximating the linear diffusion equation on coarse nets and found the resulting solution very sensitive to values assigned at the singular points.

Some problems are such that singularities occur at points p_i inside the integration domain. *Interior singularities* occur when one, or more than one, coefficient of the partial differential equation becomes singular. In such problems the solution will ordinarily also have a singularity at p_i, and the finite difference scheme will not be applicable without modification. Physical problems where such singularities occur are source-sink and concentrated point-load problems.

Singularities inside the region occur in a Poisson equation

$$\nabla^2 u = f(x, y) \tag{5-5}$$

if f is not well behaved. A typical source problem involves $f(x, y) = 0$, except at the origin whereupon

$$u = \frac{1}{2\pi} \log r, \qquad r^2 = x^2 + y^2. \tag{5-6}$$

Clearly, u has a singularity at the origin and we would not expect success near this point with any finite difference process.

The deflection of a plate by a concentrated point load at the origin is represented (see, for example, Timoshenko and Woinowsky-Krieger [2]) by the *biharmonic* equation

$$\nabla^4 u = f(x, y) \tag{5-7}$$

where f is zero except at the origin. The solution contains a term $r^2 \log r$ for which the *second derivatives* do not exist at $r = 0$.

In Chapter 2 we met Eqn (2-112) which has a singularity at $x = 0$ resulting from the term $x^{-1} u_x$. In the event that $\lim_{x \to 0} u_x = 0$, the indeterminate form $x^{-1} u_x$ takes the value u_{xx} as $x \to 0$.

The accepted technique in interior singularity problems is to subtract the singularity where possible, thereby generating a new problem with different boundary conditions but with a 'well-behaved' solution. In practice, this procedure works well with some linear problems. The more difficult nonlinear problems require individual treatment. The generation of 'local' solutions can sometimes be useful if the character of the singularity is known. Weinstein [3] discusses theoretical problems associated with the singular equation

$$u_{xx} + \frac{k}{y} u_y + u_{yy} = 0 \tag{5-8}$$

and its generalizations. An extensive bibliography constitutes a guide to much of the theoretical literature on linear equations such as the Tricomi equation, $u_{xx} + y^{-1} u_{yy} = 0$, and the Poisson–Euler–Darboux equation, $u_{xy} + a(x + y)^{-1}(u_x + u_y) = 0$.

Most of the effort on singularities has been concentrated on scientific problems, and currently no general theory is available. We shall examine several methods for treating singularities in some linear problems.

(a) Subtracting out the singularity

Suppose u has a discontinuity at a point p. The goal is to calculate a solution U of the same partial differential equation such that $u - U$ is well behaved at the point p. To illustrate, we suppose that the solution of Laplace's equation, $\nabla^2 u = 0$, is desired in the region $y > 0$ subject to the boundary conditions $u = f(x)$ for $x > 0$ and $u = g(x)$ for $x < 0$.† Let the discontinuities at $x = 0$ be defined as

$$f^{(n)}(0) - g^{(n)}(0) = a_n, \quad n = 0, 1, 2, 3, 4 \tag{5-9}$$

where the a_n are finite. Milne [5] constructs the *local* solution w of $\nabla^2 w = 0$ as

$$w(x, y) = \left\{ \sum_{j=0}^{4} a_j p_j(x, y) \right\} \frac{1}{\pi} \tan^{-1}(y/x) + \left\{ \sum_{j=1}^{4} a_j q_j(x, y) \right\} \frac{1}{2\pi} \ln (x^2 + y^2) \tag{5-10}$$

where

$$p_0 = 1, \quad (p_j + iq_j) = \frac{1}{j!} (x + iy)^j, \quad j = 1, 2, 3, 4$$

where p_j and q_j are found by selecting real and imaginary parts. w satisfies Laplace's equation everywhere except at the origin since it is the imaginary part of the analytic function

$$\frac{1}{\pi} \sum_{j=0}^{4} (a_j z^j \ln z)/j!$$

For $y > 0$ we now take the limit as $y \to 0$ for the two cases $x > 0$ and $x < 0$. It can be shown (Problem 5-1) that

$$w(x, 0) = \begin{cases} 0, & x > 0 \\ a_0 + a_1 x + a_2 x^2/2 + a_3 x^3/6 + a_4 x^4/24, & x < 0 \end{cases} \tag{5-11}$$

so that w has the proper jumps at the origin as specified in Eqn (5-9). The required solution is then obtained by setting $v = u - w$ and solving $\nabla^2 v = 0$, together with the boundary conditions,

$$v(x, 0) = \begin{cases} f(x), & x > 0 \\ g(x) - a_0 - a_1 x - a_2 x^2/2 - a_3 x^3/6 - a_4 x^4/24, & x < 0 \end{cases} \tag{5-12}$$

which are *continuous in v and its derivatives up to fourth order* at the singularity (the origin).

Success of this method depends strongly upon our ability to construct

---

† By a translation of the axis, the point $x = 0$ can be moved to any other boundary point. There is no loss in generality in considering only $x = 0$.

suitable solutions to the homogeneous differential equation. This may be very difficult in the nonlinear case. Sometimes local solutions may be found by asymptotic or series methods.

(b) Mesh refinement

One very common method of dealing with discontinuities is effectively to ignore them and attempt to diminish their effect by using a mesh refinement in the region near to and surrounding the singularity. A rather elegant refinement procedure is reported by Trutt *et al.* [4] for the nonlinear elliptic equation $\nabla \cdot \{F(|\nabla u|) \nabla u\} = 0$. The two interior corners (see Fig. 3-7) require the use of mesh refinement. The mesh refinement procedure has the effect of minimizing the 'area of infection' created by the singularity.

Milne [5] examines the 'area of infection' problem for the event when the boundary turns through a right angle, with the conditions $u = 0$ on the x-axis and $u = 1$ on the y-axis. For Laplace's equation he shows how the effect of the singularity decreases with distance and also shows the interval size necessary near the singularity to achieve a required precision.

The procedure of subtracting out the singularity appears to be preferable to the refined net method, if the analysis can be performed.

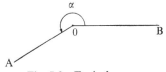

Fig. 5-2 Typical corner

(c) The method of Motz and Woods

A sudden change of direction of the boundary of a region introduces a common form of disturbance. If the angle exceeds π, the corner is called a *re-entrant corner*. Motz [6] considered Laplace's equation when $\alpha = 2\pi$, although the procedure applies to more general cases. Woods [7] uses similar methods for Poisson's equation $\nabla^2 u = f$, but his calculation requires iteration and is somewhat difficult to program for a computer. We shall discuss the approach taken by Motz for Laplace's equation.

Examination of the effect of a corner such as that of Fig. 5-2 is facilitated by writing Laplace's equation in circular cylindrical coordinates (r, θ) as

$$u_{rr} + \frac{1}{r} u_r + \frac{1}{r^2} u_{\theta\theta} = 0. \qquad (5\text{-}13)$$

By separation the elemental solutions are found to be

$$r^k \cos k\theta, \qquad r^k \sin k\theta, \qquad \text{for any } k. \qquad (5\text{-}14)$$

When u is required to vanish on the boundaries AO and OB we find the solution to be

$$u(r, \theta) = \sum_{n=1}^{\infty} a_n r^{n\pi/\alpha} \sin \frac{n\pi\theta}{\alpha}. \qquad (5\text{-}15)$$

If $\partial u/\partial n = 0$ on the boundaries, only the terms in $\cos k\theta$ are required and the solution is

$$u(r, \theta) = \sum_{n=0}^{\infty} b_n r^{n\pi/\alpha} \cos \frac{n\pi\theta}{\alpha}. \qquad (5\text{-}16)$$

For $\alpha > \pi$ it is clear in both cases that, for $n = 1$, the exponent of r is less than 1, so that $\partial u/\partial r$ has a singularity at the origin. If $\alpha \leq \pi$ this is not the case and, in this sense, the re-entrant corner is more complex.

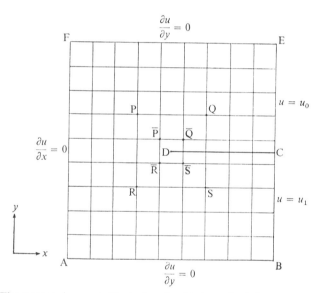

Fig. 5-3   The rectangular mesh illustrating the Motz method for a re-entrant corner

The specific problem of Motz [6] is that of finding the solution to Laplace's equation $\nabla^2 u = 0$ in the finite rectangular region of Fig. 5-3, with $\alpha = 2\pi$. The boundary conditions are $\partial u/\partial y = 0$ on AB and FE, $\partial u/\partial x = 0$ on AF, $u = u_1$ on BC, $u = u_0$ on CE, and $u_y^+ = u_y^- = 0$ (the $\pm$ are u labels above and below the line DC). At most mesh points the standard five-point star is used. However, in the neighbourhood of the singularity D, the first few terms of the appropriate series, Eqn (5-16), are used. For example, using four coefficients, we have

$$u = b_0 + b_1 r^{1/2} \cos \tfrac{1}{2}\theta + b_2 r \cos \theta + b_3 r^{3/2} \cos \tfrac{3}{2}\theta. \qquad (5\text{-}17)$$

Determination of the four coefficients in Eqn (5-17) is carried out, at the

kth iteration step in the overall solution, by using the values at four 'remote' points, say P, Q, R, and S. Thus, four linear equations of the form

$$u^k(P) = b_0 + b_1 r_P^{1/2} \cos \tfrac{1}{2}\theta_P + b_2 r_P \cos \theta_P + b_3 r_P^{3/2} \cos \tfrac{3}{2}\theta_P \quad (5\text{-}18)$$

result. These are solved for the b_i, thereby furnishing coefficients for Eqn (5-17) to be used in determining the values at the mesh points $\bar{P}$, $\bar{Q}$, $\bar{R}$, and $\bar{S}$. These special equations are *only used* at $\bar{P}$, $\bar{Q}$, $\bar{R}$, and $\bar{S}$.

The Motz idea is simple; it can produce an accuracy in solutions which is unattainable by finite difference methods, and when high precision and an economic interval size is required, it is easily improved, not by iteration, but by selecting more terms in the series. Some specific investigations have been reported by Lieberstein [8] and Greenspan and Warten [9].

(d) Removal of singularity

In some problems removal of a singularity is possible by the adoption of new independent variables. The transformation is to be so devised that the singular point is *expanded into a line or curve.* Thus, a sudden change of the type, Eqns (5-4), will become a smooth change along the line or curve in the new independent variables. Similarity transformations (see Ames [33]) are often useful or suggestive of other transformations to achieve the stated goal.

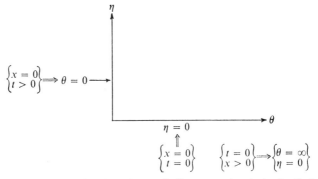

Fig. 5-4 Integration domain when the parabolic corner singularity is eliminated by a 'similarity' type transformation

As an example of the removal of a singularity, consider the diffusion equation $u_t = u_{xx}$ with a discontinuity at the origin,

$$\lim_{x \to 0} f(x) \neq \lim_{t \to 0} g(t) \quad (5\text{-}19)$$

and an integration domain as shown in Fig. 5-1. We attempt a transformation, modeled after the similarity form,

$$\theta = x^a t^b, \qquad \eta = t^c \quad (5\text{-}20)$$

with the goal of expanding the singular point into a line, and the further aim of having a relatively easy equation along this line. The selection of a, b, and c is probably not unique. One useful set is $a = 1$, $b = -\frac{1}{2}$, and $c = \frac{1}{2}$ so that Eqns (5-20) become $\theta = xt^{-1/2}$, $\eta = t^{1/2}$, and the differential equation in the new coordinates is

$$u_{\theta\theta} + \tfrac{1}{2}\theta u_\theta = \tfrac{1}{2}\eta u_\eta. \tag{5-21}$$

As a consequence of this transformation it is clear that the boundary line $x = 0$, $t > 0$ becomes the line $\theta = 0$ in the (θ, η) plane. The point of discontinuity $x = 0$, $t = 0$ becomes the line $\eta = 0$, and the line $t = 0$, $x > 0$ becomes the point at infinity on the line $\eta = 0$ (Fig. 5-4). The effect of the mapping is to expand the origin into a line, and the jump from $f(0)$ to $g(0)$ becomes a smooth change along this line.

Along $\eta = 0$ the initial values are given by the solution of the boundary value problem,

$$\frac{d^2u}{d\theta^2} + \frac{\theta}{2}\frac{du}{d\theta} = 0, \quad u(0) = g(0), \quad u(\infty) = f(0) \tag{5-22}$$

which is

$$u(\theta, 0) = [f(0) - g(0)]\pi^{-1/2} \int_0^\theta e^{-p^2/4}\, dp + g(0). \tag{5-23}$$

Having determined the initial values, Eqn (5-23), we then express Eqn (5-21) in a finite difference approximation, say by an explicit technique, and march a few steps in the η direction—sufficient to take us from the neighbourhood of the discontinuity. Rather than proceed further in the semi-infinite (θ, η) domain it is usually desirable to return to the original plane. However, for equidistant grid points in the (x, t) plane, an interpolation is usually required (at least in one direction) to obtain the values of u in the original (x, t) plane. This method, while analytically attractive, is numerically complicated.

Recent studies by Eisen [10, 11] have concerned the stability, convergence, and consistency of finite difference approximations for the n-dimensional spherically symmetric diffusion equation

$$u_t = u_{rr} + \frac{n-1}{r}\frac{\partial u}{\partial r}, \quad 0 \le t \le T, \quad 0 \le r \le R$$

with initial, regularity, and boundary conditions of the form

$$u(r, 0) = u_0, \quad u_r(0, t) = 0, \quad \alpha u(R, t) + \beta u_r(R, t) = 0$$

α and β constants. Eisen shows that the scheme,

$$u_{i,j} = u(i\,\Delta r, j\,\Delta t), \quad m = \Delta t/(\Delta r)^2,$$

$$u_{i,j+1} = u_{i,j} + m[u_{i+1,j} - 2u_{i,j} + u_{i-1,j}]$$

$$+ \frac{m(n-1)}{2i}(u_{i+1,j} - u_{i-1,j}), \tag{5-24}$$

$$u_{-1/2,j} = u_{1/2,j}, \quad u_{M+1/2,j} = 0, \quad j = \tfrac{1}{2}, \tfrac{3}{2}, \ldots, M + \tfrac{1}{2}$$

converges to the solution of the problem when n is even and $m < \frac{1}{2}$. Albasiny [12] met this limitation on m and resorted to an implicit method.

Bramble *et al.* [13] study the effect of an isolated singularity on the rate of convergence of a finite difference analog in the Dirichlet problem for Poisson's equation. Whiteman [22, 23] extends the method of Motz by using several alternative expansion techniques.

PROBLEMS

5-1 Beginning with Eqn (5-10), carry out the development of Eqn (5-11).

5-2 Generalize the method of Section 5-1(a) to apply to the Poisson equation $\nabla^2 u = f(x, y)$.

5-3 Apply the method of Motz, with three terms, to the re-entrant corner problem with $\alpha = 3\pi/2$ (Fig. 5-2). Apply the boundary conditions given in Fig. 5-3.

5-4 Suppose the wave equation $u_{tt} = u_{xx}$ has the discontinuity, Eqn (5-19), at the origin. Can the technique of Section 5-1(d) be applied?

5-2 Shocks

The motion† of an ideal gas is often characterized by curves in the (x, t) plane (or surfaces in space) across which certain of the dependent variables are discontinuous but possess one-sided limits on both sides. Examples include (see Courant and Friedrichs [15]):

 (i) **Interfaces**—boundaries separating media of different physical, chemical, and/or thermodynamic properties. For example, interfaces separate two different fluids, a liquid and solid or gas and liquid;
 (ii) **Contact discontinuity**—ρ and e are discontinuous but p and u are continuous;
 (iii) **Shock**—ρ, e, u, and p are all discontinuous;
 (iv) **Head of rarefaction wave**—only certain derivatives are discontinuous.

At discontinuities the differential equations involve *one-sided* derivatives. To these must be added *jump conditions*, which serve as internal boundary conditions, thereby insuring uniqueness. For interfaces, contact discontinuities, and rarefaction heads, the only internal boundary conditions required are the continuity of certain variables.

If the position of an interface is known, as in some problems of solid mechanics, then one can often adjust the grid so that mesh points lie on these boundaries. In such cases the basic finite difference equations can be applied,

---

† Even if the flow is initially smooth, discontinuities may develop in finite time t_1. After that time no smooth single-valued flow exists. This can also occur in fluid mechanics (see, for example, Ladyzhenskaya [14]).

as before, if care is used to change the physical parameters or use the appropriate equation of state on each side of the interface. Such an approach may lead to inaccuracies and some special treatment at interfaces is often necessary. Inaccuracies may result from large changes in some system property across the interface (e.g. density in going from a gas to liquid, conductivity from solid to gas, permeability from gas to solid, etc.). In such cases we may wish to use quite different spacing Δr on the two sides of the interface. Such a problem is seen in Fig. 3-7 where the air gap is *small* compared to the other dimensions.

Generally the position of the discontinuity does not coincide with the fixed coordinates of the Eulerian mesh. In particular problems it may be possible to choose Δt such that $\Delta r/\Delta t$ is equal to the interface velocity. The discontinuity then passes through net points. Unfortunately, this direction may lead to instability unless Δr is chosen properly, which is directly contrary to the concept of a fixed Eulerian net. One can attempt to keep track of all discontinuities and construct special schemes to handle them in terms of their closest mesh points. The complexity of the problem is usually increased by these methods. Longley [16] devised a procedure of this type using the Euler conservative form of the one-dimensional equations. His program prevents diffusion of the interface.

The calculation of fluid motion in systems with discontinuities is usually best accomplished from the Lagrangian form. In Lagrangian coordinates the paths of discontinuities are known since their spatial coordinates remain invariant in time. The mesh may therefore be introduced so that the discontinuity is always at a mesh point, halfway between mesh points, or at any other position. Modification of the schemes discussed in Section 4-15 may be necessary for increased accuracy. This is especially true for *interfaces* (see, for example, Richtmyer and Morton [17, p. 298]) where inaccuracies may arise if one uses the second of Eqns (4-178). From the Richtmyer and Morton definitition of $x = \rho_0^{-1} \int \rho(\eta) \, d\eta$ it follows that $\partial p/\partial x$ is continuous across an interface while the derivative of the pressure gradient may not be. They suggest the more accurate expression

$$\left.\frac{\partial p}{\partial x}\right|_{x=x_{\mathrm{I}}} = \frac{3(p_{\mathrm{I}+1/2,j} - p_{\mathrm{I}-1/2,j}) - \frac{1}{3}(p_{\mathrm{I}+3/2,j} - p_{\mathrm{I}-3/2,j})}{\Delta_{\mathrm{L}}x + \Delta_{\mathrm{R}}x} \qquad (5\text{-}25)$$

where the notation $i = \mathrm{I}$ represents the interface spatial index and $\Delta_{\mathrm{L}}x$ and $\Delta_{\mathrm{R}}x$ are the increments of x used on the left and right sides of the interface. Equation (5-25) is to be used in place of $(p_{\mathrm{I}+1/2,j} - p_{\mathrm{I}-1/2,j})/\Delta x$ in the second of Eqns (4-178) (the equation of motion), and may also be employed in the smooth portion of the flow.

Trulio and Trigger [18] choose the interface to be at the *center* of the mesh and define two values of ρ and e, at each time step, for each side of the interface. Values of u and r must be calculated at the interface points.

In reality the shock wave is not a discontinuity at all but a narrow zone, a few mean free paths in thickness, through which the variables change continuously, even though very rapidly. Across a shock wave the Rankine-Hugoniot conditions hold—they are discrete expressions of mass, momentum, and energy conservation.

These conditions are, respectively (see, for example, Courant and Friedrichs [15]),

mass: $\qquad\qquad \rho_1(U - u_1) = \rho_2(U - u_2) = m$

momentum: $\qquad\quad m(u_1 - u_2) = p_1 - p_2 \qquad\qquad$ (5-26)

energy: $\qquad\quad m(e_1 + \tfrac{1}{2}u_1^2 - e_2 - \tfrac{1}{2}u_2^2) = p_1u_1 - p_2u_2$

where m is the mass crossing the unit area of the shock front in unit time and subscripts 1 and 2 refer to conditions of the unshocked fluid ahead of the shock and behind the shock, respectively. U is the velocity of the shock front.

In the shock layer, heat conduction and viscosity effects are important and there is an entropy, S, increase across the shock $(S_2 > S_1)$. Consequently, the Eulerian equations for plane motion become

$$\rho_t + (\rho u)_r = 0$$

$$(\rho u)_t + (\rho u^2 + p - \bar{\mu}u_r)_r = 0$$

$$(\rho \epsilon)_t + [\rho u \epsilon + u(p - \bar{\mu}u_r) - kT_r]_r = 0 \qquad (5\text{-}27)$$

$$\rho T(S_t + uS_r) - \bar{\mu}(u_r)^2 - (kT_r)_r = 0$$

where $\epsilon = e + \tfrac{1}{2}u^2$, and $\bar{\mu}\,(=\tfrac{4}{3}\mu)$ and k are coefficients of viscosity and conductivity, respectively.

Methods of treating shocks by finite difference approximations fall basically into two areas—those which add artificial dissipative terms directly or indirectly and those which employ characteristics. We shall describe several of these procedures.

(a) Pseudoviscosity

This idea of von Neumann and Richtmyer was to introduce a purely artificial dissipative mechanism of such form and strength that there results a smooth shock transition extending over a small number of space variable intervals. The finite difference equations are constructed with the pseudoviscosity included. In a calculation the shocks appear as a narrow region across which the fluid variables change rapidly but continuously. Narrow in this context means only a few (3 or 4) Δt's in thickness, which is much larger than the thickness of the real shock.

The pseudoviscosity modified Lagrange equations ($\alpha = 0$) [see Eqns (4-152)–(4-155)] to be solved are

$$V_t = u_y, \qquad u_t = -(p + q)_y, \qquad e_t = -(p + q)V_y, \qquad p = f(e, V) \quad (5\text{-}28)$$

where $dy = \rho_0\, dx = \rho\, dr$ (y is the mass coordinate). The original form of q,

$$q = -\frac{[b(\Delta y)]^2}{V}\frac{\partial u}{\partial y}\left|\frac{\partial u}{\partial y}\right|$$

has been found to unnecessarily smear out rarefaction waves and has evolved into the limited form

$$q = \begin{cases} \dfrac{[b(\Delta y)]^2}{V}\left(\dfrac{\partial u}{\partial y}\right)^2, & \dfrac{\partial u}{\partial y} < 0 \\[2ex] 0, & \dfrac{\partial u}{\partial y} \ge 0 \end{cases} \tag{5-29}$$

The dissipative term, q, is added to the equations over the whole field so that we need not know where the shocks are, or whether they happen to be crossing each other or interfaces.

In practice, for the perfect gas, $e = pV/(\gamma - 1)$, the shock spreads out over approximately $\pi b\, \Delta y[2/(\gamma + 1)]^{1/2}$ mesh points. Satisfactory answers were found for a problem with $\gamma = 2$ when b was between 1.5 and 2.0, with the shock spread over 3 to 5 times Δy. The addition of the pseudoviscosity necessitates extra stability criteria. With very strong shocks† this method reduces the permissible Δt by about the factor $\sqrt{\gamma}/2b$, which is around one-third in practical problems. The actual criterion is

$$\sqrt{(\gamma p_f V_f)}\,\frac{\Delta t}{V_f\, \Delta y} \le \sqrt{\gamma}/2b \tag{5-30}$$

where the subscript f indicates conditions *directly behind the shock*.

Extensive calculations are reported by Richtmyer and Morton in the one-dimensional case. Brode [19, 20] has utilized the von Neumann–Richtmyer method successfully in cylindrical and spherical symmetry. His calculations concerned the determination of blast waves and explosions. More recently Schulz [21] presents arguments that in *higher dimensions* a tensor artificial viscosity is a more suitable quantity to use than the scalar extension of Eqn (5-29). He reports successful calculations by this method and asserts that the shock stability criterion of the difference equations resulting from the use of the tensor viscosity is less severe than that deduced from a generalized scalar viscosity.

Other forms of the pseudoviscosity have been introduced by various investigators (see, for example, Fox [26]), but it is now generally conceded that methods appearing in the works of Lax [24], Lax and Wendroff [25], and Godunov [27] are superior.

(b) Lax–Wendroff method

Lax [24], in his original suggestion, applied the staggered scheme [Eqn (4-167)] to hyperbolic equations. The continuation of this work led to the

---

† If an estimate of shock strength is not available, one simply assumes this case.

Lax–Wendroff system, Eqn (4-169). In turn, Lax and Wendroff [25] modified the algorithm by adding an artificial viscosity term of a particular form, thus significantly *decreasing oscillations behind the shock.*

The Lax–Wendroff method for the vector equation

$$\frac{\partial \mathbf{U}}{\partial t} + A \frac{\partial \mathbf{U}}{\partial x} = 0 \tag{5-31}$$

where $\mathbf{U}$ is a vector and A a constant matrix, is expressible [see Eqn (4-169) and Problem 4-52] as

$$(\Delta t)^{-1}(\mathbf{U}_{i,j+1} - \mathbf{U}_{i,j}) + A(2\Delta x)^{-1}(\mathbf{U}_{i+1,j} - \mathbf{U}_{i-1,j})$$

$$= \frac{\Delta t}{2} A^2 (\Delta x)^{-2}(\mathbf{U}_{i+1,j} - 2\mathbf{U}_{i,j} + \mathbf{U}_{i-1,j}) \tag{5-32}$$

The form of the right member has tempted some to regard it as an approximation to a dissipative term $\frac{1}{2}\Delta t A^2 \, \partial^2 \mathbf{U}/\partial x^2$ (see Fox [26, p. 356])—that term should therefore be included in Eqn (5-31). That this is *incorrect* results from the fact that the alleged dissipative term has the same order as the truncation error of the left member. One can then shift terms back and forth between the two sides and obtain a variety of corrections, some of which are not dissipative.

Richtmyer and Morton [17] observe that one procedure is to suppose that $\mathbf{U}$ is an exact solution of

$$\frac{\partial \mathbf{U}}{\partial t} + A \frac{\partial \mathbf{U}}{\partial x} = Q\mathbf{U} \tag{5-33}$$

where Q is to be determined so that $\mathbf{U}$ satisfies

$$\mathbf{U}_{i,j+1} - \mathbf{U}_{i,j} + \frac{1}{2} A \frac{\Delta t}{\Delta x} (\mathbf{U}_{i+1,j} - \mathbf{U}_{i-1,j})$$

$$- \frac{1}{2}\left(\frac{A\Delta t}{\Delta x}\right)^2 (\mathbf{U}_{i+1,j} - 2\mathbf{U}_{i,j} + \mathbf{U}_{i-1,j}) = O[(\Delta t)^4] \tag{5-34}$$

Without the term $Q\mathbf{U}$ the truncation error of Eqn (5-32) would be $O[(\Delta t)^3]$. Suppose Q is a constant coefficient differential operator such that $Q\mathbf{U} = O[(\Delta t)^2]$. Upon expanding the terms of Eqn (5-34) in Taylor's series about the point i, j, using Eqn (5-33), and equating to zero the sum of terms of order $(\Delta t)^3$ (Problem 5-8), there results

$$Q\mathbf{U} = -\tfrac{1}{6}A\{(\Delta x)^2 - A^2(\Delta t)^2\} \frac{\partial^3 \mathbf{U}}{\partial x^3} \tag{5-35}$$

However, Eqn (5-33), with $Q\mathbf{U}$ specified by Eqn (5-35), is *dispersive* but not *dissipative.*†

---

† That is, different Fourier components travel with different speeds but no change in amplitude takes place.

A dissipative term can be added by replacing $O[(\Delta t)^4]$ in Eqn (5-34) by $O[(\Delta t)^5]$. The Taylor's series expansion process employed in Problem 5-8 then replaces Q by $Q + Q'$, where

$$Q'\mathbf{U} = -\tfrac{1}{8}A^2\,\Delta t\{(\Delta x)^2 - A^2(\Delta t)^2\}\frac{\partial^4 \mathbf{U}}{\partial x^4} \qquad (5\text{-}36)$$

The later modification introduces dissipation into Eqn (5-33) (Problem 5-9).

It is probably preferable to study dissipative effects at the level of the difference equations. Lax and Wendroff [25] did this and found that oscillations behind the shock were reduced if the right-hand side of Eqn (4-169) had added to it an artificial dissipative term of the type

$$\frac{\Delta t}{2\Delta x}\{Q_{i+1/2}(\mathbf{U}_{i+1} - \mathbf{U}_i) - Q_{i-1/2}(\mathbf{U}_i - \mathbf{U}_{i-1})\} \qquad (5\text{-}37)$$

Here, $Q_{i+1/2} = Q_{i+1/2}(\mathbf{U}_i, \mathbf{U}_{i+1})$ is a matrix that is chosen to be negligible when $\mathbf{U}_{i+1}$ and $\mathbf{U}_i$ are nearly equal, but giving the desired dissipative effects when $\mathbf{U}_{i+1}$ and $\mathbf{U}_i$ are significantly different.

Lax and Wendroff [25] chose

$$Q_{i+1/2} = g(A_{i+1/2}) \qquad (5\text{-}38)$$

basing their arguments upon a desire to have the dissipation effective if the equations uncouple and A is constant. Secondly, for a single scalar equation, a reasonable dissipation and shock profile are obtained by selecting $Q_{i+1/2} = \eta\,|A(U_{i+1}) - A(U_i)|$, where η is a constant, $\eta = O(1)$, $A = dF/dU$, F and U scalars. Generalizing from this we select Eqn (5-38), where $A_{i+1/2} = A(\tfrac{1}{2}U_{i+1} + \tfrac{1}{2}U_i)$ in the variable coefficient case and the nth eigenvalue of $Q_{i+1/2}$ should be $\eta\,|a_{i+1}^n - a_i^n|$, where a_i^n and a_{i+1}^n are the nth eigenvalues of $A(U_i)$ and $A(U_{i+1})$. In Eqn (5-38) g is a polynomial of degree $p - 1$ (p = number of dependent variables, 3 for gas dynamics) taking the value $\eta\,|a_{i+1}^n - a_i^n|$ when its argument is equal to† $a_{i+1/2}^n$—that is, g is a Lagrange interpolation polynomial whose coefficients depend upon the components of U_i and U_{i+1}.

If A is constant, $g \equiv 0$ since $|a_{i+1}^n - a_i^n| = 0$ for each n. For the Lagrangian form, Problem 4-54, the eigenvalues of A are 0, $\pm c'$, $c' = cV_0/V$. Since c depends upon V and ϵ, $c' = c'(\mathbf{U})$. Then g reduces to the square term only ($p - 1 = 2$) and

$$Q_{i+1/2} = \eta\,\frac{|c'(\mathbf{U}_{i+1,j}) - c'(\mathbf{U}_{i,j})|}{\{c'(\mathbf{U}_{i+1/2,j})\}^2}\,(A_{i+1/2})^2 \qquad (5\text{-}39)$$

An ingenious method for one-dimensional shocks has been given by Godunov [27], based upon the Lagrangian equations in conservation form, Problem 4-51. Shocks can also be calculated by Hartree's hybrid method

---

† This is the nth eigenvalue of $A_{i+1/2}$.

(Section 4-12) and generalizations. Stein [28] carried out calculations of this type for a spherical blast wave and Keller *et al.* [29] for a shallow water bore problem.

PROBLEMS

5-5 Incorporate a pseudoviscosity term in the Lagrangian equations with spherical symmetry ($\alpha = 2$).

5-6 Express the pseudoviscosity q in terms of $\partial V/\partial t$ instead of $\partial u/\partial y$.

5-7 Extend the finite difference approximation [Eqns (4-178)] to include q. (*Hint*: Use a central difference for u_y, centered at $i + \frac{1}{2}, j$.)

5-8 Establish the indicated form for QU, Eqn (5-35).

5-9 Show that Eqn (5-33), with Q replaced by $(Q + Q')$, is dissipative by demonstrating that the Fourier component $\exp[(-1)^{1/2}kx]$ decreases as $\exp[-\text{const.}\ k^4 t]$, assuming that $(\Delta x)^2 - (\lambda \Delta t)^2 > 0$ for every eigenvalue of A. Why is the last condition necessary?

5-10 Apply an analysis similar to that of Problem 5-9 if the von Neumann–Richtmyer pseudoviscosity is employed in Eqn (5-33).

Ans: In lowest order correction $(O[(\Delta t)^2])$, it is both dispersive and dissipative.

5-11 Establish the form, Eqn (5-39), of the artificial viscosity term for the Lagrangian equations.

5-12 Write the 'smooth solution' Lax–Wendroff technique, Eqn (4-169), for $u_t + uu_r = 0$. What form does the artificial dissipative term $Q_{i+1/2}$ [Eqn (5-38)] take in this case?

5-3 Eigenvalue problems

Our discussions on elliptic equations were primarily concerned with boundary value problems. Quite frequently physical problems generate eigenvalue problems, which are closely related to equilibrium problems. Here the general problem† is to find one or more constants, λ, and corresponding functions, u, such that the differential equation

$$L_{2m}(u) = \lambda N_{2n}(u) \tag{5-40}$$

is satisfied in a domain D and the boundary conditions

$$B_i(u) = \lambda C_i(u), \quad i = 1, 2, \ldots, m \tag{5-41}$$

are satisfied on the boundary of D. The operators L_{2m} and N_{2n} are *linear*

---

† We shall consider only *linear* eigenvalue problems.

homogeneous differential operators of order $2m$ and $2n$ respectively, with $m > n$.

The constant λ need not appear in the boundary conditions. When it does not, the eigenvalue problem [Eqns (5-40) and (5-41)] is said to be *self-adjoint* if for any two functions v, w, which satisfy the boundary conditions but are otherwise arbitrary, both

$$\int_D v L_{2m}(w)\, dD = \int_D w L_{2m}(v)\, dD$$

$$\int_D v N_{2n}(w)\, dD = \int_D w N_{2n}(v)\, dD$$

$(5\text{-}42)$

are true. If, for any such function v,

$$\int_D v L_{2m}(v)\, dD \geq 0 \qquad (5\text{-}43)$$

L_{2m} is said to be *positive*. If the equality holds only for $v \equiv 0$, the operator is said to be *positive-definite*.† When the constant λ does not appear in the boundary conditions, the problem is said to be *special* if N_{2n} has the form

$$N_{2n}(u) = g \cdot u \qquad (5\text{-}44)$$

where g, a function of the coordinates of D, is positive throughout D.

Eigenvalue problems abound in engineering and the physical sciences. We shall mention two of these problems.

(i) *The membrane problem* (Courant and Hilbert [30]): A uniform elastic membrane has mass per unit area m and is stretched with surface tension T over a rigid frame whose boundary consists of piecewise smooth simple closed curves enclosing the domain D. The eigenvalue problem for the natural transverse vibrations is to determine u^k not identically zero and numbers λ^k (k is a superscript) with

$$-\nabla^2 u^k = \lambda^k u \text{ in } D, \qquad u^k = 0 \text{ on } C. \qquad (5\text{-}45)$$

Here $\lambda^k = m\omega_k^2/T$, where ω_k is the desired frequency.

(ii) *Buckling of a plate* (Timoshenko and Gere [31]): Let a thin elastic rectangular plate be hinged along all four edges and subject to a uniform longitudinal compression. We seek the buckling load—that is, the *smallest* (critical) compression, p, per unit length—for which the unbent configuration ceases to be physically stable. In dimensionless form the transverse deflection $w(x, y)$ must satisfy

$$\nabla^4 w = \frac{\partial^4 w}{\partial x^4} + 2\frac{\partial^4 w}{\partial x^2\, \partial y^2} + \frac{\partial^4 w}{\partial y^4} = -\lambda \frac{\partial^2 w}{\partial x^2} \qquad (5\text{-}46)$$

† One demonstrates the truth or falsity of these by employing integration by parts.

within the region D: $-1 < x < 1$, $-\frac{1}{2} < y < \frac{1}{2}$, and satisfy the following boundary conditions: $w = 0$, $w_{xx} = 0$ on $x = \pm 1$; $w = 0$, $w_{yy} = 0$ on $y = \pm\frac{1}{2}$.

We confine our discussion to eigenvalue problems of the form of Eqns (5-40) and (5-41), in which $C_i \equiv 0$ for all i, possessing sufficient continuity and smoothness conditions to ensure that real solutions do exist (see, for example, Courant and Hilbert [30]). In fact, there are an infinite number of solutions, each consisting of a scalar (eigenvalue) λ^k and a corresponding eigenfunction u^k satisfying

$$L_{2m}(u^k) = \lambda^k N_{2n}(u^k) \quad \text{and} \quad B_i(u^k) = 0, \quad i = 1, 2, \ldots, m.$$

The trivial solution, $u = 0$, is ruled out as being improper. If L_{2m} is *positive-definite, all eigenvalues are greater than zero*, while if it is only positive then $\lambda = 0$ is an eigenvalue. A very useful survey of exact and approximate eigenvalue methods is contained in Crandall [32].

For the most part the discretization methods discussed for elliptic equations apply also to eigenvalue problems. However, here the algebraic equations are homogeneous and one must estimate a discretization error. Since the eigenvalue is only a single number, the discretization error is the error in that scalar. Our initial discussion will employ the well-known *membrane eigenvalue problem* as a vehicle to introduce some of the basic ideas.

(a) The membrane eigenvalue problem† [Eqn (5-45)]

For this problem it is known that $L_2 = -\nabla^2$ is positive-definite. Consequently, all the eigenvalues are positive—further, there are a countable infinity of eigenvalues with no finite limit point. Numbering them, $0 < \lambda_1 \leq \lambda_2 \leq \lambda_3 \leq \ldots$, there corresponds to each an orthonormal eigenfunction, u^p—that is, for all p, l,

$$\iint_D u^p u^l \, dx \, dy = \begin{cases} 1, & p = l \\ 0, & p \neq l \end{cases} \tag{5-47}$$

Let D be a rectangular region so that there are no irregular interior mesh points. Upon approximating the Cartesian form of Eqn (5-45) with the five-point molecule, Eqn (3-14), we have

$$-h^{-2}(u_{i+1,j} - 2u_{i,j} + u_{i-1,j}) - k^{-2}(u_{i,j+1} - 2u_{i,j} + u_{i,j-1}) = \lambda u_{i,j} \tag{5-48}$$

where, as usual, $h = \Delta x$, $k = \Delta y$, $u_{i,j} = u(ih, jk)$. Upon writing out Eqn (5-48) at each mesh point, and using the boundary conditions, $u \equiv 0$, one obtains the linear homogeneous algebraic equations (Problem 5-14)

$$Au = \lambda u. \tag{5-49}$$

† Of course, this formulation applies to a variety of other field problems such as magnetic wave guides, ideal fluid mechanics, steady-state diffusion and conduction, and so forth (see, for example, Moon and Spencer [34]).

Here u now denotes the n-vector of $u_{i,j}$ values at the mesh points and A is a symmetric $n \times n$ matrix (n is the number of interior mesh points). Since the domain has been discretized with n mesh points, there are only n eigenvalues instead of an infinity of them.

In the event that our differential equation is complicated, we may wish to use a low-order discrete approximation, such as the five-point molecule used herein, for Laplace's equation. For the case at hand, we can use more accurate formulae, but these will generally *contain λ in a nonlinear form*. For example, since $\nabla^2 u = -\lambda u$, we have $\nabla^4 u = -\lambda \nabla^2 u = +\lambda^2 u$, from which the nine-point formula (Problem 5-15), with $h = k$,

$$4S_1 + S_2 = (20 - 6h^2\lambda + \tfrac{1}{2}h^4\lambda^2)u_{i,j} + O(h^6) \tag{5-50}$$

follows. For simplicity we have employed the notation

$$S_1 = u_{i+1,j} + u_{i-1,j} + u_{i,j+1} + u_{i,j-1}$$
$$S_2 = u_{i+1,j+1} + u_{i-1,j+1} + u_{i+1,j-1} + u_{i-1,j-1} \tag{5-51}$$
$$S_3 = u_{i+2,j} + u_{i-2,j} + u_{i,j+2} + u_{i,j-2}$$

The actual determination of λ can be accomplished by setting $\mu = -6h^2\lambda + \tfrac{1}{2}h^4\lambda^2$, obtaining μ by solving the resulting eigenvalue problem of the form of Eqn (5-49) and then by solving a quadratic. Mann *et al.* [35] and Collatz [36] develop and investigate more elaborate and higher-order truncation error formulas for Eqn (5-45), but all these are hardly valuable unless comparable accuracy is achieved near the boundary. Of course, the methods for irregular boundaries of Section 1-9 are applicable to eigenvalue problems without any significant modification.

An extrapolation technique, called 'h^m extrapolation,' due to Richardson and Gaunt [37], is often helpful here, as well as in other finite difference calculations. Suppose the discretization error of our finite difference algorithm is $O(h^m)$. Let u_1 and u_2 be the solutions, generated by that algorithm, at the end of an interval using $h = h_1$ and at the end of the *same* interval using $h = h_2$. The extrapolation,

$$u_E = \frac{u_1 h_2^m - u_2 h_1^m}{h_2^m - h_1^m}, \tag{5-52}$$

of these two values gives an improved approximation, provided the total round-off error is negligible and both interval sizes are sufficiently small that the majority of the discretization error remains $O(h^m)$. One can employ three numerical values and obtain the extrapolation formula

$$u_E = \frac{[(h_2^m - h_3^m)/h_1^m]u_1 + [(h_3^m - h_1^m)/h_2^m]u_2 + [(h_1^m - h_2^m)/h_3^m]u_3}{[(h_2^m - h_3^m)/h_1^m] + [(h_3^m - h_1^m)/h_2^m] + [(h_1^m - h_2^m)/h_3^m]}. \tag{5-53}$$

(b) Some methods of computation

The number of algorithms for developing eigenvalues for $Au = \lambda u$ directly and iteratively is very large (see, for example, Bodewig [38] and Wilkinson

[39]). If A is symmetric and n^2 is sufficiently small to hold the whole matrix A in high-speed storage, some method, such as that of Givens [40], can be employed. In this procedure A is reduced by orthogonal transformations to a similar tridiagonal† matrix from which the eigenvalues are obtained in various ways. In most practical problems, it is not feasible to employ the direct methods, and our problem is then that of solving $Au = \lambda u$, iteratively, thereby requiring the storage of *approximately n elements*.

In many physical problems only the largest or smallest eigenvalue, and corresponding eigenvectors, are required. For this end eigenvalue, the *power method* is attractive (see Bodewig [38]). Beginning with an initial trial vector $u^{(0)}$, form successively $u^{(k)}$ by means of

$$u^{(k)} = Au^{(k-1)} \tag{5-54}$$

until the direction of $u^{(k)}$ settles down.‡ If $u^{(k)}$ is actually an eigenvector, then $Au^{(k)} = \lambda u^{(k)}$—that is, the elements of $Au^{(k)}$ will be proportional to $u^{(k)}$. Each element will be a scalar multiple of the corresponding element of $u^{(k)}$. *That scalar multiple is the eigenvalue.* But when, and to what eigenvalue will this process converge?

To understand the behavior of the sequence, $u^{(k)}$, generated by means of Eqn (5-54), let us suppose that $u^{(0)}$ has been expanded in terms of the (true) eigenvectors u^j (j a superscript)

$$u^{(0)} = \sum_{j=1}^{n} a_j u^j. \tag{5-55}$$

Then

$$u^{(1)} = Au^{(0)} = \sum_{j=1}^{n} a_j Au^j = \sum_{j=1}^{n} a_j \lambda_j u^j$$

and generally

$$u^{(k)} = Au^{(k-1)} = \cdots = A \ldots Au^{(0)} \ (k \text{ times})$$

$$= \sum_{j=1}^{n} a_j (\lambda_j)^k u^j. \tag{5-56}$$

Denoting§ $\lambda_N = \max_j |\lambda_j|$, and if $a_N \neq 0$, we can write

$$u^{(k)} = a_N \lambda_N^k \left\{ u^N + \sum_{j=1}^{n}{}' \frac{a_j}{a_N} \left(\frac{\lambda_j}{\lambda_N}\right)^k u^j \right\} \tag{5-57}$$

where the notation $\sum'$ means the term $j = N$ is omitted. Since $|\lambda_j/\lambda_N| < 1$, $j \neq N$, the summation in Eqn (5-57) goes to zero as k increases and hence

---

† Earlier methods often employed a *diagonalization* process rather than the less 'expensive' triangularization method.

‡ If the problem is $Au = \lambda Bu$, it may be expressed in this form by writing $B^{-1}Au = Hu = \lambda u$.

§ For this demonstration we suppose that the eigenvalues are distinct. It is a minor matter to lift this restriction.

$u^{(k)}$ approaches a multiple of the eigenvalue u^N. Moreover, the scalar ratio of corresponding elements of $u^{(k)}$ and $u^{(k-1)}$ approaches λ_N. Thus the iteration process, known as the *power method, yields convergence to the eigenvector corresponding to the eigenvalue of largest absolute value.*

It is immediately obvious from Eqn (5-57) that the number of iterations required to obtain a desired accuracy will be decreased if the $|a_j/a_N|$ ($j \neq N$) and/or $|\lambda_j/\lambda_N|$ ($j \neq N$) are small. The first of these is somewhat controllable by the analyst who should utilize any physical or other available information in choosing his initial vector. If the eigenvalues are well separated, convergence will be fairly rapid but, of course, the mode for which $|\lambda_j/\lambda_N|$ is largest will be most resistant to size reduction.

In some buckling problems the eigenvalue of primary interest is the one of largest absolute value since the largest λ corresponds to the smallest critical load. However, for the natural frequency problems, typified by the membrane problem, the largest eigenvalue corresponds to the highest natural frequency, which is seldom of interest. To find the smallest frequency we can write $Au = \lambda u$ in the form

$$\mu u = Bu$$

where $\mu = 1/\lambda$, $B = A^{-1}$ and apply the power method. Convergence is then to the eigenvalue of smallest absolute value.

A second method, that of *inverse iteration* (perhaps first introduced by Wielandt [41]), has many good features. This has somewhat more flexibility if we are interested in one particular eigenvalue somewhere in the spectrum. For the more general eigenvalue problem,

$$Au = \lambda Bu \qquad (5\text{-}58)$$

the determination of that λ nearest to a particular number p, by inverse iteration, is accomplished by solving successively

$$(A - pB)u^{(k+1)} = Bu^{(k)}. \qquad (5\text{-}59)$$

The convergence properties of *inverse iteration* have been studied by Crandall [42] and Ostrowski [43]. When $p = 0$ the convergence of the ratio of the components of $u^{(k)}$ to $u^{(k+1)}$ is to the smallest value of λ; otherwise to $\lambda - p$ where λ is the desired root. Generally Eqn (5-59) is solved by iteration processes (see, for example, Chapter 3).

More difficult eigenvalue problems have been studied by various authors. Copley [44] and Forsythe and Wasow [45] have addressed themselves to eigenvalue problems possessing re-entrant corners; that of the second authors is the L-shaped membrane problem. Small fluid oscillations in a tank generate an eigenvalue problem with an eigenvalue in a boundary condition. Several finite difference procedures for determining the smallest eigenvalue are described by Ehrlich *et al.* [46]. Wilkinson [47] develops *a priori* error bounds for the eigenvalues, computed by several methods, employing orthogonal transformations.

(c) A nonlinear eigenvalue problem (Motz [48])

Nonlinear eigenvalue problems are appearing with increasing frequency. The present example is that concerned with the calculation of a plasma configuration confined by a radio-frequency field in a resonant cavity. The problem of Motz [48] is the eigenvalue problem

$$\frac{1}{r}\frac{\partial}{\partial z}\left(\frac{1}{K}\frac{\partial\phi}{\partial z}\right) + \frac{\partial}{\partial r}\left(\frac{1}{Kr}\frac{\partial\phi}{\partial r}\right) + \lambda\phi = 0$$

where $\lambda = w^2/c^2$. The problem is nonlinear since K depends upon $\phi = \phi(r, z)$. The numerical process employed by Motz is composed of an *inner–outer iteration* process.

For the *outer iteration* we suppose K is known at the nth step of the method and solve, for λ_{n+1} and ϕ_{n+1}, the linear eigenvalue problem

$$\frac{1}{r}\frac{\partial}{\partial z}\left(\frac{1}{K_n}\frac{\partial\phi_{n+1}}{\partial z}\right) + \frac{\partial}{\partial r}\left(\frac{1}{K_n r}\frac{\partial\phi_{n+1}}{\partial r}\right) + \lambda_{n+1}\phi_{n+1} = 0. \tag{5-60}$$

At each step of the outer iteration the linear eigenvalue problem, Eqn (5-60), is solved by an *inner iteration*, say, inverse iteration. If the smallest eigenvalue is desired, the inverse iteration process is represented by

$$\frac{1}{r}\frac{\partial}{\partial z}\left(\frac{1}{K_n}\frac{\partial\phi_{n+1}^{(s+1)}}{\partial z}\right) + \frac{\partial}{\partial r}\left(\frac{1}{K_n r}\frac{\partial\phi_{n+1}^{(s+1)}}{\partial r}\right) = -\phi^{(s)}. \tag{5-61}$$

Upon completion of the computation of ϕ_{n+1} a new estimate, K_{n+1}, is obtained and the outer iteration continued.

PROBLEMS

5-13 By employing *separation of variables* show that the wave equation, $u_{xx} + u_{yy} = (1/c^2)u_{tt}$, for the vibration of a membrane with initial conditions $u(x, y, 0) = f(x, y)$, $u_t(x, y, 0) = g(x, y)$, and boundary conditions $u(\pm 1, y, t) = u(x, \pm 1, t) = 0$, is exactly reducible to the eigenvalue problem of Eqn (5-45).

5-14 It is required to set up the algebraic problem, $Au = \lambda u$, for $-(u_{xx} + u_{yy}) = \lambda u$ in the rectangular region $D: 0 < y < 1, 0 < x < 2$. Let $u = 0$ on the boundary of D and $h = \frac{1}{2}, k = \frac{1}{3}$. Find the explicit form for A and verify that it is symmetric and diagonally dominant.

5-15 Develop Eqn (5-50).

5-16 The biharmonic equation, $\nabla^4 u = f(x, y)$, is 'elliptic' in the sense that, for well-posed problems, the domain is closed and two boundary conditions are specified at each boundary point. Develop the thirteen-point finite difference molecule

$$\frac{1}{h^4}(20u_{i,j} - 8S_1 + 2S_2 + S_3) + O(h^2)$$

for $\nabla^4 u$, with $h = k$.

5-17   Apply the finite difference approximation of Problem 5-16 for a rectangular domain, $|x| < a$, $|y| < b$, on the boundaries of which are specified the values of the function and the normal derivatives.

5-18   Apply the power method to Problem 5-14 and find the largest and the smallest eigenvalue.

5-19   Let $B = I$ (identity matrix) in Eqn (5-59). Write the inverse iteration method in terms of the power method.

5-20   Apply the inverse iteration method to Problem 5-14.

5-21   Write the inverse iteration, Eqn (5-61), in matrix form.

5-4   Parabolic equations in several space variables

The presently available finite difference methods for parabolic equations in several space variables (the transient Navier–Stokes equations are discussed in Section 5-9) fall into basically two categories. In the first one finds generalizations of the elementary methods and, in the second, alternating direction methods which have no single space variable analog. The treatment of two space variables is typical, and since the algebraic manipulations are relatively simpler, we restrict our attention to this case.

(a) Generalizations of the elementary methods

Consider the diffusion equation,

$$u_t = u_{xx} + u_{yy} \tag{5-62}$$

to be solved in $0 < t \leq T$ and a connected region R of the (x, y) plane. We suppose $u(x, y, 0) = f(x, y)$ and $u(x, y, t) = g(x, y, t)$ on the boundary of R, say B. In what follows we write $u_{i,j,n} = u(i \Delta x, j \Delta y, n \Delta t)$ and employ the notation

$$\Delta_x u_{i,j,n} = (u_{i+1,j,n} - u_{i-1,j,n})/2\Delta x$$
$$\delta_x^2 u_{i,j,n} = (u_{i+1,j,n} - 2u_{i,j,n} + u_{i-1,j,n})/(\Delta x)^2 \tag{5-63}$$

which is slightly at variance with that previously used—note that now we employ the symbols to denote *divided* differences.

The *explicit forward difference scheme* [Eqn (2-8)] generalizes to

$$u_{i,j,n+1} = u_{i,j,n} + \Delta t\{\delta_x^2 + \delta_y^2\}u_{i,j,n} \tag{5-64}$$

providing that each spatially neighboring net point is in either R or on B. If B is made up entirely of segments of lines $x = $ constant and $y = $ constant, then the convergence analysis employed for one space variable generalizes. The stability restriction (Problem 5-22) is

$$\Delta t[(\Delta x)^{-2} + (\Delta y)^{-2}] \leq \tfrac{1}{2} \tag{5-65}$$

and if $\Delta x = \Delta y$, $r = \Delta t/(\Delta x)^2 \leq \frac{1}{4}$. Such a severe restriction makes this method of doubtful practicality. With q space variables and equal space increments the stability restriction becomes

$$r \leq \frac{1}{2q}. \tag{5-66}$$

A discussion of the modifications necessary when the boundary is not so simple as that above is given by Douglas [49].

The backward difference, Eqn (2-33), can also be generalized to permit an implicit finite difference approximation which has *no stability restriction*, but the resulting linear equations are no longer tridiagonal (Problem 5-24). Elimination methods require excessive calculation if the mesh sizes are small leaving the field open to iteration. ADI procedures have been popular (see Douglas [50]).

The Crank–Nicolson scheme, Eqn (2-52), can be extended to the two-dimensional probem as

$$\frac{u_{i,j,n+1} - u_{i,j,n}}{\Delta t} = \tfrac{1}{2}(\delta_x^2 + \delta_y^2)(u_{i,j,n+1} + u_{i,j,n}). \tag{5-67}$$

It is always stable and is locally second-order correct in time and space (compare Section 2-5) for rectangular regions, but in the curvilinear case the methods used by Douglas [49] reduce the local error to $\mathrm{O}[\Delta x + (\Delta t)^2]$. The three-level formula, Eqn (2-135), obviously generalizes (Problem 5-27). Additional details are available in Fox [26, Chapters 29 and 30].

(b) Alternating direction methods

Implicit difference schemes are sought for a number of reasons, the primary one being the desire for unconditional stability. One also seeks to improve accuracy and from a practical viewpoint the generated algebraic system should be easily solvable. The implicit methods for a single space variable (see Chapter 2) satisfied these desires quite well, but the generalizations of the elementary methods were lacking in the 'ease of computation' category. The ADI methods, previously extensively discussed in Chapter 3, are intended to simplify the solution of the algebraic equations while preserving the stability and accuracy requirements. The foundations of ADI, for parabolic equations, were laid by Douglas [51], Peaceman and Rachford [52], Douglas and Rachford [53], and Douglas [54]. Originally devised for parabolic equations in two dimensions, the process has been reformulated and extended to any number of dimensions.

The main feature of the method is a reformulation of the finite difference equations so that the algebraic problem consists of a set of linear equations possessing a tridiagonal matrix. We then solve this set of equations in each coordinate direction in turn, so that we do three net sweeps in a three-dimensional problem. This comprises a single iteration. We illustrate the basic concepts for the diffusion equation, Eqn (5-62), on the unit square.

If the algebraic equations resulting from the implicit finite difference approximations for Eqn (5-62) are to be tridiagonal, then *only one* of the space derivatives can be evaluated at the advanced time level, $n + 1$. This restriction leads to the equation (say the difference is taken in x at the advanced time level)

$$\delta_x^2 u_{i,j,n+1} + \delta_y^2 u_{i,j,n} = (\Delta t)^{-1}(u_{i,j,n+1} - u_{i,j,n}).$$ (5-68)

The stability condition of Eqn (5-68) is easily obtained since it is linear. The eigenfunctions† for the discrete problem are

$$v_{i,j,n}^{p,q} = \rho_n \sin{(p\pi i \Delta x)} \sin{(q\pi j \Delta y)}$$ (5-69)

and one easily determines (Problem 5-28)

$$\frac{\rho_{n+1}}{\rho_n} = \frac{1 - 4r \sin^2{(q\pi \Delta x/2)}}{1 + 4r \sin^2{(p\pi \Delta x/2)}}$$ (5-70)

where $r = \Delta t/(\Delta x)^2$ and $\Delta x = \Delta y$. If r is large, the absolute value of the ratio can be made large by taking $q = M - 1$ ($\Delta x = 1/M$) and $p = 1$. Thus, *unconditional stability is not* one of the properties of Eqn (5-69). The use of

$$\delta_x^2 u_{i,j,n+1} + \delta_y^2 u_{i,j,n+2} = (\Delta t)^{-1}(u_{i,j,n+2} - u_{i,j,n+1})$$ (5-71)

would lead to a stability ratio with the positions of p and q interchanged. The crucial and intriguing part of the agument follows: *If we take one time step with Eqn* (5-68) (a tridiagonal matrix) *followed by one using Eqn* (5-71) (another tridiagonal matrix) then the stability ratio for the double step is

$$\frac{\rho_{n+2}}{\rho_n} = \left\{\frac{1 - 4r \sin^2{(p\pi \Delta x/2)}}{1 + 4r \sin^2{(p\pi \Delta x/2)}}\right\} \left\{\frac{1 - 4r \sin^2{(q\pi \Delta x/2)}}{1 + 4r \sin^2{(q\pi \Delta x/2)}}\right\}$$ (5-72)

which is bounded, in absolute value, by one for any time step! The effect of using two possibly unstable methods alternately, is to produce a stable equation. We are really only interested in the solution after the double step. Thus, let Δt be the double step and introduce the notation $w_{i,j,n+1/2}$ for the intermediate value. Then, the difference equations become

$$\delta_x^2 u_{i,j,n+1/2} + \delta_y^2 u_{i,j,n} = \frac{u_{i,j,n+1/2} - u_{i,j,n}}{\frac{1}{2}\Delta t}$$

$$\delta_x^2 u_{i,j,n+1/2} + \delta_y^2 u_{i,j,n+1} = \frac{u_{i,j,n+1} - u_{i,j,n+1/2}}{\frac{1}{2}\Delta t}$$ (5-73)

where each half represents tridiagonal algebraic systems which are simple.

---

† The initial value problem $u_{xx} + u_{yy} = u_t$ is easily separated and with vanishing boundary values the continuous eigenfunctions are $\sin{p\pi x} \sin{q\pi y}$. This motivates Eqn (5-69) although they can be obtained directly by separation.

Upon eliminating the intermediate values from the pair of Eqns (5-73), one finds, for a rectangle,

$$\tfrac{1}{2}(\delta_x^2 + \delta_y^2)(u_{i,j,n+1} + u_{i,j,n})$$

$$= \frac{u_{i,j,n+1} - u_{i,j,n}}{\Delta t} + \frac{\Delta t}{4}\delta_x^2\delta_y^2(u_{i,j,n+1} - u_{i,j,n}). \quad (5\text{-}74)$$

If $u \in C^5$, it is clear that Eqn (5-74) is locally *second order in both space and time*.

When Eqns (5-73) are extended to three space variables, the lack of symmetry in the stability ratio leads to a method which is unstable for any useful value of r. Douglas [54] introduced a modification which removed the aforementioned limitation. His argument is based upon the following observations: (i) Eqn (5-74) is a perturbation of the Crank–Nicolson scheme, Eqn (5-67); (ii) Eqns (5-73) and (5-74) are equivalent on a rectangle. Thus, any alternating direction scheme that leads to a perturbation of the Crank–Nicolson equation [like Eqn (5-74)] is equally satisfactory. With this as our goal, let us begin with the Crank–Nicolson method, Eqn (5-67), and impose an alternating direction modification upon it. First, evaluate the x derivative at $n + \tfrac{1}{2}$ and obtain a first approximation u_{n+1}^* at time $n + 1$ from

$$\tfrac{1}{2}\delta_x^2(u_{i,j,n+1}^* + u_{i,j,n}) + \delta_y^2 u_{i,j,n} = \frac{u_{i,j,n+1}^* - u_{i,j,n}}{\Delta t} \quad (5\text{-}75a)$$

and then move the evaluation of the y derivative ahead by means of

$$\tfrac{1}{2}\delta_x^2(u_{i,j,n+1}^* + u_{i,j,n}) + \tfrac{1}{2}\delta_y^2(u_{i,j,n+1} + u_{i,j,n}) = \frac{u_{i,j,n+1} - u_{i,j,n}}{\Delta t}. \quad (5\text{-}75b)$$

Equation (5-75b) is very nearly the Crank–Nicolson equation. If the intermediate value u_{n+1}^* is eliminated, then Eqn (5-74) is satisfied for a rectangular region. Consequently, Eqns (5-73) and (5-75) are equivalent on a rectangle although the intermediate values differ.

In three (or more) dimensions, the generalization of Eqns (5-75) is

$$\tfrac{1}{2}\delta_x^2(u_{n+1}^* + u_n) + \delta_y^2 u_n + \delta_z^2 u_n = \frac{u_{n+1}^* - u_n}{\Delta t}$$

$$\tfrac{1}{2}\delta_x^2(u_{n+1}^* + u_n) + \tfrac{1}{2}\delta_y^2(u_{n+1}^{**} + u_n) + \delta_z^2 u_n = \frac{u_{n+1}^{**} - u_n}{\Delta t} \quad (5\text{-}76)$$

$$\tfrac{1}{2}\delta_x^2(u_{n+1}^* + u_n) + \tfrac{1}{2}\delta_y^2(u_{n+1}^{**} + u_n) + \tfrac{1}{2}\delta_z^2(u_{n+1} + u_n) = \frac{u_{n+1} - u_n}{\Delta t}$$

where the obvious spatial indices are omitted. Both two- and three-dimensional forms are locally second-order correct in space and time, are unconditionally stable, and each of the algebraic systems is tridiagonal. Brian [55] obtained a related technique.

The original method of Douglas–Rachford [53], for three space variables, was developed from the backward difference equation. The method presented herein should be superior. Douglas *et al.* [56] has introduced a multistage alternating direction method. The convergence of these methods has been the subject of research by Douglas and Pearcy [57], in the presence of singular operators, and Pearcy [58].

Fairweather and Mitchell [59, 60] obtain finite difference formulae for the solution of the three space dimensions diffusion equation. Their method is fourth-order correct in space and second-order in time. Rounding errors in alternating direction methods for parabolic equations are examined by Rachford [61]. Stone [62] utilizes the sparseness of the coefficient matrix in a new method which proves superior, in some examples, to previously employed methods. The analysis of Stone's method is incomplete at this writing.

Alternating direction methods have proved useful on many more complex problems. Lees [63] has developed convergence proofs for alternating direction methods when applied to the nonlinear parabolic equation

$$u_t = [a(x, y, t)u_x]_x + [b(x, y, t)u_y]_y + f(x, y, t, u, u_x, u_y).$$

PROBLEMS

5-22 Let the region R of Eqn (5-62) be $0 < x < a$, $0 < y < b$. Establish the stability restriction, Eqn (5-65).

5-23 If a, b, c do not involve partial derivatives of u, but are otherwise unrestricted, the equation $u_t + au = \nabla \cdot (b\nabla u) + c$ is quasilinear. Describe an explicit algorithm analogous to Eqn (5-64) in the two space dimensions case.

5-24 By employing backward differences generate an implicit algorithm for Eqn (5-62) and show the linear system is *not* tridiagonal.

5-25 Can Eqn (2-35) be generalized to two space dimensions?   *Ans:* Yes

5-26 Can Eqn (2-129) be generalized to two space dimensions? (*Hint:* With what do we replace $u_{xxxx} + u_{yyyy}$?)   *Ans:* Yes: $(u_{xx} + u_{yy})_t - 2u_{xxyy}$.

5-27 Generalize Eqn (2-135) to two space dimensions.

5-28 Establish the ratio, Eqn (5-70).

$$(Hint: \quad \delta_x^2 \sin p\pi x = -\frac{4}{(\Delta x)^2} \sin^2\left(\frac{p\pi \Delta x}{2}\right) \sin p\pi x)$$

5-29 Compare Eqn (5-74) with the Crank–Nicolson form, Eqn (5-67).

5-5 Additional comments on elliptic equations

In this section we shall discuss some extensions of the methods of Chapter 3 as they have been applied to nonlinear and three-dimensional problems.

(a) Nonlinear over-relaxation

When finite difference approximations are applied directly to a nonlinear elliptic equation, nonlinear algebraic equations result. Several methods have been proposed to solve these systems—perhaps the earliest is an extended Liebmann method, which is a direct generalization of the (linear) Gauss–Seidel procedure (see Bers [64] and Schechter [65]). Schechter discusses a number of iteration methods for nonlinear problems. While the extended Liebmann method has a sound theoretical basis, the following method of nonlinear over-relaxation has only numerical experiments and semitheoretical arguments to justify it as this book is being written.

In many of the previous methods the solution of the large nonlinear algebraic system was obtained by an outer iteration (Newton's method, say) which linearizes, followed by some iterative technique (say, SOR). This process is repeated, thus constructing a cascade of outer (Newton) iterations alternated with a large sequence of inner linear iterations.

The complexity of the strategy, discussed above, suggests that alternatives would be highly desirable. One direct and simple method which is particularly well adapted for solving algebraic systems associated with nonlinear elliptic equations is due to Lieberstein [66, 67]—a method called *nonlinear over-relaxation* (NLOR).

Consider a system of k algebraic equations each having continuous first derivatives

$$f_p(x_1, x_2, \ldots, x_k) = 0, \quad p = 1, 2, \ldots, k. \tag{5-77}$$

For convenience set $f_{pq} = \partial f_p / \partial x_q$. The basic idea is the introduction of a relaxation factor ω so that

$$x_1^{n+1} = x_1^n - \omega \frac{f_1(x_1^n, x_2^n, \ldots, x_k^n)}{f_{11}(x_1^n, \ldots, x_k^n)}$$

$$x_2^{n+1} = x_2^n - \omega \frac{f_2(x_1^{n+1}, x_2^n, \ldots, x_k^n)}{f_{22}(x_1^{n+1}, x_2^n, \ldots, x_k^n)} \tag{5-78}$$

$$x_3^{n+1} = x_3^n - \omega \frac{f_3(x_1^{n+1}, x_2^{n+1}, x_3^n, \ldots, x_k^n)}{f_{33}(x_1^{n+1}, x_2^{n+1}, x_3^n, \ldots, x_k^n)}$$

and so on.

This method has a feature of the Gauss–Seidel method in that it uses corrected results immediately upon becoming available. In addition, if the f_p are linear functions of the x_j, this method reduces to SOR (Problem 5-30).

The convergence criteria for NLOR can be shown to be the same as those for SOR (Chapter 3) with the coefficient matrix A replaced by the Jacobian of Eqn (5-77), $J(f_{pq}^n)$. This is accomplished by use of the Taylor series so that the vector form for $x^{n+1} - x^n$ is

$$x^{n+1} - x^n \approx -\omega[D_n^{-1}L_n e^{n+1} + (I + D_n^{-1}U_n)e^n] \tag{5-79}$$

where e^n stands for the error vector, $e^n = x^n - x$ and L_n, D_n, and U_n are lower triangular, diagonal, and upper triangular matrices formed from

$$J(f_{pq}^n) = L_n + D_n + U_n. \qquad (5\text{-}80)$$

From Eqn (5-79) it follows that

$$e^{n+1} \approx L_{\omega_n} e^n = -\left[\frac{1}{\omega} D_n + L_n\right]^{-1}\left[\left(1 - \frac{1}{\omega}\right)D_n + U_n\right]e^n \qquad (5\text{-}81)$$

which is exactly the same form as the error matrix in SOR (Chapter 3). For convergence, the Jacobian $J(f_{pq}^n)$, at each stage of the iteration, must have the properties required for A in SOR. To make the process most efficient, an ω_{opt} should be calculated minimizing the spectral norm of L_{ω_n} at *each* iteration n. This may be expensive and unrewarding since one does better to overestimate ω_{opt} than to underestimate. Further, it is suggested that for small systems one usually runs with $\omega = 1$, and for large systems with a constant ω slightly less than 2.

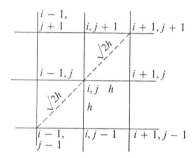

Fig. 5-5   Alternative computational net for the partial derivative u_{xy}

Consider an application of this method to a boundary value problem for the nonlinear elliptic equation in a closed domain,

$$F(x, y, u, u_x, u_y, u_{xx}, u_{xy}, u_{yy}) = 0 \qquad (5\text{-}82)$$

where F has at least one continuous derivative with respect to each of its eight arguments. At an interior point, we use a nine-point molecule (Fig. 5-5) and write for each interior point of the region

$$F\left\{x_i, y_j, u_{ij}, \frac{u_{i+1,j} - u_{i-1,j}}{2h}, \frac{u_{i,j+1} - u_{i,j-1}}{2h}, \frac{u_{i+1,j} + u_{i-1,j} - 2u_{i,j}}{h^2}, \right.$$

$$\left. \frac{u_{i+1,j+1} + u_{i-1,j-1} - u_{i+1,j-1} - u_{i-1,j+1}}{2h^2}, \frac{u_{i,j+1} + u_{i,j-1} - 2u_{i,j}}{h^2}\right\} = 0. \qquad (5\text{-}83)$$

Equation (5-83) furnishes the equation

$$f[x_i, y_j, u_{i,j}, u_{i+1,j}, u_{i-1,j}, u_{i,j+1}, u_{i,j-1},$$

$$u_{i-1,j+1}, u_{i-1,j-1}, u_{i+1,j+1}, u_{i+1,j-1}] = 0 \qquad (5\text{-}84)$$

which must be satisfied at every point (i, j) of a square net with interval size h. These are the nonlinear algebraic equations, Eqn (5-77).

To compute using NLOR, read in values of the initial guess $u_{i,j}^0$ and scan the mesh replacing the value of $u_{i,j}$ at each point by

$$u_{i,j}^0 - \omega \frac{f}{\partial f/\partial u_{ij}}$$

where f and $\partial f/\partial u_{i,j}$ are evaluated, as in Eqns (5-78), by using corrected results immediately upon becoming available.

If a *consistent* ordering is chosen (e.g. scanning the grid from left to right and up) and u_{xy} is not present, then the Jacobian $J(f_{i,j}^n)$ will have Property (A). Otherwise it does not, and we cannot use the molecule of Fig. 5-5 and still generate a system with Property (A). In view of the knowledge that SOR is useful on some matrices without property (A), it is probably also true that NLOR is useful for a large class of systems not having Property (A).

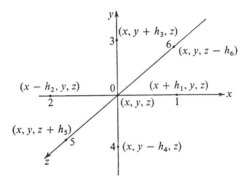

Fig. 5-6   Point configuration in three dimensions

Several numerical experiments (Lieberstein [67] and Greenspan and Yohe [68]) have been reported in which excellent results were obtained with NLOR. Greenspan and Yohe compare NLOR to an extended Liebmann method (using a generalized Newton–Raphson procedure) for the three-dimensional problem

$$u_{xx} + u_{yy} + u_{zz} = e^u \tag{5-85}$$

in the spherical sector G bounded by $x^2 + y^2 + z^2 = 1$; $x = 0$; $y = 0$; $z = 0$ and with $u = x + 2y + z^2$ on the boundary of G. With $h = 0.1$ the NLOR took only *one-half* the computation time of the extended Liebmann method.

A problem of interest in gas dynamics (see Wise and Ablow [69]) is the Dirichlet problem for the nonlinear elliptic equation

$$\nabla^2 u = u^2. \tag{5-86}$$

Iteration methods for this problem have previously been discussed in Problem 3–42. The Newton method, discussed in Section 3-9 and employed by Bellman *et al.* [70] in two dimensions, is applied by Greenspan [71] to the *three-dimensional problem* of Eqn (5-86). The outer iteration for Eqn (5-86) is

$$\nabla^2 u^{(k)} - 2u^{(k-1)}u^{(k)} = -\{u^{(k-1)}\}^2. \tag{5-87}$$

When the *seven-point* three-dimensional molecule (Fig. 5-6) is employed, we obtain the finite difference simulation

$$-[u_0^{(k-1)} + (h_1 h_2)^{-1} + (h_3 h_4)^{-1} + (h_5 h_6)^{-1}]u_0^{(k)}$$
$$+ [h_1(h_1 + h_2)]^{-1}u_1^{(k)} + [h_2(h_1 + h_2)]^{-1}u_2^{(k)}$$
$$+ [h_3(h_3 + h_4)]^{-1}u_3^{(k)} + [h_4(h_3 + h_4)]^{-1}u_4^{(k)}$$
$$+ [h_5(h_5 + h_6)]^{-1}u_5^{(k)} + [h_6(h_5 + h_6)]^{-1}u_6^{(k)}$$
$$= -\tfrac{1}{2}[u_0^{(k-1)}]^2 \tag{5-88}$$

Greenspan [71] reports four successful calculations of this three-dimensional problem.

PROBLEMS

5-30 Demonstrate that the NLOR of Lieberstein, Eqns (5-78), reduces to the (linear) SOR method if Eqns (5-77) are linear.

5-31 Mackenroth and Fisher [72] examine the flow field of an ideal axisymmetric jet impinging on a plane surface by interchanging the spatial variables r, z with the potential and stream functions, ϕ and ψ (a method probably first used by Woods [73]). The result is the nonlinear elliptic equation

$$(r^{-1}r_\phi)_\phi + (rr_\psi)_\psi = 0. \tag{5-89}$$

This method is advantageous since the position of a free streamline does not have to be calculated and since the (ψ, ϕ) integration domain is rectangular. The authors solved their problem by underrelaxation. Formulate the problem employing the NLOR scheme.

(b) Some alternative procedures

In two papers Gunn [74, 75] discusses a *semi-explicit technique* for solving the quasilinear elliptic equation in m dimensions

$$\sum_{i=1}^{m} (a_i u_{x_i})_{x_i} + \sum_{i=1}^{m} b_i u_{x_i} + f(x_i, u) = 0.$$

The results of numerical experiments for the Poisson equation, $\nabla \cdot (a\nabla u) = f$, on a cube, are compared with the Brian–Douglas alternating direction method of Section 5-4 and with the SOR process. While the new method was asymptotically faster than SOR, it actually took more time because of machine limitations.

Some useful methods for accelerating the convergence of iterative algorithms have been obtained and applied to nonlinear elliptic equations by Erdelyi *et al.* [76] and Ahamed [77].

Lastly, we recommend Wachspress [78] for his excellent discussion of iterative methods for elliptic systems.

(c) Some questions associated with numerical weather prediction

It is trite to note that there is great practical interest in weather prediction. Some significant assistance is being made to this interpretative art by numerical weather prediction—although considerable time must elapse before one can call this 'art' quantitative.

The basic considerations and equations of meteorology are given in Holmboe *et al.* [79]. Brief summaries of the state of the art to 1961 are contained in Forsythe and Wasow [45, p. 386] and Fox [26, p. 478]. Richtmyer and Morton [17] discuss the more limited problem of 'front' movement. Herein, we wish to discuss calculations associated with the so-called *balance equation* (see, for example, Charney and Phillips [80] or Fox [26, p. 480])

$$2[\psi_{xx}\psi_{yy} - \psi_{xy}^2] + f\nabla^2\psi + \nabla f \cdot \nabla\psi = g\nabla^2 h \qquad (5\text{-}90)$$

which is of *Monge–Ampere* type (see, for example, Ames [33] and Pogorelov [83]). The name, balance equation, stems from the fact that it gives the nondivergent wind, **V**, exactly balancing the pressure forces. The variables ψ, f, g, and h of Eqn (5-90) are, respectively, velocity potential ($\mathbf{V} = \mathbf{k} \times \nabla\psi$, $\mathbf{k}$ a unit vector), Coriolis parameter ($f = 2\alpha \sin \phi$, where α is earth's angular velocity and ϕ is latitude), gravitational constant, and height of a pressure surface. To numerically integrate this equation, we suppose that h is known from observations.

The condition that the balance equation be elliptic is that

$$g\nabla^2 h - \nabla f \cdot \nabla\psi + f^2/2 > 0$$

which cannot be evaluated since it contains ψ. A very good approximation is obtained by noting that $f\nabla\psi \sim g\nabla h$, so the *approximate* condition for ellipticity is

$$g\nabla^2 h - \frac{g}{f}\nabla f \cdot \nabla h + f^2/2 > 0.$$

This form allows the test to be made since f and h are known. In what follows the ellipticity of Eqn (5-90) is assumed.

Various numerical methods for solving the balance equation have been proposed. The computational molecule used is the nine-point molecule shown in Fig. 5-5. The occurrence of ψ_{xy} makes such a molecule advisable.

The simplest *outer iteration* would probably have the form

$$\nabla^2\psi^{(n+1)} = \frac{1}{f}\{g\nabla^2 h - \nabla f \cdot \nabla\psi^{(n)} - 2[\psi_{xx}^{(n)}\psi_{yy}^{(n)} - (\psi_{xy}^{(n)})^2]\} \qquad (5\text{-}91)$$

but the convergence is questionable in some cases (see, for example, Bolin [81]). The inner iteration can be any of those previously discussed. The difficulty here appears to be in the *nature of the nonlinearity*. Previous cases included nonlinearities involving the first derivative or function. *Here the 'Monge-Ampere' term has the same order as the Laplacian*!

The possible divergence of the above simple outer iteration has spurred other investigations of the balance equation. One of these is due to Arnason [82]. The Arnason procedure uses an alternative linearization yielding the outer iteration

$$\psi_{xx}^{(n+1)}\psi_{yy}^{(n)} + \psi_{yy}^{(n+1)}\psi_{xx}^{(n)} - 2\psi_{xy}^{(n+1)}\psi_{xy}^{(n)}$$
$$+ f\nabla^2\psi^{(n+1)} + \nabla f \cdot \nabla\psi^{(n+1)} = g\nabla^2 h. \quad (5\text{-}92)$$

The matrices of the finite difference approximation, obtained from Eqn (5-92), are functions of the iteration number n. The test of a solution, ψ, is carried out by treating Eqn (5-90) as a Poisson equation for h and comparing the computed values and the original values of h.

Special problems occur in the inner iteration of the balance equation which lead to a systematic error in the calculation of ψ. Bolin [81] noted that this systematic error was due to the use of two different grid lengths in computing derivatives. To see this we refer to Fig. 5-5 and note that

$$\psi_{xx} = \frac{\psi_{i+1,j} - 2\psi_{i,j} + \psi_{i-1,j}}{h^2} + O(h^2)$$

with a similar expression for ψ_{yy} in the other direction where both use the same grid length h. On the other hand, ψ_{xy} is approximated using a grid length of $2h$—that is,

$$\psi_{xy} = \frac{1}{(2h)^2}\{\psi_{i+1,j+1} - \psi_{i-1,j+1} - \psi_{i+1,j-1} + \psi_{i-1,j-1}\} + O(h^2).$$

Theoretically, *both* these approximations have a truncation error of $O(h^2)$ but, in the sparse meteorological fields, ψ_{xy} is underestimated compared with ψ_{xx} and ψ_{yy}. The end result is an underestimate of the values of ψ.

This systematic error can be eliminated by computing ψ_{xx}, ψ_{yy}, and ψ_{xy} of the 'Monge-Ampere' term $\psi_{xx}\psi_{yy} - \psi_{xy}^2$ with an interval length of $\sqrt{(2)}h$ (Fig. 5-5) so that

$$\psi_{xx}\psi_{yy} - \psi_{xy}^2 \approx \frac{1}{2h^2}\{\psi_{i+1,j+1} - 2\psi_{i,j} + \psi_{i-1,j-1})$$

$$(\psi_{i-1,j+1} - 2\psi_{i,j} + \psi_{i+1,j-1}) - (\psi_{i+1,j} + \psi_{i-1,j} - \psi_{i,j+1} - \psi_{i,j-1})^2\}. \quad (5\text{-}93)$$

It would seem appropriate to compute $\nabla^2\psi$ in the same way, thus obtaining

$$\psi_{xx} + \psi_{yy} \approx \frac{1}{2h^2}\{\psi_{i+1,j+1} + \psi_{i-1,j+1} + \psi_{i+1,j-1} + \psi_{i-1,j-1} - 4\psi_{i,j}\}. \quad (5\text{-}94)$$

This representation has the disadvantage that $\psi_{i,j}$ is only related to $\psi_{i+1,j}$, $\psi_{i-1,j}$, $\psi_{i,j+1}$, and $\psi_{i,j-1}$ through the (usually) weak link ψ_{xy} and not through the Laplacian. If Eqn (5-94) is not used for $\nabla^2\psi$, but an alternative field with mesh side $\sqrt{(2)}h$ is used, then the result is a checkerboard pattern of *two* fields for ψ. The usual compromise is to use Eqn (5-93) with the standard five-point molecule for $\nabla^2\psi$—that is,

$$\nabla^2\psi \approx [\psi_{i+1,j} + \psi_{i-1,j} + \psi_{i,j+1} + \psi_{i,j-1} - 4\psi_{i,j}]/h^2.$$

5-6 Hyperbolic equations in higher dimensions

(a) Characteristics in three independent variables

The method of characteristics (Chapter 4) cannot be easily generalized because of its special applicability to *two* equations in *two* independent variables. Some of the desirable features are lost during generalization. Several computational generalizations have been proposed with more or less success. We shall discuss some of the more promising ones.

The theory of characteristics in more than two independent variables, as it applies to simultaneous quasilinear first-order hyperbolic equations, is given in Courant and Hilbert [84]. As before, this is no restriction because higher order equations may be transformed to a larger first-order set. A *characteristic* in n variables is an $(n-1)$-dimensional subspace at which derivative discontinuities can occur. In general, a characteristic is the space generated by a wave front.

Thornhill [85] designed a computational method aimed at minimizing the amount of interpolation. A network of points is chosen in three dimensions by taking the intersections of three families of characteristic surfaces. This is a direct generalization of the Massau method (Section 4-3) for two variables. The method can be used for plane unsteady gas flow. If the *entropy S is constant*, no interpolation is required. The method is *explicit* and the local error is $O(h^2)$, which is difficult to reduce.

An *implicit* method is given by Coburn and Dolph [86]. The net is constructed using two families of characteristic surfaces drawn through each of a family of curves on the initial surface. The vertices of these 'tents' generate the new surface and we begin again. The net points are chosen on these curves, but only a one-dimensional array is handled simultaneously. Some interpolation is required, while the method is $O(h^2)$.

Butler [87] describes a third method which makes *no attempt to construct a characteristic net*. In this respect it is more akin to Hartree's scheme (Section 4-12). Use is made of the fact that an infinity of characteristics passes through each point instead of a finite number in the two-variable case. The solution at each point is computed without referring to data outside its domain of dependence. Butler [87] worked the case of unsteady plane flow.

Bruhn and Haack [88] used the method of Coburn and Dolph for starting flow in a nozzle, where no shocks occur. Talbot [89] studies a shock-thermal layer interaction problem, while Richardson [90] and Ansorge [91] look at the general theory.

One can generally assert that the method of characteristics suffers from the disadvantages that it necessitates a great deal of coding and entails an increase in the operating time per step. On the other hand, the method keeps close to the physical model upon which the equations are based and gives reasonable precision.

(b) Finite difference methods

For gas flows which undergo large distortions and for gas flows in two or more space variables, Lagrangian calculations lead to inaccuracies because of the distortion of the Lagrangian mesh as time advances. A new net can be introduced, from time to time, to restore rectangularity, but this leads to severe interpolation problems. Using this method, some good results have been obtained by Kolsky [92], Blair *et al.* [93], and Schultz [94].

The Eulerian formulation encounters problems in flows containing non-rigid boundaries or interfaces between fluids of differing properties since there is no simple way of distinguishing the type of fluid present at a net point at any particular instant. The Lax–Wendroff equations offer the best solution at present, and we shall discuss them briefly (see, for example, Burstein [95–97]).

In Lagrangian calculations, the motion of the net lines, if not carefully monitored, produces distorted zones which eventually lead to errors and even instabilities as net points or lines become crowded locally. Mixed formulations should help and at least two have been suggested (Frank and Lazarus [98] and Noh [99]). Noh's CEL (Coupled Eulerian–Lagrange Code) is composed first of a Lagrangian calculation followed by the mapping of the Lagrange net onto the fixed Eulerian grid. Noh presents (graphically) several successful applications in one and two space dimensions. Generally the coupled methods are advantageous when the character of the flow is radically different in two distinct directions.

In multidimensional problems there are some serious mathematical questions. One of these, which is often unclear, is whether the physical problem one wishes to solve is *properly posed* (see, for example, Morawetz [100] and Richtmyer and Morton [17]). Nonlinear instability is even more of a hazard in two dimensions than in one. It appears that we are far from any universal methods for solving multidimensional problems, and it is quite probable that this situation will continue to exist for some time.†

The work of Lax [24] and Lax and Wendroff [25] has stimulated research

† Richtmyer and Morton [17, p. 353] believe that the need in multidimensional problems is *not* computing machines of vastly increased speed and capacity, but improved mathematical (including numerical) analysis and methods!

aimed at generalizations of their methods. For the smooth part of the flow in Cartesian coordinates, x, y, the Eulerian equations in conservation form are

$$\frac{\partial \mathbf{U}}{\partial t} + \frac{\partial}{\partial x} \mathbf{F}(\mathbf{U}) + \frac{\partial}{\partial y} \mathbf{G}(\mathbf{U}) = 0 \tag{5-95}$$

where

$$\mathbf{U} = \begin{bmatrix} \rho \\ m \\ n \\ e \end{bmatrix}, \qquad \mathbf{F}(\mathbf{U}) = \begin{bmatrix} m \\ p + (m^2/\rho) \\ mn/\rho \\ (e + p)m/\rho \end{bmatrix}, \qquad \mathbf{G}(\mathbf{U}) = \begin{bmatrix} n \\ mn/\rho \\ p + (n^2/\rho) \\ (e + p)n/\rho \end{bmatrix} \tag{5-96}$$

Here $m = \rho u$, $n = \rho v$ are x and y components of momentum, e is total energy per unit volume, and p and ρ are pressure and density. With u, v, and ϵ the velocity components and internal energy per unit volume,

$$e = \rho[\epsilon + (u^2 + v^2)/2] \qquad \text{and} \qquad p = f(\epsilon, v)$$

becomes

$$p = f\left(\frac{e}{\rho} - \frac{m^2 + n^2}{2\rho^2}, \frac{1}{\rho}\right).$$

The generalization of Richtmyer [101], of the two-step method [Eqns (4-170) and (4-171)] to two Cartesian space variables, x and y, is first calculate

$$\mathbf{U}_{i,j,n+1} = \tfrac{1}{4}(\mathbf{U}_{i+1,j,n} + \mathbf{U}_{i-1,j,n} + \mathbf{U}_{i,j+1,n} + \mathbf{U}_{i,j-1,n})$$

$$- \frac{\Delta t}{2\Delta x}(\mathbf{F}_{i+1,j,n} - \mathbf{F}_{i-1,j,n}) - \frac{\Delta t}{2\Delta y}(\mathbf{G}_{i,j+1,n} - \mathbf{G}_{i,j-1,n}) \tag{5-97}$$

and the final value, at $(n + 2)\,\Delta t$, is then obtained from

$$\mathbf{U}_{i,j,n+2} = \mathbf{U}_{i,j,n} - \frac{\Delta t}{\Delta x}(\mathbf{F}_{i+1,j,n+1} - \mathbf{F}_{i-1,j,n+1})$$

$$- \frac{\Delta t}{\Delta y}(\mathbf{G}_{i,j+1,n+1} - \mathbf{G}_{i,j-1,n+1}). \tag{5-98}$$

Fractional indices have been avoided by considering the basic cell of the mesh to have dimensions $2\Delta x$, $2\Delta y$, and $2\Delta t$. The quantities $\mathbf{U}$, $\mathbf{F}(\mathbf{U})$, and $\mathbf{G}(\mathbf{U})$ come from the vector equation [Eqn (5-95)] and, in the case of gas dynamics, are given by Eqns (5-96). $\mathbf{F}_{i,j,n}$ is an abbreviation for $\mathbf{F}(\mathbf{U}_{i,j,n})$. Equations (5-97) and (5-98) are not coupled between the set of space–time mesh points having even values of $i + j + n$ and the set having odd values (Problem 5-32). Half the net points can be omitted, if desired.

As in one-space variable, these equations have a local truncation error $O(\Delta^3)$ and the linearized stability condition, with $\Delta x = \Delta y$, is

$$[\sqrt{(u^2 + v^2)} + c]\frac{\Delta t}{\Delta x} < \frac{1}{\sqrt{2}}. \tag{5-99}$$

Modifications of Eqns (5-97) and (5-98), for shock calculations, are discussed by Burstein [95–97], Lapidus (see Richtmyer and Morton [17]), and Houghton *et al.* [102].

Gourlay and Morris [103] describe explicit and implicit *second-order* correct two-dimensional schemes for Eqn (5-95).

(c) Singularities in solutions of nonlinear hyperbolic equations

Initial value problems of nonlinear hyperbolic equations may develop singularities in their solutions after only a finite elapsed time. At these singularities the derivatives become unbounded and the solution ceases to exist in its differentiable form. These singularities are not necessarily attributable to discontinuities in the intial data. Mathematical examples (see, for example, Zabusky [104] and Ames *et al.* [105]) exist in which singularities occur *even* when the prescribed initial data are analytic. Singularities usually represent the appearance of important physical phenomena. Examples involving systems of hyperbolic first-order partial differential equations include *gas shock formation* (Courant and Friedrichs [106], Jeffrey and Taniuti [107]), *breaking of water waves* (Stoker [108], Jeffrey [109]), formation of *transverse shock waves* from finite amplitude plane shear waves in an incompressible perfectly elastic material (Chu [110], Bland [111]), and *velocity jump phenomena* in traveling threadlines (Ames *et al.* [105]).

Estimates of the time to *occurrence* of singularities in nonlinear hyperbolic equations, of order greater than 1, have been carried out by Zabusky [104] and Kruskal and Zabusky [112] for their special problems. More general techniques, together with examples, have evolved from the initial work of Lax [113] to the more general results of Jeffrey [114, 115] and Ames [157].

We include herein some of the results of Jeffrey [115] because of the observation by Zabusky [104] that a *convergent finite difference approximation for a nonlinear problem may not exhibit the singularities of the exact solution.*†
The derivation of the following results will be omitted.

The evolution of discontinuities in solutions of homogeneous nonlinear hyperbolic equations from smooth initial data‡ will be developed for the *reducible* system

$$\mathbf{U}_t + A\mathbf{U}_x = 0 \qquad (5\text{-}100)$$

in which
$$U = \begin{bmatrix} u_1 \\ u_2 \end{bmatrix}, \qquad A = \begin{bmatrix} a_{11} & a_{12} \\ a_{21} & a_{22} \end{bmatrix}$$

where the a_{ij} are functions only of the dependent variables u_1 and u_2. Let

---

† That is, the finite difference solution is smooth. This is quite a different situation from that in shocks where an artificial viscosity is required.

‡ We discuss only the pure initial value problem. Combined initial boundary value problems must be converted to a pure initial value problem.

$\lambda^{(1)}$ and $\lambda^{(2)}$ be the distinct eigenvalues of $|A - \lambda I| = 0$, so that the characteristic curves are $C^{(1)}: dx/dt = \lambda^{(1)}$; $C^{(2)}: dx/dt = \lambda^{(2)}$. If $l^{(1)}$ and $l^{(2)}$ are the *left* eigenvectors—that is, $\lambda^{(i)}l^{(i)} = l^{(i)}A$—and if the $C^{(1)}$ characteristic is parameterized by $\beta(t)$ and the $C^{(2)}$ characteristic by $\alpha(t)$, α and β montone differentiable functions, then

$$l^{(1)}\mathbf{U}_\alpha = 0 \quad \text{along } C^{(1)} \text{ characteristics} \tag{5-101}$$

$$l^{(2)}\mathbf{U}_\beta = 0 \quad \text{along } C^{(2)} \text{ characteristics.} \tag{5-102}$$

If we write $l^{(i)} = [l_1^{(i)}, l_2^{(i)}]$, Eqns (5-101) and (5-102) become

$$
\begin{aligned}
l_1^{(1)}u_{1\alpha} + l_2^{(1)}u_{2\alpha} = 0 \quad \text{along } C^{(1)} \text{ characteristics} \\
l_1^{(2)}u_{1\beta} + l_2^{(2)}u_{2\beta} = 0 \quad \text{along } C^{(2)} \text{ characteristics.}
\end{aligned}
\tag{5-103}
$$

Alternative forms of Eqns (5-103) are determined if we multiply by the integrating factors q_1 and q_2, if necessary, so that they become exact differentials. Upon integrating with respect to α and β, we obtain

$$\int q_1 l_1^{(1)} \, du_1 + \int q_2 l_2^{(1)} \, du_2 = r(\beta)$$

$$\int q_2 l_1^{(2)} \, du_1 + \int q_2 l_2^{(2)} \, du_2 = s(\alpha)$$

where r and s are the Riemann invariants. Thus

$$
\begin{aligned}
0 = \frac{dr}{d\alpha} = \frac{\partial r}{\partial t} + \lambda^{(1)} \frac{\partial r}{\partial x} \quad \text{along } C^{(1)} \text{ characteristics} \\
0 = \frac{ds}{d\beta} = \frac{\partial s}{\partial t} + \lambda^{(2)} \frac{\partial s}{\partial x} \quad \text{along } C^{(2)} \text{ characteristics}
\end{aligned}
\tag{5-104}
$$

are replacements for Eqns (5-103).

The actual value, t_C, of the *time of existence of a solution* of the original system, Eqn (5-100), satisfies the inequality

$$t_{\text{inf}} < t_C < t_{\text{sup}}$$

where the solution is bounded if $t < t_{\text{inf}}$ and unbounded if $t > t_{\text{sup}}$. Along the initial line let $r_0(x) = r(x, 0)$ and $s_0(x) = s(x, 0)$. If the invariant initial values, $r_0(x)$ and $s_0(x)$, differ only slightly from the constant values $\hat{r}_0$ and $\hat{s}_0$, then $t_{\text{sup}} - t_{\text{inf}}$ is small and the bounds provide a good estimate† of t_C. The quantity t_{inf} is the least positive of the two quantities

$$\frac{-1}{\max\limits_{r,s}\left[\left(\frac{\partial \lambda^{(1)}}{\partial r}\right) \exp\{g_1(\hat{r}_0, \hat{s}_0) - g_1(\hat{r}_0, s)\}\right] \max\left(\frac{\partial r}{\partial x}\right)_{t=0}}$$

and

$$\frac{-1}{\max\limits_{r,s}\left[\left(\frac{\partial \lambda^{(2)}}{\partial s}\right) \exp\{g_2(\hat{r}_0, \hat{s}_0) - g_2(r, \hat{s}_0)\}\right] \max\left(\frac{\partial s}{\partial x}\right)_{t=0}}$$

(5-105)

† See Jeffrey [115] for the general case.

and t_{sup} is the least positive of the two quantities

$$\frac{-1}{\min_{r,s}\left[\left(\dfrac{\partial \lambda^{(1)}}{\partial r}\right)\exp\{g_1(\hat{r}_0,\hat{s}_0)-g_1(\hat{r}_0,s)\}\right]\max\left(\dfrac{\partial r}{\partial x}\right)_{t=0}}$$

(5-106)

and

$$\frac{-1}{\min_{r,s}\left[\left(\dfrac{\partial \lambda^{(2)}}{\partial s}\right)\exp\{g_2(\hat{r}_0,\hat{s}_0)-g_2(r,\hat{s}_0)\}\right]\max\left(\dfrac{\partial s}{\partial x}\right)_{t=0}}.$$

Here

$$g_1(r,s) = \int \frac{1}{\lambda^{(1)}-\lambda^{(2)}}\frac{\partial \lambda^{(1)}}{\partial s}\,ds$$

(5-107)

$$g_2(r,s) = \int \frac{1}{\lambda^{(2)}-\lambda^{(1)}}\frac{\partial \lambda^{(2)}}{\partial r}\,dr$$

are auxiliary functions. If they are represented approximately by the first two terms of their Taylor series—that is,

$$g_1(\hat{r}_0,s) = g_1(\hat{r}_0,\hat{s}_0) + (s-\hat{s}_0)\frac{\partial g_1}{\partial s}\Big|_0 + O[(s-\hat{s}_0)^2],$$

then, from Eqns (5-107),

$$g_1(\hat{r}_0,\hat{s}_0) - g_1(\hat{r}_0,s) = \frac{\hat{s}_0-s}{\lambda_0^{(1)}-\lambda_0^{(2)}}\left(\frac{\partial \lambda^{(1)}}{\partial s}\right)_0 + O[(s-\hat{s}_0)^2] \quad (5\text{-}108)$$

where 0 refers to initial values. Insertion of this and the corresponding result for g_2 simplifies Eqns (5-105) and (5-106) in the case that r and s differ only slightly from constant values $\hat{r}_0$ and $\hat{s}_0$. From this it follows (Problem 5-35) that the solution will only break down due to $C^{(1)}$ characteristics when $\max\,(\partial r/\partial x)_{t=0} > 0$ and to $C^{(2)}$ characteristics when $\max\,(\partial s/\partial x)_{t=0} > 0$.

Example: As an example, we consider the mixed initial boundary value problem (see Ludford [116]) for gas motion in a closed tube ($x = 0$ to $x = l$). Thus,

$$\mathbf{U}_t + A\mathbf{U}_x = 0$$

in which

$$\mathbf{U} = \begin{bmatrix} \rho \\ u \end{bmatrix} \qquad A = \begin{bmatrix} u & \rho \\ c^2/\rho & u \end{bmatrix}$$

where ρ, u, $c^2 = \partial p/\partial \rho$ are density, velocity, and the square of the sound speed, respectively. We shall assume the gas is polytropic—that is, $p = A\rho^\gamma$ where A and γ are constants.

From Section 4-5 we have the following results:

(1) $\lambda^{(1)} = u + c, \qquad \lambda^{(2)} = u - c$

$$l^{(1)} = \left[\frac{c}{\rho}, 1\right], \qquad l^{(2)} = \left[\frac{c}{\rho}, -1\right]$$

(2)   along $C^{(1)}$ characteristics we have the Riemann invariant,

$$u + \frac{2c}{\gamma - 1} = -r$$

along $C^{(2)}$ characteristics we have the Riemann invariant,

$$u - \frac{2c}{\gamma - 1} = -s$$

where the minus signs are introduced to make $\partial\lambda^{(1)}/\partial r$ and $\partial\lambda^{(2)}/\partial s$ negative, as required.

In Eqns (5-105) and (5-106) we need $\partial\lambda^{(1)}/\partial r$ and $\partial\lambda^{(2)}/\partial s$. From (1), $\lambda^{(1)} = u + c$, so that

(3) $\dfrac{\partial\lambda^{(1)}}{\partial r} = \dfrac{\partial u}{\partial r} + \dfrac{dc}{d\rho}\dfrac{\partial\rho}{\partial r}.$

Since the gas is polytropic, $c^2 = A\gamma\rho^{\gamma-1}$ and $dc/d\rho = (c/2\rho)(\gamma - 1)$. Adding and subtracting the invariants, we have

(4) $r + s = -2u$

(5) $r - s = \dfrac{-4c}{\gamma - 1}.$

Consequently,

(6) $\dfrac{\partial u}{\partial r} = -\dfrac{1}{2}$

and

(7) $1 = \left(\dfrac{-4}{\gamma - 1}\right)\dfrac{dc}{d\rho}\dfrac{\partial\rho}{\partial r}.$

Combining Eqns (3), (6), and (7), we have

(8) $\dfrac{\partial\lambda^{(1)}}{\partial r} = -\left(\dfrac{\gamma + 1}{4}\right)$

and by similar reasoning (Problem 5-36), we find

(9) $\dfrac{\partial\lambda^{(1)}}{\partial s} = \dfrac{\gamma - 3}{4}.$

Both are independent of r and s!

To apply the estimates [Eqns (5-105) and (5-106)] or the simplified forms of Problem 5-35, we must convert the initial boundary value problem to a *pure initial value problem.* The initial values of u and ρ, specified on $0 < x < l$, determine the initial values

$$r_0(x) = r(x, 0) = -u(x, 0) - \frac{2c(x, 0)}{\gamma - 1}$$

and

$$s_0(x) = s(x, 0) = -u(x, 0) + \frac{2c(x, 0)}{\gamma - 1}$$

on $0 < x < l$. The boundary conditions are $u(0, t) = u(l, t) = 0$ for all $t > 0$. Thus, by Eqn (4)

$$r_0(0) + s_0(0) = r_0(l) + s_0(l) = 0.$$

We then extend $r_0(x) + s_0(x)$ to the interval $-l \le x \le l$ and thence to the entire initial line, $-\infty < x < \infty$, by defining it to be an *even* function in $-l \le x \le l$ which is periodic of period $2l$. $r_0(x) - s_0(x)$ is similarly extended as an *odd* function in $-l \le x \le l$ which is periodic of period $2l$. Since the extension converts the problem to a pure initial value problem, the boundary conditions are disregarded.

When $\max (\partial r/\partial x)_{t=0}$ and $\max (\partial s/\partial x)_{t=0}$ are both positive, t_{inf} (from Problem 5-35) is the lesser of the two numbers

(10)

$$\frac{4}{(\gamma + 1) \max_s \left[\exp \left\{ \frac{(\gamma - 3)(\hat{s}_0 - s)}{8c_0} \right\} \right] \max \left(\dfrac{\partial r_0}{\partial x} \right)}$$

$$\frac{4}{(\gamma + 1) \max_r \left[\exp \left\{ \frac{(3 - \gamma)(\hat{r}_0 - r)}{8c_0} \right\} \right] \max \left(\dfrac{\partial s_0}{\partial x} \right)}$$

Similarly t_{sup} is the lesser of the two numbers

(11)

$$\frac{4}{(\gamma + 1) \min_s \left[\exp \left\{ \frac{(\gamma - 3)(\hat{s}_0 - s)}{8c_0} \right\} \right] \max \left(\dfrac{\partial r_0}{\partial x} \right)}$$

$$\frac{4}{(\gamma + 1) \min_r \left[\exp \left\{ \frac{(3 - \gamma)(\hat{r}_0 - r)}{8c_0} \right\} \right] \max \left(\dfrac{\partial s_0}{\partial x} \right)}$$

(Problem 5-37).

If r and s of Eqns (10) and (11) are replaced by their constant values r_0 and s_0, then t_{inf} and t_{sup} coincide and we obtain

(12) $\quad t_c = \dfrac{4}{(\gamma + 1)\beta}$

where

$$\beta = \max \left\{ \max \left(\frac{\partial r_0}{\partial x} \right), \max \left(\frac{\partial s_0}{\partial x} \right) \right\}$$

Ames [157] has given an alternative method for calculating the time to breakdown. When applicable his method is simpler.

PROBLEMS

5-32 Verify that Eqns (5-97) and (5-98) are not coupled between the set of space–time mesh points having even values and the set having odd values of $i + j + n$.

5-33 In Eqns (5-97) and (5-98), let $F(U) = AU$ and $G(U) = BU$ where A and B are constant matrices. Show that the resulting equation is similar to that of Problem 4-52 with $U_{i,j,n+2}$ in place of $U_{i,j+1}$.

5-34 (a) In the equation of Problem 5-33, how many spatial points are there? (b) Draw the space portion of the molecule and indicate how many times $\partial/\partial x$ is evaluated and where? Carry out the same discussion for $\partial/\partial y$ and $\partial^2/\partial y^2$.

Ans: (a) 9; (b) $\partial/\partial x$ is evaluated four times and averaged

5-35 Insert Eqn (5-108) and its analog for g_2 into Eqns (5-105) and (5-106), then show that the conclusions following Eqn (5-108) are correct.

Ans: t_{inf} is the least positive of the two quantities

$$\frac{-1}{\max\limits_{r,s}\left[\left(\frac{\partial\lambda^{(1)}}{\partial r}\right)\exp\left\{\left(\frac{\hat{s}_0 - s}{\lambda_0^{(1)} - \lambda_0^{(2)}}\right)\left(\frac{\partial\lambda^{(1)}}{\partial s}\right)_0\right\}\right]\max\left(\frac{\partial r}{\partial x}\right)_{t=0}}$$

$$\frac{-1}{\max\limits_{r,s}\left[\left(\frac{\partial\lambda^{(2)}}{\partial s}\right)\exp\left\{\left(\frac{\hat{r}_0 - r}{\lambda_0^{(2)} - \lambda_0^{(1)}}\right)\left(\frac{\partial\lambda^{(2)}}{\partial r}\right)_0\right\}\right]\max\left(\frac{\partial s}{\partial x}\right)_{t=0}} \tag{5-109}$$

5-36 Find $\partial\lambda^{(1)}/\partial s$, $\partial\lambda^{(2)}/\partial r$, and $\partial\lambda^{(2)}/\partial s$ for the gas flow example.

5-37 Complete the evaluation of Eqns (10) and (11) of the gas flow example.

5-38 Apply the Jeffrey–Lax theory to shallow water wave theory $U_t + AU_x = 0$, where

$$U = \begin{bmatrix} u \\ c \end{bmatrix} \qquad A = \begin{bmatrix} u & 2c \\ \frac{1}{2}c & u \end{bmatrix}$$

Here u is fluid velocity, $c = \sqrt{(gy(x))}$ is the wave propagation speed, $y(x)$ is the water depth, and g is the gravitational constant.

Ans: For $r \to \hat{r}_0$, $s \to \hat{s}_0$, and β as in the text example, $t_c = 4/3\beta$

5-7 Mixed systems

Problems which incorporate ideal fluid motion and some other transport process, such as heat transfer, have mathematical models which are coupled equations of mixed parabolic–hyperbolic type. One of the consequences of mixed parabolic–hyperbolic equations is that there are *two time constants*. In the case of practical magneto-gas dynamics (Jeffrey and Taniuti [107]) problems the time constant for the hyperbolic equations (that is, the gas-dynamic equations of Section 4-13) is considerably smaller than that for the parabolic equations. This occurs since hydromagnetic shocks and other phenomena are relatively quick as compared with diffusion (the energy equation is parabolic) which is relatively slow. We cannot, of course, neglect the diffusion because it counteracts the confinement of the plasma. What we must be careful about is *not* to let the diffusion be swamped by errors of the finite difference process.

There are many physical systems in which parabolic equations are coupled to hyperbolic equations so that two (or more) transport phenomena must be calculated simultaneously. In addition to the already mentioned magneto-gas dynamics, such phenomena as exploding wires, strong shocks, and the initial phases of blasts must include coupled gas flow and (radiation) diffusion.

The coupling effect occurs for infinitesimal or acoustic vibrations and this is the problem to which we first turn (see Morimoto [117]). Let p_0, V_0, and ϵ_0 be ambient values of pressure $p + p_0$, specific volume $V + V_0$, and specific internal energy $\epsilon + \epsilon_0$, where $p \ll p_0$, $V \ll V_0$, and $\epsilon \ll \epsilon_0$. Let u be the velocity, $c = \sqrt{(p_0 V_0)}$ the isothermal sound speed, and $\epsilon_0 + \epsilon = (p_0 + p)$ $(V_0 + V)/(\gamma - 1)$ the equation of state. In terms of u, $w = cV/V_0$, and $e = \epsilon/c$, the equations, to first order in the small quantities u, w, and e, are

$$u_t = c[w - (\gamma - 1)e]_x$$

$$w_t = cu_x \tag{5-110}$$

$$e_t = \sigma e_{xx} - cu_x$$

where σ is the ratio of thermal conductivity to specific heat at constant volume. An *explicit* system for Eqns (5-110) with $u_{i,j} = u(i\,\Delta x, j\,\Delta t)$ is

$$\frac{u_{i,j+1} - u_{i,j}}{\Delta t} = c\,\frac{w_{i+1/2,j} - w_{i-1/2,j} - (\gamma - 1)(e_{i+1/2,j} - e_{i-1/2,j})}{\Delta x}$$

$$\frac{w_{i+1/2,j+1} - w_{i+1/2,j}}{\Delta t} = c\,\frac{u_{i+1,j+1} - u_{i,j+1}}{\Delta x}$$

$$\frac{e_{i+1/2,j+1} - e_{i+1/2,j}}{\Delta t} = \sigma\,\frac{e_{i+3/2,j} - 2e_{i+1/2,j} + e_{i-1/2,j}}{(\Delta x)^2} \tag{5-111}$$

$$- c\,\frac{u_{i+1,j+1} - u_{i,j+1}}{\Delta x}.$$

The advanced values of the velocity, u, have been used in the second and third equations (compare Eqns (4-127) without diffusion). Equations (5-111) are effectively explicit since the first equation can be solved first for $u_{i,j+1}$ and $u_{i+1,j+1}$. These values are then employed in the other two.

If the sound waves and diffusion of heat were uncoupled, the stability conditions would be (Problem 5-39), respectively,

$$\sqrt{(\gamma)}c\,\Delta t/\Delta x < 1$$

$$\sigma\,\Delta t/(\Delta x)^2 < \tfrac{1}{2}. \tag{5-112}$$

In an actual calculation, one should choose Δx and Δt so that both conditions of Eqns (5-112) are satisfied. They are, of course, necessary.† This system has been found satisfactory but restrictive in the time step size.

To avoid the small time step imposed by Eqns (5-112) a simulation often applied is the use of an explicit method for the gas flow and an implicit method for the diffusion. In this way one keeps the first two of Eqns (5-111) but replaces the third with the implicit form

$$\frac{e_{i+1/2,j+1} - e_{i+1/2,j}}{\Delta t} = \sigma \frac{e_{i+3/2,j+1} - 2e_{i+1/2,j+1} + e_{i-1/2,j+1}}{(\Delta x)^2}$$

$$- c \frac{u_{i+1,j+1} - u_{i,j+1}}{\Delta x}. \quad (5\text{-}113)$$

For this system Morimoto [117] proved that Richtmyer's conjectured stability criterion, $c\,\Delta t/\Delta x < 1$, was true. As $\sigma \to 0$ the condition is the more stringent inequality $\sqrt{(\gamma)}c\,\Delta t/\Delta x < 1$ (see Richtmyer and Morton [17]).

To take full advantage of the attractive features of implicit methods, Rouse [118] uses implicit methods for both gas flow and diffusion.

Consider the Lagrangian form of the gas-dynamic equations in slab geometry (see Section 4-13)

$$V_t = Vu_r, \qquad u_t = -Vp_r, \qquad e_t = -pV_t + Q_t \qquad (5\text{-}114)$$

where the quantity Q represents a specific energy input. Let T be temperature and let the equations of state be

$$p = p(T, V), \qquad e = e(T, V), \qquad V = \rho^{-1}. \qquad (5\text{-}115)$$

Using these equations of state we rewrite $u_t = -Vp_r$ as

$$\frac{\partial u}{\partial t} = -V \frac{\partial p}{\partial T}\bigg|_V \frac{\partial T}{\partial r} - V \frac{\partial p}{\partial V}\bigg|_T \frac{\partial V}{\partial r}. \qquad (5\text{-}116)$$

In addition, the third of Eqns (5-114) becomes

$$\frac{\partial e}{\partial T}\bigg|_V \frac{\partial T}{\partial t} + \frac{\partial e}{\partial V}\bigg|_T \frac{\partial V}{\partial t} = -p \frac{\partial V}{\partial t} + \frac{\partial Q}{\partial t} \qquad (5\text{-}117)$$

which is transformed, by introducing $V_t = Vu_r$, into

$$\frac{\partial e}{\partial T}\bigg|_V \frac{\partial T}{\partial t} = -\left\{ p + \frac{\partial e}{\partial V}\bigg|_T \right\} V \frac{\partial u}{\partial r} + \frac{\partial Q}{\partial t}. \qquad (5\text{-}118)$$

With radiation diffusion *and* thermal conduction included, we have

$$\frac{\partial Q}{\partial t} = V \frac{\partial}{\partial r} \left\{ \lambda \frac{\partial}{\partial r} \left(\frac{a}{3} T^4 \right) + k \frac{\partial T}{\partial r} \right\} \qquad (5\text{-}119)$$

---

† As Δt and $\Delta x \to 0$, the second of conditions (5-112) always implies the first and it is generally conjectured that it is the stability condition (Problem 5-40).

where λ is the mean free path and $a = \frac{4}{3}\sigma$, σ = Stefan–Boltzmann constant. Setting this result into Eqn (5-118), one obtains

$$\frac{\partial e}{\partial T}\bigg|_V \frac{\partial T}{\partial t} = -\left\{ p + \frac{\partial e}{\partial V}\bigg|_T \right\} V \frac{\partial u}{\partial r} + V \frac{\partial}{\partial r} \left\{ \left[\frac{4a\lambda}{3} T^3 + k\right] \frac{\partial T}{\partial r} \right\} \quad (5\text{-}120)$$

With the state equations [Eqns (5-115)], Eqns (5-116) and (5-120) are prepared for numerical solution, except for the term containing V_r [Eqn (5-116)], which is converted into a function of u by the following changes. For the moment we ignore this and the implicit difference equation [Eqn (5-116), with T = constant] to obtain

$$u_{i,j+1} - u_{i,j} = -\frac{\Delta t}{\rho\,\Delta r}\frac{\partial p}{\partial V}\bigg|_{i,j+1/2} [\tfrac{1}{2}(V_{i+1/2,j} - V_{i-1/2,j})$$

$$+ \tfrac{1}{2}(V_{i+1/2,j+1} - V_{i-1/2,j-1})] \quad (5\text{-}121)$$

where $\rho\,\Delta r$ is the mass in zones (assumed constant), and the notation

$$V_{i+1/2,j} = \tfrac{1}{2}[V_{i,j} + V_{i+1,j}]$$

$$V_{i,j+1/2} = \tfrac{1}{2}[V_{i,j} + V_{i,j+1}]$$

As in all implicit techniques let us assume a knowledge of the solution at the level $t = j\,\Delta t$. On the right-hand side of Eqn (5-121), $V_{(\ ),j+1}$ and $(p_V)_{(\ ),j+1}$ are unknown. The latter is obtained by iteration and the former by using a difference approximation for $V_t = Vu_r$ at $i + \frac{1}{2}$ and $i - \frac{1}{2}$,

$$V_{i+1/2,j+1} - V_{i+1/2,j} = \frac{\Delta t}{\rho\,\Delta r}(u_{i+1,j+1/2} - u_{i,j+1/2})$$

$$V_{i-1/2,j+1} - V_{i-1/2,j} = \frac{\Delta t}{\rho\,\Delta r}(u_{i,j+1/2} - u_{i-1,j+1/2}).$$

When the second expression is subtracted from the first, we see that

$$V_{i+1/2,j+1} - V_{i-1/2,j+1} = V_{i+1/2,j} - V_{i-1/2,j}$$

$$+ \frac{\Delta t}{\rho\,\Delta r}[u_{i+1,j+1/2} - 2u_{i,j+1/2} + u_{i-1,j+1/2}].$$

Upon substituting this into Eqn (5-121), we find the relation

$$u_{i,j+1} - u_{i,j} = -\frac{\Delta t}{\rho\,\Delta r}\frac{\partial p}{\partial V}\bigg|_{i,j+1/2} [V_{i+1/2,j} - V_{i-1/2,j}$$

$$+ \frac{t}{2\rho\,\Delta r}(u_{i+1,j+1/2} - 2u_{i,j+1/2} + u_{i-1,j+1/2})]$$

which is *independent of V in the time level $j + 1$*.

The calculation of Eqn (5-120), in implicit form, is carried out as in our discussion of parabolic systems. The computations are completed from the information $u_{i,j+1}$, $T_{i,j+1}$ obtained from the remaining Lagrange equations [Eqns (4-152)–(4-154)], $\rho_0\,dx = (1/V)\,dr$, $u = r_t$, and $p = p(e, V)$.

PROBLEMS

5-39 Establish the uncoupled stability conditions, Eqns (5-112), for the finite difference simulation, Eqns (5-110).

5-40 Verify the truth of the statement 'the second condition always implies the first in the limit as Δt and $\Delta x \to 0$.' [Eqn. (5-112)].

5-41 Describe a fully implicit algorithm, analogous to that of Rouse, for Eqns (5-110).

5-8 Higher-order equations in elasticity and vibrations

Many of the equations of science and engineering are of order higher than 2. In plane elasticity one meets the (fourth-order) biharmonic equation, $\nabla^4 u = 0$; in the vibration of a thin beam, $u_{xxxx} + u_{tt} = 0$; in shell analysis, equations of eighth order; in dynamics of three-dimensional traveling strings, sixth order. As a consequence of this technological interest and application, a trickle of research papers and books is beginning to appear. One of the recent volumes, Soare [119], is devoted entirely to the application of finite differences to shell analysis. Finite difference approximations and efficient computational algorithms have been the goal of a number of researchers. In this section we shall briefly discuss problems of plane elasticity and the vibrations of thin beams.

(a) Elasticity

Southwell and his co-workers [120–122], before the digital computer era, utilized finite difference approximations for plane elasticity and other problems and applied Southwell's 'relaxation' method to obtain numerical solutions of those problems. This *noncyclic* or *free-steering* method, using human insight as it did, was very difficult to adapt efficiently for use in high-speed computers. The cyclic methods of Chapter 3—the variants of SOR and ADI—have been applied to the numerical solution of plane elasticity problems with some success. Motivated by a paper of Heller [123], Varga [124] and Parter [125, 126] applied a *two-line variant* of SOR to the numerical solution of $\nabla^4 u = f$ having clamped boundary conditions ($u = \partial u/\partial n = 0$ on the boundary) for a rectangle with uniform mesh spacings. Conte and Dames [127, 128] applied the Douglas–Rachford variant of ADI (Chapter 3) to the same problem with simply supported boundary conditions (that is, the ends are free of bending moments and are free to move in the horizontal direction but not in the vertical direction). These Conte–Dames results are theoretically restricted to rectangular domains, as shown by Birkhoff and Varga [129]. Stiefel *et al.* [130] investigated the practical application of SOR, Gaussian elimination, and conjugate gradient techniques to a range of typical biharmonic problems. They observe that SOR, with properly chosen relaxation factors, is a very promising method. Griffin and Varga [131, 132], Callaghan *et al.* [133], and Griffin [134] have developed and applied an *efficient method using nonuniform meshes* that is suitable for general, irregularly shaped regions. The case of simply connected regions is discussed herein, while the problem in multi-connected regions is described in Griffin and Varga [131].

Let R be a plane simply connected region with boundary B. In terms of the Airy stress function (see, for example, Sokolnikoff [135]), $\phi(x, y)$, defined in terms of the normal stresses $\sigma^x = \phi_{yy} + V$, $\sigma^y = \phi_{xx} + V$, and shear stress $\tau^{xy} = -\phi_{xy}$, the equilibrium equations of plane infinitesimal elasticity are satisfied identically. The compatibility of strain requires that

$$\nabla^4\phi = -\beta\nabla^2 V - \gamma\nabla^2 T = q(x, y) \tag{5-122}$$

where T is the temperature distribution, (x, y) is in R, $\beta = 1 - \nu$, $\gamma = \alpha E$ for plane *stress*, or $\beta = (1 - 2\nu)/(1 - \nu)$, $\gamma = \alpha E/(1 - \nu)$ for *plane strain*, where α is the coefficient of thermal expansion, ν is Poisson's ratio, and E is the modulus of elasticity. Typical boundary conditions are those which arise when forces are specified on B: $\phi_x = g_1(x, y)$, $\phi_y(x, y) = g_2(x, y)$, (x, y) on B, or $\phi = g_3(x, y)$, $\phi_n(x, y)\dagger = g_4(x, y)$, (x, y) on B; if there is a line of physical symmetry, the boundary conditions may be $\phi_n(x, y) = 0$; $(\nabla^2\phi)_n = 0$, (x, y) on B.

For the purpose of illustrating the development of finite difference aproximations for Eqn (5-122), on a nonuniform grid, we shall refer to Fig. 5-7.

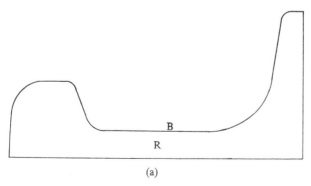

(a)

Fig. 5-7(a) Typical structural member

† n refers to the outward-pointing normal.

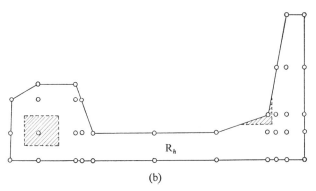

(b)

Fig. 5-7(b) Polygonal boundary approximation and nonuniform computational mesh

First, the boundary B is approximated by a polygonal boundary B_h. Second, a rectangular mesh of horizontal and vertical mesh lines, in general *not* uniformly spaced, is placed on R in such a manner that the intersections of the mesh lines coincide with the intersections of the boundary segments of B_h. The totality of mesh points, a finite set S, are shown as small circles in Fig. 5-7(b). Corresponding to each *mesh point*, P_i, is a *mesh region*, r_i, defined so that its boundaries fall halfway between the mesh lines (cross-hatched regions in Fig. 5-7(b)).

We have previously used Taylor series expansions in developing finite difference approximations. An alternative technique is based upon *integration*. It is closely related to the derivation of difference approximations using the variational formulation (see, for example, Forsythe and Wasow [45, p. 182] or Babuska *et al.* [136, p. 262]). The integration technique is very convenient for approximating differential equations for *irregularly shaped regions using a nonuniform mesh*. Equation (5-122) is integrated over each mesh region r_i, giving

$$\iint_{r_i} \nabla^4 \phi \, dx \, dy = - \iint_{r_i} (\beta \nabla^2 V + \gamma \nabla^2 T) \, dx \, dy \qquad (5\text{-}123)$$

which is equivalent to Eqn (5-122) if such integrations are taken over all possible mesh regions r_i of R_h. By Green's theorem, each area integral can be replaced by a line integral, so that

$$\int_{\gamma_i} \frac{\partial}{\partial n} (\nabla^2 \phi) \, ds = - \int_{\gamma_i} \left(\beta \frac{\partial V}{\partial n} + \gamma \frac{\partial T}{\partial n} \right) ds \qquad (5\text{-}124)$$

where γ_i are the boundaries of each mesh region r_i.

We seek approximations to ϕ, say ϕ_i at the mesh points i of the set S $(i = 1\,2, \ldots, N)$. Some of the values ϕ_i may be known directly from boundary conditions. At interim points, or boundary points where derivatives are specified, the values of ϕ_i are related by a set of linear difference equations deduced from Eqn (5-124) by approximating portions of the line integral in the following manner (Fig. 5-8). The line integral from a to c is approximated by

$$\int_a^c \frac{\partial}{\partial n} (\nabla^2 \phi) \, ds \approx \left(\frac{\partial}{\partial n} \nabla^2 \phi \right)_b \cdot l_{ac},$$

where b is on the mesh line and the normal derivative at b is approximated by central differences—that is,

$$\frac{\partial}{\partial n} (\nabla^2 \phi)_b \approx \frac{\delta^2 \phi_1 - \delta^2 \phi_0}{h_{01}}.$$

At a typical point 0 the difference approximation of Eqn (5-124) becomes

$$\sum_{i=1}^{4} (\delta^2 \phi_i - \delta^2 \phi_0) \frac{l_{0i}}{h_{0i}} = - \sum_{i=1}^{4} [\beta(V_i - V_0) + \gamma(T_i - T_0)] \frac{l_{0i}}{h_{0i}} \quad (5\text{-}125)$$

where $\delta^2 \phi_i$ is a difference approximation of $\nabla^2 \phi$ at i (Chapter 3), h_{0i} is the mesh distance between the mesh lines passing through 0 and i, and l_{0i} is the length of the side of the mesh region between points 0 and i (i.e. $l_{01} = l_{ac}$).

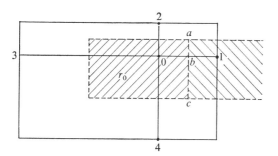

Fig. 5-8 Mesh point region in a nonuniform mesh

Lastly, we specify the differences $\delta^2 \phi_i$. At any point 0 we have

$$(\nabla^2 \phi)_0 \approx \frac{1}{A_0} \int \int_{r_0} \nabla^2 \phi \, dx \, dy = \frac{1}{A_0} \int_{\gamma_0} \frac{\partial \phi}{\partial n} \, ds \quad (5\text{-}126)$$

where A_0 is the area of r_0. Thus, we define

$$\delta^2 \phi_0 = \frac{1}{A_0} \sum_{i=1}^{4} (\phi_i - \phi_0) \frac{l_{0i}}{h_{0i}} \quad (5\text{-}127)$$

and, clearly, $A_0 = l_{01} \cdot l_{02} = l_{03} \cdot l_{04}$.

These approximations, based upon mesh lengths (and, redundantly, the areas), can be readily calculated in a computer program. We leave it as an exercise (Problem 5-42) to show that for the case of uniform spacings the approximations, Eqn (5-127), when substituted into Eqn (5-125), reduce to the 13-point biharmonic molecule

$$
\begin{array}{ccccc}
 & & 1 & & \\
 & 2 & -8 & 2 & \\
1 & -8 & 20 & -8 & 1 \\
 & 2 & -8 & 2 & \\
 & & 1 & &
\end{array}
\tag{5-128}
$$

whose local truncation error is $O(h^2)$ on a square mesh.

At boundary points the derivation of difference equations proceeds in a manner similar to that for interior points. When 0 is on a line of symmetry (say the line $2 \to 0 \to 4$ in Fig. 5-8), the line integrals are taken around the half mesh region to the left of the line of symmetry. Since $\partial/\partial n\,(\nabla^2\phi) = 0$ on the line of symmetry, $\int \partial/\partial n\,(\nabla^2\phi)\,ds = 0$ on the side of the mesh region coinciding with that line. Thus, the difference approximation for Eqn (5-124) becomes

$$
\sum_{i=2}^{4} (\delta^2\phi_i - \delta^2\phi_0)\frac{l_{0i}}{h_{0i}} = -\sum_{i=2}^{4} [\beta(V_i - V_0) + \gamma(T_i - T_0)]\frac{l_{0i}}{h_{0i}}. \tag{5-129}
$$

Of course, l_{02}, l_{04} refer to sides of the half-mesh region, while l_{03} is as before. The corresponding modifications in Eqn (5-127) are left for Problem 5-43. This method applies equally well at points on B at which the boundary conditions $\phi = g_3$, $\partial\phi/\partial n = g_4$ are specified (see Griffin and Varga [132] for details).

Some attention is being paid to higher-order elliptic equations. Zlamal [137, 138] has concerned himself with error estimates for finite difference approximations to those problems. Second-order error methods are developed for *general regions*. Bramble [139] has also developed a second-order method for biharmonic equations, and Smith [140] discusses the so called *coupled equation approach*—to solve $\nabla^4\phi = q(x, y)$ where ϕ and $\partial\phi/\partial n$ are known on the boundary set $v = c\nabla^2\phi$, $\nabla^2 v = cq$ with the same boundary conditions. Two discrete Poisson equations must then be solved, usually by an "inner-outer" iteration method. At the time of writing, the technique has been applied only to rectangular regions.

PROBLEMS

5-42 When the mesh spacing is uniform in both directions, show that the substitution of Eqn (5-127) into Eqn (5-125) leads to the 13-point molecule for the biharmonic.

5-43 Carry out the line of symmetry boundary modifications corresponding to Eqn (5-129).

5-44 Apply the integral method to obtain the finite difference approximation on the boundary when ϕ and $\partial\phi/\partial n$ are specified there.

5-45 Using the method described, reduce the 'wrench' domain of Fig. 5-9 to a polygonal domain and set up a mesh. Identify a typical mesh region.

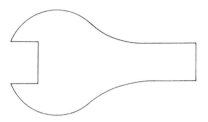

Fig. 5-9 'Wrench'-shaped domain for Problem 5-45

(b) Vibrations of a thin beam

If a thin beam of length L is supported or clamped at the ends and set into vibration, the linearized equation of motion is

$$\rho \frac{\partial^2 y}{\partial t^2} = - \frac{\partial^2}{\partial x^2}\left(EI \frac{\partial^2 y}{\partial x^2}\right). \tag{5-130}$$

Here $y = y(x, t)$ is the transverse displacement from an equilibrium position, ρ is mass per unit length, and EI is the flexural rigidity. The assumptions leading to linearity are $y \ll L$ and $\partial y/\partial x \ll 1$. In what follows we assume EI and ρ to be constant, although extension to the variable coefficient case is immediate (see Chapter 2 and Problem 5-46). When the beam is clamped at its ends, the boundary conditions are

$$y = 0, \quad y_x = 0 \quad \text{at } x = 0 \quad \text{and} \quad y = 0, \quad y_x = 0 \quad \text{at } x = L \tag{5-131}$$

Let us further assume that a static load $w(x) = w_0 \sin(\pi x/L)$ is applied to the beam so that the initial configuration is

$$y(x, 0) = \frac{w_0 L^4}{EI\pi^4}\left\{\sin \pi \frac{x}{L} - \pi \frac{x}{L}\left(1 - \frac{x}{L}\right)\right\} \tag{5-132}$$

and initial velocity is

$$y_t(x, 0) = 0. \tag{5-133}$$

A dimensionless formulation is obtained by setting

$$x' = \frac{x}{L}, \quad t' = \frac{t}{L^2\sqrt{(\rho/EI)}}, \quad \psi = \frac{y}{w_0 L^4/EI\pi^4}$$

whereupon Eqns (5-130) through (5-133) become (we have dropped the primes):

$$\psi_{xxxx} + \psi_{tt} = 0;$$

$$\psi = 0, \quad \psi_x = 0 \quad \text{at } x = 0; \qquad \psi = 0, \quad \psi_x = 0 \quad \text{at } x = 1\dagger$$

$$\psi = \sin \pi x - \pi x (1 - x) \quad \text{at } t = 0, \quad 0 < x < 1,$$

$$\psi_t = 0 \quad \text{at } t = 0, \quad 0 < x < 1.$$

(5-134)

One can develop a difference equation simulation either for $\psi_{xxxx} + \psi_{tt} = 0$ or for two equations of first order in time (thereby suggesting the parabolic nature of this problem) obtained by setting, say, $v = \psi_t$ and $w = \psi_{xx}$. Then

$$v_t = -w_{xx}$$

and

$$w_t = v_{xx}.$$

(5-135)

We describe an explicit difference approximation for the two second-order equations, Eqns (5-135), and leave the direct explicit simulation of $\psi_{xxxx} + \psi_{tt} = 0$ for Problem 5-48. If forward differences are used in both equations, the resulting system is unstable unless $\Delta t \to 0$ as $(\Delta x)^3$, which is therefore unsatisfactory (see Problem 5-49). On the other hand, if one forward and one backward time difference are employed, we find the system

$$\frac{v_{i,j+1} - v_{i,j}}{\Delta t} = -\frac{w_{i+1,j} - 2w_{i,j} + w_{i-1,j}}{(\Delta x)^2}$$

$$\frac{w_{i,j+1} - w_{i,j}}{\Delta t} = \frac{v_{i+1,j+1} - 2v_{i,j+1} + v_{i-1,j+1}}{(\Delta x)^2}$$

(5-136)

which has truncation error $O[(\Delta t)^2 + (\Delta x)^2]$ and has the sufficient condition, $\Delta t/(\Delta x)^2 < \frac{1}{2}$, for stability.

In fourth-order form one *implicit* approximation can be obtained by using, in $\psi_{xxxx} + \psi_{tt} = 0$, the average of approximations to ψ_{xxxx} in the $(j - 1)$ and $(j + 1)$ rows in place of the approximation at the jth row. Thus we find the unconditionally stable (Crandall [141]) simulation

$$\psi_{i,j+1} + \frac{1}{2}\left\{\frac{\Delta t}{(\Delta x)^2}\right\}^2 (\psi_{i+2,j+1} - 4\psi_{i+1,j+1} + 6\psi_{i,j+1} - 4\psi_{i-1,j+1} + \psi_{i-2,j+1})$$

$$= 2\psi_{i,j} - \psi_{i,j-1} - \frac{1}{2}\left\{\frac{\Delta t}{(\Delta x)^2}\right\}^2$$

$$\times (\psi_{i+2,j-1} - 4\psi_{i+1,j-1} + 6\psi_{i,j-1} - 4\psi_{i-1,j-1} + \psi_{i-2,j-1}). \quad (5\text{-}137)$$

A single application of Eqn (5-137) generates one equation for five unknown values in the $(j + 1)$ row in terms of known values in the $(j - 1)$ and jth rows. The local truncation error is the same as that for the explicit scheme.

---

† If the condition is *simple support*, lateral motion is allowed, but no turning moments are exerted. That is, at a simple support, $\psi = 0, \psi_{xx} = 0$.

The nonlinear problem concerning the large amplitude oscillations of thin elastic beams has been explored in depth by Woodall [142] using finite difference methods. The numerical results agreed with Galerkin's method in the proper parameter and variable range.

PROBLEMS

5-46 Extend the explicit finite difference approximation of Eqn (5-130) to the case where ρ and EI are functions of x.

5-47 Extend the implicit finite difference approximation of Eqn (5-130) to the case where ρ and EI are functions of x.

5-48 Develop an explicit finite difference approximation directly from $\psi_{xxxx} + \psi_{tt} = 0$, using differences centered at $x = i\,\Delta x$, $t = j\,\Delta t$ for both operators. Show that a sufficient result for stability is $\Delta t/(\Delta x)^2 < \frac{1}{2}$ (Collatz [143]). Draw the computational molecule. Compare the finite difference form obtained here with Eqns (5-136) and show that they are equivalent.

Ans: $\psi_{i,j+1} = 2\psi_{i,j} - \psi_{i,j-1} - \left[\dfrac{\Delta t}{(\Delta x)^2}\right]^2$

$$\times\ (\psi_{i+2,j} - 4\psi_{i+1,j} + 6\psi_{i,j} - 4\psi_{i-1,j} + \psi_{i-2,j})$$

5-49 Explore the consequences of writing Eqns (5-136) in the following ways:

(a) In the second equation approximate the right-hand side at the level j;
(b) In the first equation approximate the right-hand side at the level $j + 1$.

5-50 Draw the computational molecule for Eqn (5-137) and employ the implicit technique to solve the problem of Eqns (5-134). Because of symmetry only one half of the beam need be considered. Let $\Delta x = \frac{1}{8}$ and $r = \frac{1}{2}$. Write out the linear equations which must be solved simultaneously at each level, $j + 1$.

5-51 Replace the boundary condition $\partial\psi/\partial x$ at $x = 0$ and $x = 1$ of Eqns (5-134) with $\psi_{xx} = 0$ at $x = 0$ and $x = 1$. Show how the explicit scheme can be used to solve this problem.

5-9 Fluid mechanics: the Navier–Stokes equations

One of the most challenging theoretical problems is that of understanding the mathematical properties of, and extracting solutions for, the flow of viscous fluids. It is only through the essential nonlinearities of the equations that many of the most interesting observed properties of fluid mechanics can be explained. Of course, those necessary nonlinearities render the equations analytically difficult, except in special cases. A number of the known exact solutions are presented in the *Handbuch der Physik* article of Berker [158]. Here we present an exact solution obtained by Wessel [159] which was suggested by an expression in a paper by Townsend [160]. This solution is particularly useful for checking numerical methods.

The Navier–Stokes equations and the energy equation in the Boussinesq approximation (incompressibility and density variation neglected except for buoyancy effects) are

$$u_t = -(u^2)_x - (uv)_y - \sigma_x + \nu \nabla^2 u \tag{5-138}$$

$$v_t = -(uv)_x - (v^2)_y + \beta(T - T_0) - \phi_y + \nu \nabla^2 v \tag{5-139}$$

$$T_t = -(uT)_x - (vT)_y + \sigma \nabla^2 T \tag{5-140}$$

$$u_x + v_y = 0. \tag{5-141}$$

Here u and v are fluid velocity components in the x and y directions, respectively, T is temperature, ν is the coefficient of viscosity, and σ is the coefficient of heat conductivity. The product of the coefficient of volume expansion and gravitational acceleration g is called β, and $\phi = p/\rho_0 + gy$ where p is the total pressure and ρ_0 is average fluid density. The exact solution of (5-138)–(5-141) is

$$u = u_0 \exp(-\lambda t)[\sin(ax - \omega t)\sin by]$$

$$v = v_0 \exp(-\lambda t)[\cos(ax - \omega t)\cos by]$$

$$T = T_0 + cy + \tau_0 \exp(-\lambda t)\sin(ax - \omega t)\cos by$$

$$\phi = \tfrac{1}{2}\beta cy^2 + (u_0\omega/a)\exp(-\lambda t)\sin(ax - \omega t)\sin by$$

$$\quad + \tfrac{1}{4}u_0^2 \exp(-2\lambda t)\cos(2ax - 2\omega t)$$

$$\quad - \tfrac{1}{4}v_0^2 \exp(-2\lambda t)\cos 2by$$

where

$$\lambda = \nu(a^2 + b^2), \qquad \omega = [a^2\beta c/(a^2 + b^2)]^{1/2}$$

$$v_0 = au_0/b, \qquad \tau_0 = v_0c/\omega$$

and u_0, a, b, and c are arbitrary constants. In this solution it is assumed that $\nu = \sigma = $ constant in the equations.

The complexities of Eqns (5-138)–(5-141) make it easy to understand the accelerating usage of finite difference and finite element (see Chapter 6) methods in solving problems of fluid mechanics. While there has been continuing research, analytic methods and error analysis has not kept pace with the large number of problems implemented on computers during the period 1951–1976. An updated and supplemented literature survey on numerical methods for partial differential equations has been compiled by Giese [161]. Of the more than 8000 separate entries over 2000 concern

various computational aspects of fluid mechanics! Many doctoral disserta-
tions have been written in computational fluid dynamics—perhaps a dis-
proportionate number. One reference work, that of Roache [162], addresses
not only the algorithms but the 'messy' problems associated with obtaining
numerical results. We shall have occasion to discuss some of his work later.

Early computations of steady viscous flows are very limited in number.
The first use of finite differences was begun by Thom in 1928. Thom's [144]
paper of 1933 studied the wake associated with steady laminar flow past a
circular cylinder. For Reynolds numbers 10 and 20, solutions were obtained
with a desk calculator! More recently, Payne [145, 146] developed a numerical algorithm for nonsteady flows and calculated 'starting' flows in the wake
behind a circular cylinder. Building upon the preceding results Fromm [147]
used a formulation of the equations usually called the *stream function-
vorticity method*. Various examples of its application are available in Fromm
[148] and in Fromm and Harlow [149]. A conformal mapping modification
has been used by Keller and Takami [150] in numerical studies of steady
viscous flow about cylinders. Greenspan *et al.* [151] examine some questions
raised by the method and give additional references to the early ideas (to
1964). While some very difficult problems have been computed by this
method, the technique is essentially restricted to transient and steady two
space variable problems.

The second general category of simulations is based upon *primitive variables*.
This formulation is receiving increased attention because it is applicable to
transient problems in three dimensions. One of these, essentially due to
Chorin [152, 153], uses a decomposition (splitting) of Samarskii [154] (see
also Yanenko [163]) into essentially one-dimensional problems. Padmanabhan *et al.* [155] report successful computation for transient two-dimensional
wake collapse problems. Pujol [164] used Chorin's method and compared it
to a stream function–vorticity method for a two-dimensional viscous problem. All things considered, Pujol (see also Giaquinta [167]) found the stream
function–vorticity method preferable. Velocity and pressure boundary
conditions are usually simpler to apply for this method.

The three-dimensional counterpart of the stream function–vorticity formulation is the *vector-potential formulation*. The importance of this form to
transient three-dimensional problems is discussed by Aziz and Hellums [156]
who report a number of successful applications to problems of convection.

A detailed survey, presenting these and additional ideas, is given by
Ames [165, 166].

(a) Stream function–vorticity method

This first group of numerical methods has the common feature that the stream
function and vorticity are used as the dependent variables. In Cartesian
coordinates the time-dependent flow of a two-dimensional viscous incom-

pressible fluid is usually modeled by the dimensionless momentum equations (Navier–Stokes)

$$u_t + uu_x + vu_y = -p_x + \frac{1}{R}(u_{xx} + u_{yy}) \tag{5-142}$$

$$v_t + uv_x + vv_y = -p_y + \frac{1}{R}(v_{xx} + v_{yy}) \tag{5-143}$$

and continuity

$$u_x + v_y = 0. \tag{5-144}$$

The dependent variables are the velocity components u and v in the x and y directions, respectively, and the pressure p. $R = V_0 L/v$ is the Reynolds number with viscosity v and reference velocity V_0 and length L.

A stream function ψ and vorticity ω are defined by means of the relations

$$u = \psi_y, \qquad v = -\psi_x, \qquad \omega = v_x - u_y \tag{5-145}$$

whereupon one form of the vorticity equation becomes

$$\omega_t + (u\omega)_x + (v\omega)_y = \frac{1}{R}\nabla^2\omega. \tag{5-146}$$

Here ∇^2 is used to denote the Laplace operator in the appropriate coordinate system—here it is $\omega_{xx} + \omega_{yy}$. An alternative form of (5-146) is

$$\omega_t + \psi_y\omega_x - \psi_x\omega_y = \frac{1}{R}\nabla^2\omega. \tag{5-147}$$

From the definition of ω in (5-145) it follows that

$$\nabla^2\psi = -\omega. \tag{5-148}$$

A knowledge of the pressure is often useful in understanding the flow features. A suitable pressure equation is found by computing the x derivative of (5-142), the y derivative of (5-143), and summing the results, thereby generating

$$\nabla^2 p = -[(u^2)_{xx} + 2(uv)_{xy} + (v^2)_{yy}].$$

Probably Emmons [168] was the first to use the stream function–vorticity form in a digital computer calculation. To study turbulence he solved (5-147) and (5-148) at a Reynolds number ($V_0 L/v$) of 4000 by an explicit finite difference discretization constructed from a forward difference in time and a standard five-point molecule for each Laplacian. Payne [146] employs essentially the same ideas in his calculation of nonsteady flow of the wake behind a circular cylinder.

Clearly, the vorticity transport equation (5-146) or (5-147) is parabolic while the stream function equation is elliptic. Of course they are coupled, and it is well to note here that in any numerical solution there is numerical interaction. For example, an inaccurate treatment of a boundary condition in one equation can cause a drift of the solution in the other. Since these equations are usually solved cyclically, increases in the allowable time steps of the vorticity transport equation will be offset by increases in the number of iterations required in the numerical solution of the stream function Poisson equation (5-148). The final choice of the method(s) depends upon many factors, such as boundary conditions, problem geometry, problem type (steady or transient), additional information required, and the knowledge of the researcher. *The researcher is cautioned to think about the methods rather than just programming them!* At this point in time (1976) there still seems to be nearly as many numerical methods as problems. *We are still some distance from universal methods. Thus the uninformed should beware!* Here and in the later subsections we shall sketch some of the successful algorithms.

First, an algorithm will be described in which the vorticity equation is solved explicitly and the stream function equation is solved implicitly. Following this, an alternative method will be discussed in which both equations are solved implicitly. An example problem due to Giaquinta [167] will be a convenient carrier. Consider an internal separated laminar flow into the geometry of Fig. 5-10 which is impulsively started from rest and developed to steady state. The geometry is a two-dimensional sudden expansion into a simply connected rectangular domain. At the inlet HA a constant, uniform velocity U_0 is imposed so that a unit flow rate continuously

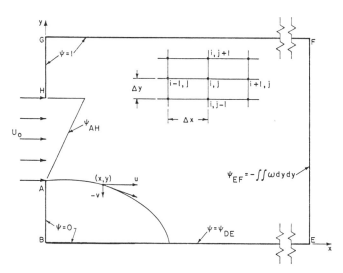

Fig. 5-10 Flow geometry

enters the expansion. Continuity requires that a unit flow rate must leave the expansion at the exit plane EF.

The fundamental no-slip boundary condition for viscous flow† is expressed as

$$u = \psi_y = 0, \qquad v = -\psi_x = 0 \qquad (5\text{-}149)$$

along boundaries AB, BE, HG, and GF, for all time. In two-dimensional flow the difference in the values of the stream function for any two stream lines must equal the discharge between them. Consequently, the values of the stream function for the expansion boundaries ABE and HGF must differ by one. Choosing the value of ψ as zero on ABE means that its value must be one on HGF. Thus

$$\psi_{AB} = \psi_{BE} = 0, \qquad \psi_{HG} = \psi_{GF} = 1. \qquad (5\text{-}150)$$

Since the velocity distribution across AH is uniform, the stream function across the inlet will be the linear function

$$\psi_{AH} = (y - y_A)/(y_H - y_A) \qquad (5\text{-}151)$$

obtained by integrating (5-145).

The boundary condition for the stream function at the exit EF cannot be specified a priori since the flow is developing with time. The condition must be predictable from previously computed or known information. This is done through the stream function equation (5-148) as

$$\psi_{EF} = -\iint \omega_{EF} \, dy \, dy \qquad (5\text{-}152)$$

where it has been assumed that $v_x = \psi_{xx} = 0$, i.e. the vertical velocity component is not a function of x. Boundary values for the vorticity are computed from the stream function distribution at each time step by (5-148) except for the initial time in which the definition $\omega = v_x - u_y$ is used. For time steps following the initial one, the vorticity along BE and GF, due to the no-slip boundary condition, is given by

$$\omega = -\psi_{yy}.$$

At time $t = 0^+$ the flow is started impulsively, the vorticity is initially zero, and the governing stream function equation is Laplace's equation, together with the previously described boundary conditions *with one change*. Initially the outlet boundary condition is assumed to be a linear function of ψ with the value 0 at BE and 1 at GF. The Laplace equation with the stated boundary conditions is then used as the stream function solution at the impulsive start (see Thoman and Szewczyk [169]).

---

† Along stationary, impermeable boundaries the fluid velocity is zero.

If (5-148) were used to compute the boundary vorticity, the result would be zero. To *initiate* the vorticity production at the boundaries, the boundary values of velocity are imposed and velocities computed throughout the field by means of $u = \psi_y$, $v = -\psi_x$, and the vorticity is calculated from the velocities by means of $\omega = v_x - u_y$. *Thus, the initial field consists of the irrotational distribution and the imposed boundary values of ψ, the computed and the imposed boundary values of velocity, and the values of vorticity on the boundaries and in the interior which are computed from the velocity.*

First, the vorticity-explicit, stream function-implicit method is described. With $\omega_{i,j}^n = \omega(i \, \Delta x, j \, \Delta y, n \, \Delta t)$ the explicit finite difference approximation for advancing the interior vorticity to time $n + 1$ is

$$\omega_{i,j}^{n+1} = \omega_{i,j}^n + \frac{\Delta t}{R(\Delta x)^2} [\omega_{i+1,j}^n - 2\omega_{i,j}^n + \omega_{i-1,j}^n]$$

$$+ \frac{\Delta t}{R(\Delta y)^2} [\omega_{i,j+1}^n - 2\omega_{i,j}^n + \omega_{i,j-1}^n]$$

$$- \frac{\Delta t}{2 \, \Delta x} [(u\omega)_{i+1,j}^n - (u\omega)_{i-1,j}^n]$$

$$- \frac{\Delta t}{2 \, \Delta y} [(v\omega)_{i,j+1}^n - (v\omega)_{i,j-1}^n] \qquad (5\text{-}153)$$

which is second-order accurate in space and first-order in time. The stability of a linearized version of (5-153) can be analyzed by von Neuman Fourier analysis (Chapter 2) to provide qualitative information about the original system. Without viscosity the restriction (see Roache [162]) from the convective terms

$$\Delta t \leq \frac{(\Delta x)^2 (\Delta y)^2}{(2/R)[(\Delta x)^2 + (\Delta y)^2] + \Delta x \, \Delta y \, (u \, \Delta y + v \, \Delta x)} \qquad (5\text{-}154)$$

must be imposed. The dynamic stability criterion based upon the viscous terms is

$$\Delta t \leq \frac{R}{2} \frac{(\Delta x)^2 (\Delta y)^2}{(\Delta x)^2 + (\Delta y)^2} \qquad (5\text{-}155)$$

In (5-154) u and v are local velocity components. A global condition is obtained by using their largest values on the entire field. These restrictions act as a guide in the numerical experimentation—but sometimes they are too strict and in other situations not strict enough.

The computation proceeds as follows:

(I) Compute the interior vorticity $\omega_{i,j}^{n+1}$ from (5-153) with 'upstream differencing'

$$(u\omega)_x \sim \begin{cases} [(u\omega)_{i+1,j} - (u\omega)_{i,j}]/\Delta x, & u < 0 \\ [(u\omega)_{i,j} - (u\omega)_{i-1,j}]/\Delta x, & u > 0 \end{cases}$$

$$(v\omega)_y \sim \begin{cases} [(v\omega)_{i,j+1} - (v\omega)_{i,j}]/\Delta y, & v < 0 \\ [(v\omega)_{i,j} - (v\omega)_{i,j-1}]/\Delta y, & v > 0 \end{cases} \qquad (5\text{-}156)$$

and a central difference if u or v equals zero, replacing the convective terms if desired (see Roache [162] for these arguments). Of course that replacement reduces the order of the spatial truncation error to one.

(II) Compute the vorticity at the downstream boundary ($i = N$) $\omega_{N,j}^{n+1}$ by extrapolation from the interior using a linear form

$$\omega_{N,j}^{n+1} = 2\omega_{N-1,j}^{n+1} - \omega_{N-2,j}^{n+1}$$

or alternative.

(III) Solve for the downstream boundary values of the stream function $\psi_{N,j}^{n+1}$. To do this assume that the velocity component v has negligible longitudinal variation along EF—i.e. $v_x = 0$. Thus we compute the boundary values from

$$(\psi_{N,j+1}^{n+1} - 2\psi_{N,j}^{n+1} + \psi_{N,j-1}^{n+1})/(\Delta y)^2 = -\omega_{N,j}^{n+1} \qquad (5\text{-}157)$$

for each downstream boundary point. Since the boundary values of ψ are known, the solution of the tridiagonal system of equations generated by (5-157) is easily carried out by the Thomas algorithm (2-46).

(IV) Compute the interior stream function values $\psi_{i,j}^{n+1}$ from a discretization of (5-148). SOR or ADI or any other iterative method can be employed. Convergence of the iteration should be checked.

(V) Calculate the velocity components in the interior and along the downstream boundary $u_{i,j}^{n+1}$, $v_{i,j}^{n+1}$ from the stream function by a finite difference form of $u = \psi_y$, $v = -\psi_x$.

(VI) Update the vorticity along the walls by means of the definitions. Along the horizontal walls $\omega = -\psi_{yy}$ and along the vertical walls $\omega = \psi_{xx}$. There is a minor controversy over this point (see Roache [162, p. 141]). Roache recommends that $\omega_{\text{wall}} = \omega_w$ be computed by means of a Taylor's expansion out from the wall. Thus for BE

$$\omega_{i,1}^{n+1} = \frac{2(\psi_{i,2}^{n+1} - \psi_{i,1}^{n+1})}{(\Delta y)^2} + O(\Delta y)$$

$$= \frac{2\psi_{i,2}}{(\Delta y)^2} + O(\Delta y)$$

and for GF

$$\omega_{i,N}^{n+1} = \frac{2(\psi_{i,N-1}^{n+1} - 1)}{(\Delta y)^2} + O(\Delta y).$$

Similar expressions apply for the other walls. Second-order approximations are recommended by others. For vorticity along BE, one form is (Roache [162])

$$\omega_{i,1}^{n+1} = -[8\psi_{i,2}^{n+1} - \psi_{i,3}^{n+1}]/2(\Delta y)^2 + O(\Delta y)^2$$

and for GF the backward difference equation is

$$\omega_{i,N}^{n+1} = -[-7 + 8\psi_{i,N-1}^{n+1} - \psi_{i,N-2}^{n+1}]/2(\Delta y)^2 + O(\Delta y)^2.$$

Additional features such as predictor–correctors can be employed.

Instead of (5-153) Fromm [147, 148] uses a DuFort–Frankel molecule [see (2-73)], so with $\Delta x = \Delta y = a$ the algorithm for advancing the interior vorticity to a new time is the explicit scheme

$$\omega_{i,j}^{n+1} = \frac{1}{1 + 4\nu\,\Delta t/a^2}$$

$$\times \left\{ \omega_{i,j}^{n-1} + \frac{2\,\Delta t}{a} [(u\omega)_{i-1/2,j}^n - (u\omega)_{i+1/2,j}^n + (v\omega)_{i,j-1/2}^n - (v\omega)_{i,j+1/2}^n] \right.$$

$$\left. + \frac{2\nu\,\Delta t}{a^2} [\omega_{i+1,j}^n + \omega_{i-1,j}^n + \omega_{i,j+1}^n + \omega_{i,j-1}^n - 2\omega_{i,j}^{n-1}] \right\}. \quad (5\text{-}158)$$

Experience indicates (see Harlow [170], Fromm [171], Hung and Macagno [172], and Torrance [173]) that the stability conditions when (5-158) is used are very nearly the same as those previously given (5-154), (5-155).

Harlow and Fromm [174] use the DuFort–Frankel approach for the two-dimensional problem of heat transfer from a rectangular cylinder into a surrounding fluid. Wilkes [175] solved a thermal convection problem in a two-dimensional cell using ADI. Though the method is unconditionally stable for the linear diffusion equation, Wilkes found it to be unstable for his problem at a Grashof number of 200,000. This instability may be due to the boundary condition approximation for the vorticity, which always lagged one time step behind the rest of the field. *Proper boundary condition treatment is especially important in these fluid problems.*

The Douglas [175a, b] method has been applied, with certain alterations, by Alonso [176], Pujol [164], and Aziz and Hellums [156] to the solution of the vorticity transport equation (5-146). The method can be roughly described as an ADI which is a perturbation of the Crank–Nicolson (Section 2-3) procedure. The use of Crank–Nicolson averaging results in a scheme that has second-order truncation error in both space and time. This accuracy improvement coupled with the linearized system's unconditional stability makes the method attractive.

Initiation of the solution is the same as before. To advance one time step, (5-146) is solved by an ADI method with the insertion of the stream function equation solution between each sweep. Further details must be left for the references.

Pearson [177] solved the unsteady, axisymmetric viscous flow generated by the differential rotation of two infinite parallel disks using an adaptation of the Crank–Nicolson centered scheme for the vorticity transport equation. The resulting algebraic problem was solved by successive overrelaxation when the flow was almost linear or by a modified alternating direction implicit method (ADI) of Peaceman and Rachford when the nonlinear terms are not small. Additional stability is achieved by employing a smoothing process at boundary points.

Keller and Takami [150] study the steady viscous incompressible flow about circular cylinders. They introduce new Cartesian coordinates

$$\xi + i\eta = \frac{1}{\pi} \ln (x + iy)$$

which map the exterior of the unit circle in the (x, y) plane onto a semi-infinite strip in the (ξ, η) plane. The transformed vorticity–stream function equations are discretized and an 'extrapolated line Liebmann' method is used, resulting in coupled tridiagonal linear systems which are easily solved by the tridiagonal algorithm.

A large number of additional modifications of the basic stream function vorticity concept are given in the 1969 IUTAM Symposium on High Speed Computing in Fluid Mechanics, edited by Frankiel and Stewartson (see [169]), and the 1971 conference edited by Holt [178]. Additional work continues to appear.

Hamielec *et al.* [179] and Rimon and Cheng [180] have solved the axisymmetric stream function and vorticity formulation for viscous incompressible flow in logarithmically contracted spherical polar coordinates $(z = \ln r, \theta)$. Different numerical procedures of second-order accuracy are employed. The novel feature here is the use of a varying (exponential) step size in the radial direction with the small step size near the boundary of the sphere. Cheng [169, pp. 34–41] discusses the accuracy of these results and questions the claimed accuracy.

Alonso [176] also employs a graded mesh structure in his solutions of the time-dependent confined rotating flow of an incompressible viscous fluid. Small cells are employed in regions of high velocity and vorticity gradients, intermediate cells in regions of intermediate gradients, and large cells in regions of small gradients. As Fromm, he finds a circular cylindrical form of (5-146) to be more convenient. The tangential momentum and vorticity transport equations are discretized by a Crank–Nicolson centered scheme for the z (axial) derivatives and an explicit one for the r derivatives. The discretized linear equations are subjected to ADI resulting in two tridiagonal systems. The stream function is also obtained by ADI.

Fromm [169, pp. 1–12, 113–119] amplifies his earlier work and stresses numerical dispersion effects. Fourth-order methods are shown to have better phase properties, so that distortions resulting from dispersion are reduced.

(b) Primitive variable methods

The use of the Navier–Stokes and continuity equations in the primitive variables u, v, p in two dimensions (u, v, w, p in three), in the form discussed here, was undertaken by Harlow and Welch [181] and Welch *et al.* [182] in their now famous MAC (marker and cell) method. We shall describe their ideas in two dimensions first.

The conservation† form of the dimensionless momentum equations is easily obtained as

$$u_t + (u^2)_x + (uv)_y = -p_x + R^{-1}(u_{xx} + u_{yy}) \qquad (5\text{-}159)$$

$$v_t + (uv)_x + (v^2)_y = -p_y + R^{-1}(v_{xx} + v_{yy}). \qquad (5\text{-}160)$$

A Poisson equation for pressure is obtained by differentiation and addition of (5-159) and (5-160):

$$\nabla^2 p = -[(u^2)_{xx} + (uv)_{xy} + (v^2)_{yy}] - D_t + \frac{1}{R}[D_{xx} + D_{yy}] \equiv F_p$$

$$(5\text{-}161)$$

where the term D is defined as

$$D = u_x + v_y.$$

In the continuum $D \equiv 0$, of course; but due to approximations or incomplete iteration of the Poisson equation error accumulates and $D_{ij} \not\equiv 0$. The result of omitting D in (5-161) is not only inaccuracy but an instability in the momentum equations. Inclusion of D can eliminate the instability. The term D_t is to be evaluated by whatever scheme is used for u_t and v_t, by forcing $D_{i,j}^{n+1} = 0$. This instability has been observed by others using the MAC method. Hirt and Harlow [183], Donovan [184], and Putre [185] have generalized the concept.

The no-slip boundary conditions $u = v = 0$ along a wall are very simple here, which is a great advantage for implicit methods since no iteration is required for the advance boundary condition. Aziz and Hellums [156] found implicit methods not useful because of an instability of the pressure term. Roache [162] conjectures that this is curable by retaining $\partial D / \partial t$ as in (5-161).

The MAC method net structure is shown in Fig. 5-11. Pressure is defined at nodal locations in the cell centers while the velocity components are

---

† Experience indicates that conservative systems do generally give more accurate results, although not everyone agrees (see Roache [162, p. 32]); but conservation does not imply accuracy.

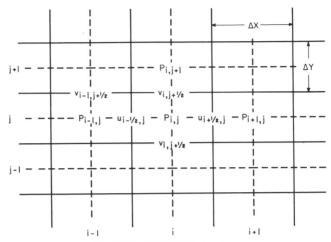

Fig. 5-11   MAC Net structure

defined along the cell boundaries as illustrated. The finite differences are taken as

$$u_t|_{i+1/2,j} = [u_{i+1/2,j}^{n+1} - u_{i+1/2,j}^n]/\Delta t$$

$$p_x^n|_{i+1/2,j} = [p_{i+1,j}^n - p_{i,j}^n]/(\Delta x)$$

$$u_{xx}^n|_{i+1/2,j} = [u_{i+3/2,j}^n - 2u_{i+1/2,j} + u_{i-1/2,j}]/(\Delta x)^2.$$

The evaluation of

$$(u^2)_x|_{i+1/2,j} = [u_{i+1,j}^2 - u_{i,j}^2]/\Delta x$$

requires the definition of u at the node points. These are obtained by averaging

$$u_{i+1,j} = \tfrac{1}{2}(u_{i+1/2,j} + u_{i+3/2,j})$$

while the product terms are evaluated as the product of the averages

$$(uv)_{i+1/2,j+1/2} = \tfrac{1}{4}(u_{i+1/2,j} + u_{i+1/2,j+1})(v_{i+1,j+1/2} + v_{i,j+1/2}).$$

The assembled algorithm for (5-159), with the superscript n understood on the right-hand side, is

$$u_{i+1/2,j}^{n+1} = u_{i+1/2,j} + \Delta t \left\{ -\frac{u_{i+1,j}^2 - u_{i,j}^2}{\Delta x} - \frac{p_{i+1,j} - p_{i,j}}{\Delta x} \right.$$

$$- \frac{(uv)_{i+1/2,j+1/2} - (uv)_{i+1/2,j-1/2}}{\Delta y}$$

$$+ \frac{1}{R} \left(\frac{u_{i+3/2,j} - 2u_{i+1/2,j} + u_{i-1/2,j}}{(\Delta x)^2} \right.$$

$$\left. \left. + \frac{u_{i+1/2,j+1} - 2u_{i+1/2,j} + u_{i+1/2,j-1}}{(\Delta y)^2} \right) \right\}.$$

$$(5\text{-}162)$$

There is a corresponding result for $v_{i,j+1/2}^{n+1}$. Equation (5-161) is treated with the same type of differencing, but with $D_{i,j}^{n+1}$ set equal to zero. Thus

$$D_t^n|_{i,j} = [D_{i,j}^{n+1} - D_{i,j}^n]/\Delta t = -D_{i,j}^n/\Delta t$$

so that the algorithm becomes

$$-F_p|_{i,j} = \frac{u_{i+1,j}^2 - 2u_{i,j}^2 + u_{i-1,j}^2}{(\Delta x)^2}$$

$$+ \frac{2}{\Delta x\,\Delta y}\left[(uv)_{i+1/2,j+1/2} - (uv)_{i+1/2,j-1/2}\right.$$

$$\left. - (uv)_{i-1/2,j+1/2} + (uv)_{i-1/2,j-1/2}\right]$$

$$+ \frac{v_{i,j+1}^2 - 2v_{i,j}^2 + v_{i,j-1}^2}{(\Delta y)^2} - \frac{D_{i,j}}{\Delta t}$$

$$- \frac{1}{R}\left[\frac{D_{i+1,j} - 2D_{i,j} + D_{i-1,j}}{(\Delta x)^2}\right.$$

$$\left. + \frac{D_{i,j+1} - 2D_{i,j} + D_{i,j-1}}{(\Delta y)^2}\right]$$

where $D_{i,j} = [u_{i+1/2,j} - u_{i-1/2,j}]/\Delta x + [v_{i,j+1/2} - v_{i,j-1/2}]/\Delta y$.

Boundary conditions on u and v are given in accordance with the averaging definitions at nodal points. A no-slip wall along $j = $ wall in Fig. 5-11 gives $u_w = 0$ so

$$u_{i+1/2,w} = 0, \qquad u_{i-1/2,w} = 0$$

while $v_w = 0$ implies

$$v_{i,w} = \tfrac{1}{2}(v_{i,w+1/2} + v_{i,w-1/2}) = 0$$

or a fictitious value of

$$v_{i,w-1/2} = -v_{i,w+1/2}$$

inside the wall is determined.

The marker particles used in MAC are massless, move with the advection field, but do not participate in the computation. By tracing and plotting the position of the marker particles one obtains a 'streakline' pattern. The positions of each particle (x_p^n, y_p^n) (Lagrangian coordinates) are obtained by numerical integration of

$$x_p^n = x_p^0 + \int_0^t u_p\,dt$$

$$y_p^n = y_p^0 + \int_0^t v_p\,dt$$

(5-163)

where (x_p^0, y_p^0) is the initial particle position at $t = 0$. Consistent with the numerical technique just described, (5-163) is marched ahead with

$$x_p^{n+1} = x_p^n + u_p \, \Delta t, \qquad y_p^{n+1} = y_p^n + v_p \, \Delta t$$

where various interpolation schemes are employed to evaluate the velocities u_p, v_p in the Eulerian net at the time location of the particle. For further details and modifications, the reader is urged to consult Roache [162], Chan et al. [186], Amsden and Harlow [187], and other references contained therein.

Chorin [152, 188] designed a semi-implicit method as an alternative to MAC. In his first paper he studied the thermal convection of a fluid heated from below (Bénard problem) and introduced an artificial compressibility into the continuity equation; the pressure became a function of an artificial equation of state. The velocity field was advanced in time using an ADI scheme on the complete Navier–Stokes equations, with the pressure term expressed by means of the artificial equation of state. At the new time level the corresponding artificial density was obtained using a DuFort–Frankel molecule on the perturbed continuity equation. The paper includes a rigorous treatment of boundary conditions.

In his successive papers Chorin [153, 189] improves his own method. He designed an auxiliary field through which the velocity field and pressure values are projected from the old to the new time level. For this purpose, the pressure gradient does not appear explicitly in the ADI discrete simulation of the Navier–Stokes equations. At the new time level, the auxiliary field is decomposed into the velocity vector field and the pressure scalar field by using a set of algorithms in which his previous idea of an artificial compressibility has evolved into a formulation that is analogous to SOR. Several implementations of this technique exist (Padmanabhan [155], Pujol [164]), but the programming is complicated.

Three-dimensional applications of MAC in various modifications are now in progress.

(c) Vector potential methods

Aziz [190] and Pearson [177] found that ADI methods for solving the finite difference form of the parabolic vorticity equation were more accurate, converged faster, and were more stable than explicit methods such as that of DuFort–Frankel. Aziz and Hellums [156] in their study of three-dimensional laminar natural convection found that the Douglas ADI scheme (p. 254) was superior to the Peaceman–Rachford one in the sense that both two- and three-dimensional forms are unconditionally stable for the linear diffusion equation. The Douglas scheme is locally second order in space and time. While this procedure can and has been used in two-dimensional problems, its major impact is felt in three dimensions where the basic equations are recast in *vector potential form*.

The vector potential method of Aziz and Hellums [156] and Hirasaki [192] consists of a transformation of the complete Navier–Stokes equations in terms of a vorticity and a vector potential. These are discretized and solved using the Douglas ADI method for the parabolic portion of the problem and SOR for the elliptic portion. The formulation of the equations of change in terms of the vector potential was found to be an essential ingredient in this analysis. The existence of the vector potential has been known for many years. We shall briefly review the basic concepts.

A useful classification of vector fields† $\mathbf{E}$ is possible in terms of the divergence (div) and curl operators. If div $\mathbf{E} = 0$ at every point of a region R the field is said to be solenoidal in that region.‡ Physically, this means there are no sources or sinks in R. If, at every point of R, curl $\mathbf{E} = 0$, the field is said to be irrotational in R. The classification of fields is as follows:

Class I   Solenoidal and irrotational:
curl $\mathbf{E} = 0$,   div $\mathbf{E} = 0$.
Class II   Irrotational but not solenoidal:
curl $\mathbf{E} = 0$,   div $\mathbf{E} \neq 0$.
Class III   Solenoidal but not irrotational:
curl $\mathbf{E} \neq 0$,   div $\mathbf{E} = 0$.
Class IV   Neither solenoidal nor irrotational:
curl $\mathbf{E} \neq 0$,   div $\mathbf{E} \neq 0$.

The velocity field of an incompressible fluid falls into Class III, while that of a compressible fluid is of Class IV.

A necessary and sufficient condition for the existence of a scalar potential ϕ is that curl $\mathbf{E} = 0$, whereupon ϕ is defined by

$$\mathbf{E} = -\operatorname{grad}\phi \quad \text{and} \quad \operatorname{div}\mathbf{E} = -\operatorname{div}\operatorname{grad}\phi = -\nabla^2\phi.$$

Thus, in fields of Class I and II it is always possible to introduce a scalar potential ϕ defined by $\nabla^2\phi = K$ (Poisson's equation for a field of Class II).

For fields that are not irrotational (curl $\mathbf{E} \neq 0$), in some cases it is still possible to employ a scalar function and thus avoid the difficulties usually associated with the more difficult case, that of the vector potential. Suppose that curl $\mathbf{E} \neq 0$ but there exists a scalar function ρ such that curl $(\rho\mathbf{E}) = 0$. In this case a scalar quasi-potential Γ exists, defined by the equation

$$\mathbf{E} = -\rho^{-1}\operatorname{grad}\Gamma. \tag{5-164}$$

---

† In this section a vector is denoted by boldface, e.g. **E**.
‡ Much of the foundation research has been with respect to electric and electromagnetic fields, especially for the Maxwell equations. Thus the terms are reminiscent of that area.

Since $0 = \text{curl} (\rho\mathbf{E}) = \rho \, \text{curl} \, \mathbf{E} + \text{grad} \, \rho \times \mathbf{E}$ it follows that

$$0 = \mathbf{E} \cdot \text{curl} (\rho\mathbf{E}) = \mathbf{E} \cdot \rho \, \text{curl} \, \mathbf{E} + \mathbf{E} \cdot \text{grad} \, \rho \times \mathbf{E}$$

$$= \rho(\mathbf{E} \cdot \text{curl} \, \mathbf{E}) + \text{grad} \, \rho \cdot \mathbf{E} \times \mathbf{E}.$$

Consequently, a necessary and sufficient condition for the existence of a quasi-potential is that $\mathbf{E} \cdot \text{curl} \, \mathbf{E} = 0$.

From (5-164)

$$K = \text{div} \, \mathbf{E} = -\text{div} (\rho^{-1} \, \text{grad} \, \Gamma)$$

$$= -\rho^{-1} \, \nabla^2\Gamma - (\text{grad} \, \rho^{-1}) \cdot \text{grad} \, \Gamma$$

where K is zero for Class III and a function for Class IV. Thus a quasi-potential can be determined by solving

$$\nabla^2\Gamma + \rho \, (\text{grad} \, \rho^{-1}) \cdot \text{grad} \, \Gamma = -\rho K. \tag{5-165}$$

The latter equation occurs in field problems associated with inhomogeneous media.

Most solenoidal fields do not admit a potential or quasi-potential, but it is always possible to introduce a vector potential $\mathbf{A}$, which, though not as simple as a scalar potential, nevertheless behaves in a somewhat similar fashion. If $\text{curl} \, \mathbf{E} = \mathbf{B}$, $\text{div} \, \mathbf{E} = 0$, the vector potential is defined by the relation

$$\mathbf{E} = \text{curl} \, \mathbf{A}. \tag{5-166}$$

Then

$$\text{curl} \, \mathbf{E} = \text{curl curl} \, \mathbf{A} = \text{grad div} \, \mathbf{A} - *\mathbf{A} = \mathbf{B} \tag{5-167}$$

where $*$ is used to denote the vector Laplacian†

$$*\mathbf{A} = \text{grad div} \, \mathbf{A} - \text{curl curl} \, \mathbf{A}. \tag{5-168}$$

We can assign any desired value to the divergence of our new vector $\mathbf{A}$. If we take $\text{div} \, \mathbf{A} = 0$, then (5-168) becomes

$$*\mathbf{A} = -\mathbf{B} \tag{5-169}$$

which is the vector form of Poisson's equation. In Cartesian coordinates (x, y, z) it splits into three scalar Poisson equations

$$*_x\mathbf{A} = -\mathbf{B}_x, \qquad *_y\mathbf{A} = -\mathbf{B}_y, \qquad *_z\mathbf{A} = -\mathbf{B}_z.$$

In the case of a general orthogonal coordinate system (x^1, x^2, x^3) the vector Laplacian is given in Moon and Spencer [191]. To solve a problem using the vector potential we first obtain $\mathbf{A}$ from (5-169) and then find $\mathbf{E}$ from (5-166).

---

† The vector Laplacian is often written ∇^2, but this is an ambiguous notation. Except in the special case of rectangular coordinates, where each component is similar to a scalar Laplacian, the two operators are quite distinct.

For completeness, we now present the scheme used by Helmholtz in his classic study of vortex motion. In Class IV suppose E has known divergence and curl, neither of which is zero. If U is irrotational (curl $U = 0$) and V is solenoidal (div $V = 0$), we suppose

$$E = U + V. \tag{5-170}$$

Since curl $U = 0$, a scalar potential ϕ can be introduced so that

$$U = -\operatorname{grad} \phi. \tag{5-171}$$

Likewise, since div $V = 0$ there exists a vector potential A such that

$$V = \operatorname{curl} A, \qquad \operatorname{div} A = 0. \tag{5-172}$$

From (5-170) we find

$$\operatorname{div} E = \operatorname{div} U + \operatorname{div} V = -\operatorname{div} \operatorname{grad} \phi = -\nabla^2 \phi$$

and

$$\operatorname{curl} E = \operatorname{curl} U + \operatorname{curl} V = \operatorname{curl} \operatorname{curl} A = -*A.$$

Consequently, fields of Class IV are evaluated by the solution of the scalar Poisson equation

$$\nabla^2 \phi = -\operatorname{div} E$$

and the vector Poisson equation

$$*A = -\operatorname{curl} E$$

for the scalar potential ϕ and vector potential A. Using these we calculate U and V by means of (5-171) and (5-172). Finally, E is calculated by means of (5-170). With this introduction we are now in position to discuss the method of the *vector potential*.

The *vector vorticity* is defined as

$$\boldsymbol{\omega} = \operatorname{curl} V = \nabla \times V. \tag{5-173}$$

In Cartesian form,

$$\omega_x = w_y - v_z, \qquad \omega_y = -w_x + u_z, \qquad \omega_z = v_x - u_y$$

and the corresponding vorticity transport equations is a vector equation with components

$$(\omega_x)_t = -\nabla \cdot (V\omega_x) + \frac{1}{R} \nabla^2 \omega_x$$

$$(\omega_y)_t = -\nabla \cdot (V\omega_y) + \frac{1}{R} \nabla^2 \omega_y$$

$$(\omega_z)_t = -\nabla \cdot (V\omega_z) + \frac{1}{R} \nabla^2 \omega_z.$$

These are in conservation form and all two-dimensional methods, extended, can be applied.

A stream function does not exist for general three-dimensional flows, but for a 'solenoidal' vector field (see the previous discussion) there exists a *vector potential* $\boldsymbol{\psi}$ *such that*

$$\mathbf{V} = \text{curl } \boldsymbol{\psi} = \boldsymbol{\nabla} \times \boldsymbol{\psi} \qquad (5\text{-}174)$$

or

$$u = \frac{\partial \psi_z}{\partial y} - \frac{\partial \psi_y}{\partial z}$$

$$v = -\frac{\partial \psi_z}{\partial x} + \frac{\partial \psi_x}{\partial z}$$

$$w = \frac{\partial \psi_y}{\partial x} - \frac{\partial \psi_x}{\partial y}$$

From (5-173) and (5-174) there follows

$$\boldsymbol{\omega} = \text{curl curl } \boldsymbol{\psi} = \boldsymbol{\nabla} \times \boldsymbol{\Phi} \times \boldsymbol{\psi}$$

and with $\boldsymbol{\nabla} \cdot \boldsymbol{\psi} = 0$, chosen arbitrarily, $\boldsymbol{\psi}$ satisfies the vector form of Poisson's equation

$$*\boldsymbol{\psi} = -\boldsymbol{\omega} \qquad (5\text{-}175)$$

but remember that except in the special case of Cartesian coordinates the vector and scalar Laplacians are quite different! Thus at each time step there are three, three-dimensional, Poisson equations to solve.

The boundary conditions at a no-slip boundary are tricky. The $\boldsymbol{\psi}$ components are not all zero; rather, the components tangential to the surface are zero and the normal derivative of the normal component is zero. For example, for a wall along (y, z) at $x = x_a$, we have

$$\frac{\partial \psi_x}{\partial x} = 0, \qquad \psi_y = \psi_z = 0 \quad \text{at } x = x_a.$$

The no-slip wall vorticity components can be expressed in terms of their fundamental velocity component definitions. With our example of a wall along (y, z) at $x = x_a$, we have

$$\left.\begin{array}{l} \omega_x = 0 \\ \omega_y = -\partial w/\partial x \\ \omega_z = +\partial v/\partial x \end{array}\right\} \quad \text{at } x = x_a. \qquad (5\text{-}176)$$

Aziz and Hellums [156] suggest evaluating wall vorticity components directly from Eqn (5-176), using forms such as the following for a wall at $x = I \, \Delta x$

$$\omega_y(I, j, k) = -\left.\frac{\partial w}{\partial x}\right|_{I,j,k} = -\frac{4w(I + 1, j, k) - w(I + 2, j, k)}{2 \, \Delta x}$$

(d) General comments and literature

Roache [162] has advanced the premise that *computational fluid dynamics* has emerged as a distinct field. It is not simply applied numerical analysis because of the difficulties mathematical theory has in handling nonlinear partial differential equations. The methods of numerical analysis are utilized whenever possible, but this is usually restricted to simple model problems and to linearized forms of the equations. Extension to nonlinear equations of a particular method which is consistent, stable, and convergent for the linearized form often requires some physical reasoning and experimentation. On the other hand the mathematics cannot and should not be ignored. The amount of mathematics needed to develop an accurate efficient algorithm for a flow problem is considerable. The concepts of convergence, stability, consistency, dispersion, dissipation, and error analysis must be well understood. And a certain amount of knowledge about differential equations, numerical analysis, functional analysis, and linear algebra is essential. No, the subject is not one for the uneducated in the mathematical sciences, but as we have remarked just that knowledge is not enough.

The literature has become very extensive, so that comprehensive publications are valuable. In addition to those of the references we recommend:

(1) AIAA Proceedings of Computational Fluid Dynamics (1973 and odd years)

(2) International Conferences on Numerical Methods in Fluid Dynamics (1972 and even years)

(3) Taylor [193]

(4) Orszag and Israeli [194]

(5) Kreiss and Oliger [195]

(6) Roache [196]

(7) Proceedings of the AICA Symposium on Computer Methods for PDE, Lehigh University, June 17–19, 1975 [197]

5-10 Introduction to Monte Carlo methods

In the literature a *Monte Carlo* method is any procedure that involves the use of statistical sampling procedures to approximate the solution of a mathematical or physical problem. Monte Carlo methods are not known for all problems nor do specific problems admit a unique Monte Carlo procedure. There may exist different, apparently unrelated, Monte Carlo methods for a given problem. While such methods exist for certain classes of integral equations, eigenvalue problems, and other functional equations, our interest here will be confined to the elementary ideas for partial differential equations. Useful surveys of theory and applications include Brown [198], Barber and Ninham [199], Meyer [200], and Fox [26, Chapters 32, 33].

Some of the fundamental probability ideas employed here are presented in Section 1-7. Since we aim herein only for an introduction, those elementary concepts will suffice.

Consider a particle that is constrained to move on the nodal points $Q(i \, \Delta x, j \, \Delta y) = (i, j)$, i, j integers, of a lattice in the (x, y) plane. At each step the particle will move to one of the neighboring nodes directly above or below or directly to the right or left of the current position. Thus the four possible moves, from (i, j) to $(i + 1, j)$, $(i - 1, j)$, $(i, j + 1)$, or $(i, j - 1)$, are each assumed to have probability $\frac{1}{4}$. The particle describes a random path composed of such transitions in succession, with the new move randomly determined each time independently of the present position and past history of the particle. Assuming that the particle started at $(0, 0)$, there will be defined for each node (i, j) and for each positive integer s a probability $P(i, j, s)$ that after s steps the particle will be found at (i, j). In order to arrive, after $s + 1$ steps, at (i, j) the particle must have been at one of the node's four neighbors after the sth step. From each of these positions the probability is $\frac{1}{4}$ for the transition of the particle to (i, j). Thus $P(i, j, s)$ satisfies the linear difference† equation

$$P(i, j, s + 1) = \tfrac{1}{4}[P(i + 1, j, s) + P(i - 1, j, s)$$
$$+ P(i, j + 1, s) + P(i, j - 1, s)]. \qquad (5\text{-}177)$$

An elementary manipulation transforms (5-177) into

$$P(i, j, s + 1) - P(i, j, s) = \tfrac{1}{4}\{[P(i + 1, j, s) - 2P(i, j, s) + P(i - 1, j, s)]$$
$$+ [P(i, j + 1, s) - 2P(i, j, s) + P(i, j - 1, s)]\}$$
$$(5\text{-}178)$$

which relates the first difference of P with respect to s with the second differences with respect to i and j.

Let us now difference the diffusion equation

$$P_t = K(P_{xx} + P_{yy})$$

with a first-order forward difference in time and central differences in x and y with $\Delta x = \Delta y = a$, $P(ia, ja, s \, \Delta t) = P(i, j, s)$

$$P(i, j, s + 1) - P(i, j, s) = \frac{K \, \Delta t}{a^2}\{[P(i + 1, j, s) - 2P(i, j, s) + P(i - 1, j, s)]$$
$$+ [P(i, j + 1, s) - 2P(i, j, s) + P(i, j - 1, s)]\}$$

† More complicated functional equations will often occur here.

which is the same as (5-177) with $K \, \Delta t / a^2 = \frac{1}{4}$. This value of $K \, \Delta t / a^2$ is the stability limit in two space variables [see Section 5-4(a)] and depending upon boundary conditions may be a stable or unstable algorithm from that theory.

If instead of starting with a particle at $(0, 0)$ for $s = 0$, the process (called a *random walk*) is started with a probability distribution $P(i, j, 0)$ on the lattice, representing the probability of finding the particle at i, j initially, the function $P(i, j, s)$ is interpretable as before. We still get the same difference equation and $P(i, j, 0)$ is the arbitrary initial function generating a particular solution of (5-178).

Random behavior corresponding to the process (random walk) described here may be sampled by establishing an appropriate correspondence between ranges of random numbers and the decisions required at each step. For example, the first draw of a random number may determine the particle's initial point in the network, the second may determine the first move, etc. A sequence of such draws will determine a particular *path*, often referred to as a *particle history*. A sample of histories of this kind is used to estimate $P(i, j, s)$ for any node (i, j) and for any s to which the histories have been carried. The estimate is determined by counting the number of histories for which the corresponding particles are at (i, j) after the sth step and dividing by the sample size.

The statistical process just described is a *Monte Carlo* process and provides an alternative computational method to those of earlier chapters. The determination of random sampling numbers is usually done by generating on the computer a sequence of pseudorandom numbers with a suitable recurrence relation. A relation of the type†

$$r_{n+1} = A r_n \pmod{m}$$

is often used, where A and m are constants whose optimal choice depends upon the problem and the computer (see e.g. Meyer [200], Orcutt *et al.* [202]).

Suppose now that random walks of the type just described are begun in the *interior* of a bounded region enclosed by a specific boundary lattice of points (i_n, j_n). For purposes of this discussion, a boundary separates the interior from the exterior of the region by ensuring that any admissible walk, of the type used above, must contain a boundary point on the path between a pair of points one of which is inside and the other outside.

The random walks described above will now be altered by *terminating any walk upon its first arrival at the boundary*. The physicist says such a boundary is absorbing. It can be proved that the walks eventually terminate in the sense that the probability of remaining forever in the region is zero.

† With 35-bit binary storage Haviland [201] used $A = 2^{18} + 3$, $m = 2^{35}$.

Therefore it is appropriate to attach to any starting point (i, j) a function $P(i_n, j_n, i, j)$ which is the probability that a particle starting at (i, j) terminates its history at boundary point (i_n, j_n). If a function $U(i_n, j_n)$ is initially defined on the boundary, the expected value of U is

$$E[U] = V(i, j) = \sum_n U(i_n, j_n)P(i_n, j_n, i, j). \qquad (5\text{-}179)$$

In other words, if the value $U(i_n, j_n)$ is attached to all walks terminating at (i_n, j_n), then $V(i, j)$ is the expected terminal value for walks starting at (i, j). Of course, $V(i_n, j_n) = U(i_n, j_n)$. As before $V(i, j)$ satisfies the difference equation

$$V(i, j) = \tfrac{1}{4}[V(i + 1, j) + V(i - 1, j) + V(i, j + 1) + V(i, j - 1)]$$

which is the standard five-point molecule for Laplace's equation $V_{xx} + V_{yy} = 0$ with Dirichlet boundary conditions specified by U.

This random walk can serve as a basis for the statistical estimation of $V(i, j)$ at a single interior point by starting a number of particle histories at that point and attaching to each history the value $U(i_n, j_n)$ assigned to the boundary point (i_n, j_n) where that walk terminates. In this way the method leads to an approximate value to the solution at a single point without carrying out the usual simultaneous solution by direct methods or iteration (see Chapter 3). Let us note in passing that $P(i_n, j_n, i, j)$ is a discrete analogue of a Green's function.

Other random and restricted walks are described in Barber and Ninham [199]. Littschwager and Ames [203] have shown how Markoff processes are related to more complicated difference approximations of the diffusion equation. Monte Carlo methods have been employed on a great variety of problems and are especially useful on molecular dynamics and flow problems (Haviland [201], Gentry *et al.* [201a]). It is not particularly convenient because of the relatively long running times to obtain reasonable answers. However, it is the only feasible method for many problems in molecular flows and is therefore an essential tool in that area of physics. This author believes that the same can be said of calculations in turbulent flow.

5-11 Method of lines

The recent availability of sophisticated and reliable algorithms and computer programs for the automatic numerical computation of complicated systems of ordinary differential equations (Gear [204] and Hindmarsh [205], for example) makes the classical *method of lines* attractive for a number of problems. There is an extensive literature, primarily of Russian origin, for the method of lines which has been summarized to 1965 by Liskovets [206].

The method is simple in concept—for a given system of partial differential equations discretize all but one of the independent variables. This semi-discrete procedure yields a coupled system of ordinary differential equations which are then numerically integrated with a digital scheme or perhaps an analog computer. To aid in the presentation consider the nonlinear diffusion (parabolic) equation

$$u_t = [D(x, t, u)u_x]_x + f(x, t, u, u_x), \quad 0 < x < 1, \ 0 < t \le T$$

$$u(x, 0) = F(x), \quad 0 \le x \le 1$$

$$\alpha_1(t)u(0, t) + \alpha_2(t)u_x(0, t) = \alpha_3(t), \quad 0 < t \le T$$ (5-180)

$$\beta_1(t)u(1, t) + \beta_2(t)u_x(1, t) = \beta_3(t), \quad 0 < t \le T$$

where $0 < m \le D < M$ for stability and $-\infty < u < \infty$.

The first step is to discretize the spatial variable in the partial differential equation to obtain the system of ordinary differential equations that also include the boundary conditions. The selected finite differences in the spatial derivative are self-evident. Let $\Delta x = h = 1/N$, with the left-hand boundary at $i = 0$ and the right-hand one at $i = N$, and $D_{i \pm 1/2} = D(x_{i \pm 1/2}, t, u_{i \pm 1/2})$. Then, for $i = 0$,

$$\frac{du_0}{dt} = \begin{cases} 0 \quad \text{and} \quad u_0 = \alpha_3/\alpha_1 & \text{if } \alpha_2 = 0 \\[2mm] \frac{2}{h}\left[D_{1/2}\frac{u_1 - u_0}{h} - D_0\frac{\alpha_3 - \alpha_1 u_0}{\alpha_2}\right] + f\left(x_0, t, u_0, \frac{\alpha_3 - \alpha_1 u_0}{\alpha_2}\right) \\[4mm] \hspace{8cm} \text{if } \alpha_2 \ne 0 \end{cases}$$

for $i = 1, \ldots, N - 1$,

$$\frac{du_i}{dt} = [D_{i+1/2}(u_{i+1} - u_i) - D_{i-1/2}(u_i - u_{i-1})]/h^2$$

$$+ f\left(x_i, t, u_i, \frac{u_{i+1} - u_{i-1}}{2h}\right)$$

and for $i = N$,

$$\frac{du_N}{dt} = \begin{cases} 0 \quad \text{and} \quad u_N = \beta_3/\beta_1 & \text{if } \beta_2 = 0 \\[2mm] \frac{2}{h}\left[D_N\frac{\beta_3 - \beta_1 u_N}{\beta_2} - D_{N-1/2}\frac{u_N - u_{N-1}}{h}\right] + f\left(x_N, t, u_N, \frac{\beta_3 - \beta_1 u_N}{\beta_2}\right) \\[4mm] \hspace{8cm} \text{if } \beta_2 \ne 0 \end{cases}$$

The initial data are $u_i(0, x_i) = F(x_i)$, $i = 0, 1, 2, \ldots, N$.

For the numerical solution, the Gear–Hindmarsh integrators have both nonstiff and stiff subroutines built into the same program. Unless one knows the system to be stiff a priori, the usual procedure is to use the less-complicated nonstiff integrator initially. If it proves ineffective, the stiff integrator is then used.

A variety of problems have been solved by lines. Here we mention the paper of Chang and Madsen [207] on a two-dimensional chemical kinetics transport problem, and the papers of Madsen and Sincovec [208] who study a nonlinear diffusion problem for $u_t = [D(1 - \alpha\sqrt{u})u_x]_x$ and one-dimensional shallow water flow over an isolated obstacle. For the latter problem, they claim reproduction of the results of Houghton and Kashara [209, 210] obtained by alternative methods. Further discussion of the line method is available in Mikhlin and Smolitskiy [211]. Walter [212] discusses applications to approximate solutions and existence–uniqueness questions.

5-12 Fast Fourier transform and applications

Analogous to the Fourier series

$$f(x) = \sum_{m=-\infty}^{\infty} a_k e^{im\pi x} \tag{5-181}$$

for $f(x)$ defined on the continuum $-1 \leq x \leq 1$ and with

$$a_m = \tfrac{1}{2} \int_{-1}^{1} f(x) e^{-im\pi x} \, dx$$

there exists a discrete Fourier polynomial for a function defined on an equally spaced set of points $\{x_j\}$ for $x_j = j/J$, where $-J \leq j < J$.

If the number of points is even, consider the vectors $\{f_j\}$, for $-J \leq j < J$, which has an even number of components. The discrete Fourier series is

$$f_j = \sum_{m=-J}^{J-1} A_m e^{im\pi j/J} \tag{5-182}$$

$$A_m = \frac{1}{2J} \sum_{j=-J}^{J-1} f_j e^{-im\pi j/J}. \tag{5-183}$$

Now suppose f_j of (5-182) is defined from $f(x)$ on the continuum $-1 \leq x \leq 1$, as $f_j = f(x_j)$, $x_j = j/J$, $-J \leq j < J$. Upon setting (5-181) into (5-183) there results

$$A_m = \frac{1}{2J} \sum_{j=-J}^{J-1} \sum_{n=-\infty}^{\infty} \sum_{k=-J}^{J-1} a_{k+2Jn} e^{i(k+2Jn)\pi j/J} e^{-im\pi j/J}$$

$$= \sum_{n=-\infty}^{\infty} a_{m+2Jn}$$

from the discrete orthogonality relation. Thus when we sample the values of $f(x)$ on the discrete mesh $\{x_j\}$, the modes $e^{i(k+2Jn)\pi x}$ cannot be distinguished from the lower frequency modes $e^{ik\pi x}$. The higher frequencies are folded into the lower one. This is the *aliasing* phenomenon discussed in the book by Blackman and Tukey [213]. In discrete Fourier analysis and synthesis a *fast method* is often needed to evaluate the Fourier sum

$$X_j = \sum_{k=0}^{N-1} A_k e^{2\pi ijk/N}, \quad 0 \le j \le N - 1 \tag{5-184}$$

where X_j and A_k are complex numbers. Simple evaluation of this formula for the N components $\{X_j\}$ would require N^2 complex multiplications and $N \times (N - 1)$ additions, where evaluation of $(e^{2\pi i})^j$, $0 \le j \le N - 1$, has not been included. The *fast Fourier transform* is a method introduced by Cooley and Tukey [214] to reorder the computation so that the operational count is reduced from the $O(N^2)$ value required by the straightforward summation. The Cooley and Tukey paper shows how the reduction can be efficiently implemented by a computer program if N is a power of 2 or 4. Algorithms for other factorizations are available. Here we describe the algorithm for complex Fourier series and merely note it is possible to apply it to real Fourier sums, sine sums, or cosine sums.

Suppose $N = N_1 N_2 \cdots N_p$ where† the N_i are integers. For convenience set $N = N_1 N_1^*$, $N_1^* = N_2 \cdots N_p$. Then for all values $0 \le j \le N - 1$ and $0 \le k \le N - 1$ we can define j_0, j_1, k_0, k_1 such that

$$j = j_1 N_1^* + j_0, \quad 0 \le j_0 < N_1^*, 0 \le j_1 < N_1$$

$$k = k_1 N_1 + k_0, \quad 0 \le k_0 < N_1, 0 \le k_1 < N_1^*$$

and there are one-to-one correspondences between j and the pairs (j_1, j_0), k and the pairs (k_1, k_0). Thus it follows that

$$X_j = X(j_0, j_1) = \sum_{k_0=0}^{N_1} \sum_{k_1=0}^{N_1^*-1} A(k_0, k_1) w^{jk_1 N_1 + jk_0} \tag{5-185}$$

where $w = e^{2\pi i/N}$ has been written for convenience. From the foregoing decomposition

$$w^{jk_1 N_1} = w^{j_1 N_1^* k_1 N_1 + j_0 k_1 N_1}$$

$$= w^{j_1 k_1 N + j_0 k_1 N_1} = w^{j_0 k_1 N_1}$$

since $e^{2\pi i} = 1$. Consequently, (5-185) becomes

$$X(j_0, j_1) = \sum_{k_0=0}^{N_1} \sum_{k_1=0}^{N_1^*-1} A(k_0, k_1) w^{j_0 k_1 N_1} w^{(j_1 N_1^* + j_0)k_0}$$

$$= \sum_{k_0=0}^{N_1} A_1(j_0, k_0) w^{(j_1 N_1^* + j_0)k_0} \tag{5-186}$$

† For example, $N = 2^p$ or $N = 3 \times 5 \times 2^{p-2}$.

where

$$A_1(j_0, k_0) = \sum_{k_1=0}^{N_1^*-1} A(k_0, k_1)w^{j_0 k_1 N_1}. \tag{5-187}$$

With j_0 and k_0 ranging through all possible values it is clear that A_1 has N components. If no further reductions are carried out, but A_1 is computed directly from (5-187), then N_1^* operations are needed for each component of A_1. To compute each of the N components of X requires N_1 operations. Therefore, this single step reduction (i.e. first compute A_1, then X) requires a total of $NN_1^* + NN_1$ operations. In the special case $N = N_1 N_2$, $N_1 = N_2 = 2^s$, the total is $N^2 = 2^{4s}$ for the straightforward method and 2^{3s+1} for this reduction. For $s > 1$, this clearly pays. In fact the single step reduction is

$$N(2N^{1/2})/N^2 = 2/N^{1/2} = 2^{-s+1}.$$

Additional reductions are possible and great improvements over N^2 can be made. The extension is sketched below. Clearly Eqn (5-187), for A_1, has exactly the same form as that for X, Eqn (5-184), namely

$$A_1(j_0, k_0) = \sum_{k_1=0}^{N_1^*-1} A(k_0, k_1)(w^*)^{j_0 k_1} \tag{5-188}$$

where $w^* = w^{N_1}$ and k_0 is a parameter. From the first factorization we have $N_1^* = N_2 N_2^*$, $N_2^* = N_3 N_4 \cdots N_p$. With fixed k_0, the N_1^* values $A_1(j_0, k_0)$ can be computed using the same method for (5-188) as used for (5-184). For (5-184), $N(N_1 + N_1^*)$ operations were required. Consequently, for (5-188), $N_1^*(N_2 + N_2^*)$ operations are required for each value of k_0, and thus there are $N_1 N_1^*(N_2 + N_2^*)$ operations altogether. The calculation of A_1 directly from (5-188) requires NN_1^* operations. By carrying out this next reduction NN_1^* is replaced by $N_1 N_1^*(N_2 + N_2^*) = N(N_2 + N_2^*)$ in the operation count. The total number of operations required to compute X_j by this two-stage reduction is $N(N_1 + N_2 + N_2^*)$. By induction the operational count for $p - 1$ stages of reduction is

$$N(N_1 + N_2 + \cdots + N_p). \tag{5-189}$$

If $N = r^p$, i.e. $N_i = r$ for each i, then the operation count is pr^{p+1} which is a great improvement over $N^2 = r^{2p}$ for $p \gg 1$. If $N = 2^8 = 256$, then the ratio is $8 \cdot 2^9/2^{16} = 2^{-4} = \frac{1}{16}$ a much faster result!

Hockney's [215] *Fourier analysis-cyclic reduction (FACR) method* is a fast method for the *direct*† solution of the five-point approximation to the Poisson equation $u_{xx} + u_{yy} = f$. It owes its speed to the use of the fast

---

† No iterations are required.

Fourier transform. It cannot be used for the general elliptic equation but is applicable to Poisson's equation for Dirichlet boundary conditions $u = h$, Neuman conditions $\partial u/\partial n = 0$, and periodic conditions providing the same boundary conditions apply to opposite sides of the rectangle (cf. Hockney [216]). The method can be modified to solve other elliptic equations, e.g.

$$f_1(y)u_{xx} + (f_2(y)u_y)_y + f_3(y)u = f_4.$$

The *spectral* method of Orszag [217, 218] uses the fast Fourier transform with Galerkin's method (see Chapter 6) for numerically simulating incompressible flows within simple boundaries.

PROBLEMS

5-52 Write out the details of the fast Fourier transform for (5-187).

5-53 Carry out the induction leading to (5-189).

5-54 Apply the fast Fourier transform algorithm to the sine series

$$X_j = \sum_{k=1}^{J-1} A_k \sin \frac{\pi k j}{J}, \quad 1 \le j \le J - 1$$

$$A_k = \frac{2}{J} \sum_{j=1}^{J-1} X_j \sin \frac{\pi k j}{J}.$$

5-13 Method of fractional steps

Beginning with the fundamental work of Peaceman, Rachford, and Douglas [see Sections 3-12, 3-13, and 5-5(b)] on ADI procedures, the *method of fractional steps* has been extended and improved in the works of many American and Soviet authors. For details, the reader should consult Yanenko [163]. The method, designed for problems in more than one space dimension, reduces the multidimensional problem to a series of steps each of which involves difference approximations in only one dimension. Usually the difference scheme reduces to a product of operators, each operating in only one direction, e.g. $U^{n+1} = L(U^n)$ where $L = L_1 L_2 L_3$. If each operator L_i satisfies $\|L_i\| \le 1 + \alpha_i \, \Delta t$, then $\|L\|$ satisfies a similar condition thus ensuring stability. As previously remarked, ADI schemes are one type of fractional step method. Also like the ADI scheme, most fractional step procedures are implicit, but in only one direction at a time [compare Section 5-5(b)]. Since ADI, called the *method of stabilizing corrections* by Yanenko, has been discussed previously, here we discuss the *splitting method*. After introducing the concept a brief discussion of applications will be presented.†

---

† While numerical analysis has profited greatly from these ideas, they have not been used with any frequency in analytic studies.

Consider an initial value problem of the form

$$u_t = L_x(u) + L_y(u) + L_z(u) + f \tag{5-190}$$

with $u = u(x, y, z, t)$ and subject to appropriate initial data. The subscripts on the operator indicate that only derivatives with respect to those variables are involved, e.g. $L_x(u) = (\alpha u_x)_x$. Suppose the domain is all of R^3 or is rectangular with periodic boundary conditions, so that boundary conditions can be ignored. Let the individual operators be approximated by finite difference operators indicated with the same subscript. For example, on an equally spaced mesh (x_i, y_j, z_k), where $x_i = ih$, $y_j = jh$, $z_k = kh$ with $t_n = n \, \Delta t$, i, j, k, and n ranging through all integer values,

$$D_x(u)_{i,j,k} = [\alpha_{i+1/2,j,k}(u_{i+1,j,k} - u_{i,j,k})$$
$$- \alpha_{i-1/2,j,k}(u_{i,j,k} - u_{i-1,j,k})]/h^2 \tag{5-191}$$

where $\alpha_{i+1/2,j,k} = \alpha(x_i + h/2, y_j, z_k)$, with similar results for D_y and D_z. In general we suppose the difference operators have truncation error $O(h^p)$ ($p = 2$ in Eqn (5-191)). Let

$$A_x = I - \frac{\Delta t}{2} D_x, \qquad A_y = I - \frac{\Delta t}{2} D_y, \qquad A_z = I - \frac{\Delta t}{2} D_z$$

$$B_x = I + \frac{\Delta t}{2} D_x, \qquad B_y = I + \frac{\Delta t}{2} D_y, \qquad B_z = I + \frac{\Delta t}{2} D_z$$

and assume $f \equiv 0$. Then the splitting method consists of

$$A_x u^{n+1/3} = B_x u^n$$
$$A_y u^{n+2/3} = B_y u^{n+1/3} \tag{5-192}$$
$$A_z u^{n+1} = B_z u^{n+2/3}$$

where the subscripts have been suppressed. If we replace the A's and B's by their definitions in terms of the D's, we obtain ($D = D_x + D_y + D_z$)

$$\frac{u^{n+1} - u^n}{\Delta t} = D\left[\frac{u^{n+1} + u^n}{2}\right] - \frac{(\Delta t)^2}{4}(D_x D_y + D_x D_z + D_y D_z)$$
$$\times \left(\frac{u^{n+1} - u^n}{\Delta t}\right) + \frac{(\Delta t)^2}{8} D_x D_y D_z(u^{n+1} + u^n) \tag{5-193}$$

where we assume the operators commute (see Problem 5-55).

From (5-192) it is clear that the intermediate results do not contain approximations to all of the spatial derivatives. Therefore, the fractional steps are not even first-order approximations to the solution. When using the method the boundary conditions must be treated carefully. For proper boundary treatment, consider the two-dimensional diffusion equation (Yanenko [163])

$$u_t = u_{xx} + u_{yy}, \quad 0 < x < 1, \, 0 < y < 1, \, 0 < t$$

$$u(x, y, 0) = f(x, y), \qquad u(0, y, t) = g_1(y, t)$$

$$u(x, 0, t) = g_2(x, t), \qquad u(1, y, t) = g_3(y, t).$$

The boundary of the region is denoted by c_1 for $x = 0$, c_2 for $y = 0$, c_3 for $x = 1$, and c_4 for $y = 1$. The following splitting algorithm with proper boundary treatment produces second-order accuracy, with truncation error $O[(\Delta t)^2 + h^2]$.

Let A_x, A_y, B_x, and B_y be defined, as before, in terms of the centered second-order difference approximations D_x, D_y for $\partial^2/\partial x^2$ and $\partial^2/\partial y^2$. The splitting is

$$u^{n+1/2} = \left(I - \frac{\Delta t}{2} D_y\right) f^n \quad \text{in interior of } c_1 \text{ and } c_3$$

$$u^{n+1/2} = f^{n+1/2} \qquad \text{at the four corners}$$

$$A_x u^{n+1/2} = B_x f^n \qquad \text{in interior of } c_2 \text{ and } c_4$$

$$A_x u^{n+1/2} = B_x u^n \qquad \text{in square's interior}$$

$$u^{n+1} = f^{n+1} \qquad \text{on } c_2 \text{ and } c_4$$

$$A_y u^{n+1} = B_y u^{n+1/2} \qquad \text{on } c_1, c_3 \text{ and interior of the square.}$$

Marchuk [219] shows that the splitting method has only first-order accuracy $O(\Delta t + h^p)$ if the operators D_x etc. do not commute. His alternative, which yields second-order accuracy, is to reverse the order of the operators on every other time step. His scheme applied to $u_t = (\alpha u_x)_x + (\alpha u_y)_y$ is

$$A_x u^{n+1/2} = B_x u^n \qquad\qquad A_y u^{n+3/2} = B_y u^{n+1}$$

$$A_y u^{n+1} = B_y u^{n+1/2} \qquad\qquad A_x u^{n+2} = B_x u^{n+3/2}$$

and the operators are as previously described.

A fractional steps method known as the *method of approximate factorization* is an approximation to a Crank–Nicolson scheme (Sections 2-3, 5-4), which provides second-order accuracy with noncommutative operators. It is given here for (5-190). The Crank–Nicolson scheme is

$$\left(I - \frac{\Delta t}{2} D\right) u^{n+1} = \left(I + \frac{\Delta t}{2} D\right) u^n + \Delta t \, f^{n+1/2} \qquad (5\text{-}194)$$

where $D = D_x + D_y + D_z$ and the subscripted D's are difference approximations to the L's. With

$$A_x = I - \frac{\Delta t}{2} D_x, \qquad B_x = I + \frac{\Delta t}{2} D_x$$

etc., the *approximate factorization method* is

$$A_x A_y A_z u^{n+1} = B_x B_y B_z u^n + \Delta t f^{n+1/2}$$

which can be written in fractional time steps as

$$A_x u^{n+1/3} = B_x B_y B_z u^n + \Delta t f^{n+1/2}$$

$$A_y u^{n+2/3} = u^{n+1/3} \tag{5-195}$$

$$A_z u^{n+1} = u^{n+2/3}.$$

Equation (5-194) can be written as (Problem 5-60)

$$\left(I - \frac{\Delta t}{2} D\right) u^{n+1} = \left(I + \frac{\Delta t}{2} D\right) u^n - \frac{(\Delta t)^3}{4} (D_x D_y + D_x D_z + D_y D_z)$$

$$\times \left(\frac{u^{n+1} - u^n}{\Delta t}\right) + \frac{(\Delta t)^3}{8} D_x D_y D_z (u^{n+1} + u^n) + \Delta t f^{n+1/2}$$

which is the same as (5-193) under the assumption of commutative operators. For commutative operators, (5-192) and (5-195) are equivalent methods with similar boundary condition problems.

On the other hand the ADI equation (5-76), or one of its equivalent forms, is an $O(\Delta t)$ approximation to the solution of the differential equation at each step. Accurate insertion of the boundary conditions is thus easier in ADI.

The various methods of fractional steps are applicable to a wide variety of problems including potential problems, elasticity, and fluid dynamics. We have already remarked on applications of ADI in several inviscid flow calculations (see Yanenko [163]). Further development of the method, with corresponding analytic applications, should lead to the numerical solution of a wide variety of three-dimensional problems that presently defy adequate treatment. Perfection of the method should lead to greater competition with the traditional methods, including finite elements which we describe in the next chapter.

PROBLEMS

5-55   If $L_x(u) = (\alpha u_x)_x$, $L_y(u) = (\alpha u_y)_y$, $L_z(u) = (\alpha u_z)_z$ show that D_x, D_y, D_z will commute if α is constant. Are there any other conditions on α permitting commutativity?

5-56 Consider $u_t = K(u_{xx} + u_{yy}) + f$, K constant, $-\infty < x < \infty$, $-\infty < y < \infty$, $u(x, y, 0) = u_0(x, y)$. Set up a splitting method that is $O[(\Delta t)^2 + h^2]$, $h = \Delta x = \Delta y$.

5-57 Assume $f \neq 0$ in (5-190); modify Eqns (5-192) to obtain a second-order splitting scheme for this case. Assume the D_x, D_y, and D_z commute and obtain the analogue of (5-193).

5-58 Let

$$L_x = \alpha u_{xx} + \beta_x u_x, \qquad L_y = \alpha u_{yy} + \beta_y u_y, \qquad L_z = \alpha u_{zz} + \beta_z u_z$$

with α, β_x, β_y, β_z constants. Assume the domain is all of R^3 and that $f \equiv 0$. Analyze stability of the splitting scheme by means of the Fourier transform.

5-59 Verify (5-193).

5-60 Verify (5-195).

5-61 Write out the ADI scheme analogous to (5-76) in the present notation. Show that a more economical computation form is

$$\frac{u^{n+1/3} - u^n}{\Delta t} = D_x \frac{u^{n+1/3} + u^n}{2} + D_y u^n + D_z u^n$$

$$\frac{u^{n+2/3} - u^{n+1/3}}{\Delta t} = D_y \frac{u^{n+2/3} - u^n}{2}$$

$$\frac{u^{n+1} - u^{n+2/3}}{\Delta t} = D_z \frac{u^{n+1} - u^n}{2}.$$

REFERENCES

1. Crandall, S. H. *J. Ass. comput. Mach.*, **2**, 42, 1955.
2. Timoshenko, S. P. and Woinowsky-Krieger, S. *Theory of Plates and Shells*, 2nd edit. McGraw-Hill, New York, 1959.
3. Weinstein, A. Singular partial differential equations and their applications, in *Fluid Dynamics and Applied Mathematics* (J. B. Diaz and S. I. Pai, eds), p. 29. Gordon & Breach, New York, 1962.
4. Trutt, F. C., Erdelyi, E. A., and Jackson, R. F. *I.E.E.E. Trans. Aerospace*, **1**, 430, 1963.
5. Milne, W. E. *Numerical Solution of Differential Equations*, p. 122. Wiley, New York, 1953.
6. Motz, H. *Q. appl. Math.*, **4**, 371, 1946.
7. Woods, L. C. *Q. Jl. Mech. appl. Math.*, **6**, 163, 1953.
8. Lieberstein, H. M. Singularity occurrence and stably posed problems for elliptic equations, *Rept. No. MRC-TR-81*. University of Wisconsin Mathematics Research Center, Madison, Wisconsin, 1959.
9. Greenspan, D. and Warten, R. On the approximate solution of Dirichlet type problems with singularities on the boundaries, *Rept. No. MRC-TR-254*. University of Wisconsin Mathematics Research Center, Madison, Wisconsin, 1961.

10. Eisen, D. *Soc. ind. appl. Math., J. num. Analysis*, **4**, 545, 1966.

11. Eisen, D. *Maths Comput.*, **22**, 347, 1968.

12. Albasiny, E. L. *Q. J. Mech. appl. Math.*, **13**, 374, 1960.

13. Bramble, J. H., Hubbard, B. E., and Zlamal, M. *Soc. ind. appl. Math., J. num. Analysis*, **5**, 1, 1968.

14. Ladyzhenskaya, O. A. *The Mathematical Theory of Viscous Incompressible Flow.* Gordon & Breach, New York, 1963.

15. Courant, R. and Friedrichs, K. O. *Supersonic Flow and Shock Waves.* Wiley (Interscience), New York, 1948.

16. Longley, H. J. Methods of differencing in Eulerian hydrodynamics, *Rept. No. LA-2379.* Los Alamos Scientific Laboratory, Los Alamos, New Mexico, 1960.

17. Richtmyer, R. D. and Morton, K. W. *Difference Methods for Initial Value Problems*, 2nd edit. Wiley (Interscience), New York, 1967.

18. Trulio, J. G. and Trigger, K. R. Numerical solution of the one-dimensional Lagrangian hydrodynamic equations, *Rept. No. UCRL-6267.* Lawrence Radiation Laboratory, University of California, Berkeley, 1961.

19. Brode, H. L. *J. appl. Phys.*, **26**, 766, 1955.

20. Brode, H. L. Point source explosion in air, *Res. Memo. No. RM-1824-AEC.* Rand Corp., Santa Monica, California, 1956.

21. Schulz, W. D. *J. Math. Phys.*, **5**, 133, 1964.

22. Whiteman, J. R. Singularities due to re-entrant corners in harmonic boundary value problems, *Rept No. MRC-TR-829.* University of Wisconsin Mathematics Research Center, Madison, Wisconsin, 1967.

23. Whiteman, J. R. Treatment of singularities in a harmonic mixed boundary value problem by dual series methods, in *Q. J. Mech. appl. Math.*, 1969 (in press).

24. Lax, P. D. *Communs pure appl. Math.*, **7**, 159, 1954.

25. Lax, P. D. and Wendroff, B. *Communs pure appl. Math.*, **13**, 217, 1960; **17**, 381, 1964.

26. Fox, L. (ed.). *Numerical Solution of Ordinary and Partial Differential Equations*, pp. 353–363. Pergamon Press (London) and Addison-Wesley (Reading, Mass.), 1962.

27. Godunov, S. K. *Mat. Sb.*, **47**, 271, 1959 (*see also* Richtmyer and Morton [17]).

28. Stein, L. R. A numerical solution of a spherical blast wave utilizing a completely tabular equation of state, *Rept No. LA-2277.* Los Alamos Scientific Laboratory, Los Alamos, New Mexico, 1959.

29. Keller, H. B., Levine, D. A., and Whitham, G. B. *J. Fluid Mech.*, **7**, 302, 1960.

30. Courant, R. D. and Hilbert, D. *Methods of Mathematical Physics*, vol. 1. Wiley (Interscience), New York, 1953.

31. Timoshenko, S. P. and Gere, J. M. *Theory of Elastic Stability*, 2nd edit. McGraw-Hill, New York, 1961.

32. Crandall, S. H. *Engineering Analysis—A Survey of Numerical Procedures.* McGraw-Hill, New York, 1956.

33. Ames, W. F. *Nonlinear Partial Differential Equations in Engineering.* Academic Press, New York, 1965.

34. Moon, P. and Spencer, D. E. *Field Theory for Engineers.* Van Nostrand, Princeton, New Jersey, 1961; *Field Theory Handbook.* Springer-Verlag, Berlin, 1961.

35. Mann, W. R., Bradshaw, C. L., and Cox, J. G. *J. Math. Phys.*, **35**, 408, 1957.

36. Collatz, L. *Eigenwertaufgaben mit Technischen Anwendungen.* Akad. Verlag, Leipzig, 1949.
37. Richardson, L. F. and Gaunt, J. A. *Trans. R. Soc.,* A226, 299, 1927.
38. Bodewig, E. *Matrix calculus.* Wiley (Interscience), New York, 1956.
39. Wilkinson, J. H. Theory and practice in linear systems and the determination of characteristic values and vectors, in *Application of Advanced Numerical Analysis to Digital Computers* (J. W. Carr, ed.), pp. 41–152. University of Michigan College of Engineering, Ann Arbor, Michigan, 1959.
40. Givens, W. *J. Ass. comput. Mach.,* 4, 298, 1957.
41. Wielandt, H. Bestimmung Eigenwerte durch gebrochene Iteration, *Rept No. 44/J/37.* Bericht der Aerodyn. Versuch., Göttingen, 1944.
42. Crandall, S. H. *Proc. R. Soc.,* A207, 416, 1951.
43. Ostrowski, A. M. *Archs. ration. Mech. Analysis,* 1, 233, 1958; 4, 153, 1959.
44. Copley, D. T. A note on the solution of the wave equation in the neighborhood of a re-entrant corner, *Admiralty Rept C.V.D. MAN/1,* 1945.
45. Forsythe, G. E. and Wasow, W. R. *Finite Difference Methods for Partial Differential Equations.* Wiley, New York, 1960.
46. Ehrlich, L. W., Riley, J. D., Strang, W. G., and Troesch, B. A. *J. Soc. ind. appl. Math.,* 9, 149, 1961.
47. Wilkinson, J. H. *J. Soc. ind. appl. Math.,* 10, 162, 1962.
48. Motz, H. Self-consistent calculation of the configuration of an R.F. confined plasma, in *Ionization Phenomena in Gases* (Conference at Munich). North Holland Pub. Co., Amsterdam, 1961.
49. Douglas, J., Jr. Numerical methods for parabolic equations, in *Advances in Computers* (F. L. Alt, ed.), vol. 2. Academic Press, New York, 1961.
50. Douglas, J., Jr. *J. Soc. ind. appl. Math.,* 9, 433, 1961.
51. Douglas, J., Jr. *J. Soc. ind. appl. Math.,* 3, 42, 1955.
52. Peaceman, D. W. and Rachford, H. H. *J. Soc. ind. appl. Math.,* 3, 28, 1955.
53. Douglas, J., Jr and Rachford, H. H. *Trans. Am. math. Soc.,* 82, 421, 1956.
54. Douglas, J., Jr. *Num. Math.,* 4, 41, 1962.
55. Brian, P. L. T. *A.I.Ch.E. Jl.,* 7, 367, 1961.
56. Douglas, J., Jr, Garder, A. O., and Pearcy, C. *Soc. ind. appl. Math., J. Num. Analysis,* 3, 570, 1966.
57. Douglas, J., Jr and Pearcy, C. *Num. Math.,* 5, 175, 1963.
58. Pearcy, C. *Num. Math.,* 4, 172, 1962.
59. Fairweather, G. and Mitchell, A. R. *Num. Math.,* 6, 285, 1964.
60. Fairweather, G. and Mitchell, A. R. *J. Soc. ind. appl. Math.,* 13, 957, 1965.
61. Rachford, H. H., Jr. *Soc. ind. appl. Math., J. Num. Analysis,* 5, 407, 1968.
62. Stone, H. L. *Soc. ind. appl. Math., J. Num. Analysis,* 5, 530, 1968.
63. Lees, M. *J. Soc. ind. appl. Math.,* 10, 610, 1962.
64. Bers, L. *J. Res. natn. Bur. Std.,* 51, 229, 1953.
65. Schechter, S. *Trans. Am. math. Soc.,* 106, 179, 1962.
66. Lieberstein, H. M. Overrelaxation for nonlinear elliptic partial differential equations, *Tech. Rept No. MRC-TR-80.* University of Wisconsin Mathematics Research Center, Madison, Wisconsin, 1959.
67. Lieberstein, H. M. A numerical test case for the nonlinear overrelaxation algorithm, *Tech. Rept No. MRC-TR-122.* University of Wisconsin Mathematics Research Center, Madison, Wisconsin, 1960.
68. Greenspan, D. and Yohe, M. On the approximate solution of $\nabla^2 u = F(u)$, *Tech. Rept No. MRC-TR-384.* University of Wisconsin Mathematics Research Center, Madison, Wisconsin, 1963; see also *Communs Ass. comput. Mach.,* 6, 564, 1963.

69. Wise, H. and Ablow, C. M. *J. chem. Phys.*, **35**, 10, 1961.
70. Bellman, R. E., Juncosa, M., and Kalaba, R. Some numerical experiments using Newton's method for nonlinear parabolic and elliptic boundary value problems, *Rept. No. P-2200*. Rand Corp., Santa Monica, California, 1961.
71. Greenspan, D. Recent computational results in the numerical solution of elliptic boundary value problems, *Tech. Rept No. MRC-TR-408*. University of Wisconsin Mathematics Research Center, Madison, Wisconsin, 1963.
72. Mackenroth, E. and Fisher, G. D. Determination of the flow field of an ideal axisymmetric impinging jet. Preprint 31F, *Symp. Nonlinear Problems in Transport*, 61st AIChE Meeting, December, 1968.
73. Woods, L. C. *J. Mech. appl. Math.*, **4**, 4, 1951.
74. Gunn, J. E. *Num. Math.*, **6**, 181, 1964.
75. Gunn, J. E. *Soc. ind. appl. Math., J. Num. Analysis*, **2**, 24, 1965.
76. Erdelyi, E. A., Ahamed, S. V., and Burtness, R. D. *I.E.E.E. Trans. Power Appl. and Systems*, *PAS-84*, 375, 1965.
77. Ahamed, S. V. *Comput. J.*, **8**, 73, 1965.
78. Wachspress, E. L. *Iterative Solution of Elliptic Systems*. Prentice-Hall, Englewood Cliffs, N.J., 1966.
79. Holmboe, J., Forsythe, G. E., and Gustin, W. *Dynamic Meteorology*. Wiley, New York, 1945.
80. Charney, J. G. and Phillips, N. A. *J. Meterol.*, **71**, 1953.
81. Bolin, B. *Tellus*, **8**, 1956.
82. Arnason G. A convergent method for solving the balance equation. Internal, Report Joint Numer. Weather Prediction Unit, U.S. Weather Bureau, Washington, D.C., 1957.
83. Pogorelov, A. V. *Monge-Ampere Equations of Elliptic Type*. P. Noordhoff, Ltd, Groningen, The Netherlands, 1964.
84. Courant, R. and Hilbert, D. *Methods of Mathematical Physics*, vol. 2. Wiley (Interscience), New York, 1962.
85. Thornhill, C. K. The numerical method of characteristics for hyperbolic problems in three independent variables. *A.R.C. Rept and Mem. 2615*, 1948.
86. Coburn, N. and Dolph, C. L. *Proc. First Appl. Math. Symp. Am. Math. Soc.*, **55**, 1947.
87. Butler, D. S. *Proc. R. Soc.*, A**255**, 232, 1960.
88. Bruhn, G. and Haack, W. *Z. angew. Math. Phys.*, **9**, 173, 1958.
89. Talbot, G. P. Application of the numerical method of characteristics in three independent variables to shock-thermal layer interaction problems. Report of U.K. Royal Armament Research and Development, 1963.
90. Richardson, D. J. In *Methods in Computational Physics*, vol. 3, p. 295. Academic Press, New York, 1964.
91. Ansorge, R. *Num. Math.*, **5**, 443, 1963.
92. Kolsky, H. *Rept No. LA-1867*. Los Alamos Scientific Laboratory, Los Alamos, New Mexico, 1955.
93. Blair, A., Metropolis, N. C., von Neumann, J., Taub, A. H., and Tsingow, M. *Maths Comput.*, **13**, 145, 1959.
94. Schultz, W. D. Two-dimensional Lagrangian hydrodynamical difference equations, in *Methods in computational physics*, vol. 3, p. 1. Academic Press, New York, 1964.
95. Burstein, S. Z. Numerical calculation of multidimensional shocked flows, *Rept No. NYO-10433*. Courant Institute of Mathematical Science, New

York University, 1963; also published in *Am. Inst. Aero. Astron. Jl.*, **2**, 211, 1964.

96. Burstein, S. Z. Finite difference calculations for hydrodynamic flows containing discontinuities, *Rept No. NYO-1480-33*. Courant Institute of Mathematical Science, New York University, 1965.

97. Burstein, S. Z. High accurate difference methods in hydrodynamics, in *Nonlinear Partial Differential Equations—A Symposium on Methods of Solution* (W. F. Ames, ed.), p. 279. Academic Press, New York, 1967.

98. Frank, R. M. and Lazarus, R. B. Mixed Eulerian–Lagrangian method, in *Methods in computational physics*, vol. 3, p. 47. Academic Press, New York, 1964.

99. Noh, W. F. CEL: A time dependent, two space dimensional coupled Eulerian-Lagrange code, in *Methods in Computational Physics*, vol. 3, p. 117, Academic Press, New York, 1964.

100. Morawetz, C. S. *Communs pure appl. Math.*, **9**, 45, 1956.

101. Richtmyer, R. D. A survey of finite difference methods for nonsteady fluid dynamics, *Rept No. NCAR-TN-63-2*. National Center for Atmospheric Research, Boulder, Colorado, 1962.

102. Houghton, D., Kasahara, A., and Washington, W. *Mon. Weath. Rev. U.S. Dep. Agric.*, **94**, 141, 1966.

103. Gourlay, A. R. and Morris, J. Ll. *Maths Comput.*, **22**, 549, 1968.

104. Zabusky, N. J. *J. Math. Phys.*, **3**, 1020, 1962.

105. Ames, W. F., Lee, S. Y., and Zaiser, J. N. *Int. J. Nonlinear Mech.*, **3**, 449, 1968.

106. Courant, R. and Friedrichs, K. O. *Supersonic Flow and Shock Waves.* Wiley (Interscience), New York, 1948.

107. Jeffrey, A. and Taniuti, T. *Nonlinear wave propagation.* Academic Press, New York, 1964.

108. Stoker, J. J. *Communs pure appl. Math.*, **1**, 1, 1948.

109. Jeffrey, A. *Z. angew. Math. Phys.*, **15**, 97, 1964.

110. Chu, B. T. *J. Mech. Phys. Solids*, **12**, 45, 1964.

111. Bland, D. R. *J. Mech. Phys. Solids*, **12**, 245, 1964.

112. Kruskal, M. D. and Zabusky, N. J. *J. Math. Phys.*, **5**, 231, 1964.

113. Lax, P. D. *J. Math. Phys.*, 5, 611, 1964.

114. Jeffrey, A. *J. Math. Mech.*, **15**, 585, 1966.

115. Jeffrey, A. *J. Math. Mech.*, **17**, 331, 1967.

116. Ludford, G. S. S. *Proc. Camb. phil. Soc. Math.*, **48**, 499, 1952.

117. Morimoto, H. *Num. Math.*, **4**, 136, 1962.

118. Rouse, C. A. *J. Soc. ind. appl. Math.*, **9**, 127, 1961.

119. Soare, M. *Application of Finite Difference Equations to Shell Analysis.* Editura Acad. Repub. Soc. Romania, Bucharest, 1967.

120. Southwell, R. V. *Relaxation Methods in Engineering Science.* Oxford University Press, London, 1940.

121. Fox, L. and Southwell, R. V. *Phil. Trans. R. Soc. Camb.*, A, 15, 1941.

122. Southwell, R. V. *Relaxation Methods in Theoretical Physics.* Oxford University Press, London, 1946.

123. Heller, J. *J. Soc. ind. appl. Math.*, **8**, 150, 1960.

124. Varga, R. S. Factorization and normalized iterative methods, in *Boundary Problems in Differential Equations* (R. E. Langer, ed.), p. 121. University of Wisconsin Press, Madison, Wisconsin, 1960.

125. Parter, S. V. *Num. Math.*, **1**, 240, 1959.

126. Parter, S. V. *J. Ass. comput. Mach.*, **8**, 359, 1961.

127. Conte, S. D. and Dames, R. T. *Mathl. Tabl. natn. Res. Coun., Wash.* (now *Maths Comput.*), **12**, 198, 1958.
128. Conte, S. D. and Dames, R. T. *J. Ass. comput. Mach.*, **7**, 264, 1960.
129. Birkhoff, G. D. and Varga, R. S. *Trans. Am. math. Soc.*, **92**, 13, 1959.
130. Engli, M., Ginsburg, T., Rutishauser, H., and Stiefel, E. Refined iterative methods for computation of the solution and the eigenvalues of self-adjoint boundary value problems, *Mitterlungen aus dem Institut für angewandte Mathematik, No. 8.* Birkhauser Verlag, Basel/Stuttgart, 1959.
131. Griffin, D. S. and Varga, R. S. A numerical solution for plane elasticity problems, *Rept No. WAPD-256.* Bettis Atomic Power Laboratory, Pittsburgh, Pa., 1962.†
132. Griffin, D. S. and Varga, R. S. *J. Soc. ind. appl. Math.*, **11**, 1046, 1963.
133. Callaghan, J. B., Jarvis, P. H. and Rigler, A. K. A program for the solution of thin elastic plate equations, *Rept No. WAPD-TM-255.* Bettis Atomic Power Laboratory, Pittsburgh, Pa. 1961.†
134. Griffin, D. S. A numerical solution for plate bending problems, *Rept No. WAPD-230.* Bettis Atomic Power Laboratory, Pittsburgh, Pa., 1963.†
135. Sokolnikoff, I. S. *Mathematical Theory of Elasticity*, 2nd edit. McGraw-Hill, New York, 1956.
136. Babuska, I., Prager, M. and Vitasek, E. *Numerical Processes in differential Equations.* Wiley (Interscience), New York, 1966.
137. Zlamal, M., *Soc. ind. appl. Math., J. Num. Analysis*, **2**, 337, 1965.
138. Zlamal, M., *Soc. ind. appl. Math., J. Num. Analysis*, **4**, 626, 1967.
139. Bramble, J. H., *Num. Math.*, **9**, 236, 1966.
140. Smith, J., *Soc. ind. appl. Math., J. Num. Analysis*, **5**, 323, 1968.
141. Crandall, S. H. *J. Ass. comput. Mach.*, **1**, 111, 1954.
142. Woodall, S. R. *Int. J. Nonlinear Mech.*, **1**, 217, 1966.
143. Collatz, L. *Z. angew. Math. Mech.*, **31**, 392, 1951.
144. Thom, A. *Proc. R. Soc.*, A**141**, 651, 1933.
145. Payne, R. B. *Rept No. 3407.* Aeronautical Research Council, London, 1956.
146. Payne, R. B. *J. Fluid Mech.*, **4**, 81, 1958.
147. Fromm, J. E. A method for computing nonsteady incompressible viscous fluid flows, *Rept No. 2910.* Los Alamos Scientific Laboratory, Los Alamos, New Mexico, 1963.
148. Fromm, J. The time dependent flow of an incompressible viscous fluid, in *Methods in Computational Physics*, vol. 3, p. 345. Academic Press, New York, 1964.
149. Fromm, J. E. and Harlow, F. H. *Physics Fluids*, **6**, 975, 1963.
150. Keller, H. B. and Takami, H. Numerical studies of steady viscous flow about cylinders, in *Numerical Solution of Nonlinear Differential Equations* (D. Greenspan, ed.). Wiley, New York, 1966.
151. Greenspan, D., Jain, P. C., Manohar, R., Noble, B., and Sakurai, A. Numerical studies of the Navier–Stokes equations, *Rept No. MRC-TR-482.* University of Wisconsin, Mathematics Research Center, Madison, Wisconsin, 1964.
152. Chorin, A. J. "The numerical solution of the Navier–Stokes equations for incompressible fluids, *Rept No. NYO-1480-82.* Courant Institute of Mathematical Sciences, New York University, 1967.
153. Chorin, A. J. *Maths Comput.*, **22**, 745, 1968.

† Available from Office of Technical Services, Department of Commerce, Washington, D.C.

154. Smarskii, A. A. *U.S.S.R. Comput. Math. and Math. Phys.*, **5**, 894, 1963–1964.

155. Padmanabhan, H., Ames, W. F., and Kennedy, J. F. *J. Engrg. Math.*, **4**, 229, 1970.

156. Aziz, K. and Hellums, J. D. *Physics Fluids*, **10**, 314, 1967.

157. Ames, W. F. *Int. J. Nonlinear Mech.*, **5**, 605, 1970.

158. Berker, R. Intégration des équations du mouvement d'un fluide visqueux incompressible, *Handbuch der Physik*, Band VIII/2. Springer, Berlin–Göttingen–Heidelberg, 1963.

159. Wessel, W. R. Personal communication (*see also* [169]).

160. Townsend, A. A. *J. Fluid Mech.*, **24**, 307, 1966.

161. Giese, J. H. A bibliography for the numerical solution of partial differential equations, *BRL Mem. Rept. 2114*. Ballistic Research Laboratories, Aberdeen Proving Ground, Md., 1971. (1973 supplement and computer based searchable file also available.)

162. Roache, P. J. *Computational Fluid Dynamics*. Hermosa Publishers, Albuquerque, N.M., 1972.

163. Yanenko, N. N. *The Method of Fractional Steps*. Springer, New York–Berlin, 1971.

164. Pujol, A. Numerical experiments on the stability of Poiseuille flows of non-Newtonian fluids. Ph.D. Dissertation, University of Iowa, Iowa City, Iowa, 1971.

165. Ames, W. F. *SIAM Review*, **15**, 524, 1973.

166. Ames, W. F. *Nonlinear Partial Differential Equations in Engineering*, vol. II. Academic Press, New York, 1972.

167. Giaquinta, A. R. Numerical modeling of unsteady flow with natural and forced separation. Ph.D. Dissertation, University of Iowa, Iowa City, Iowa, 1974.

168. Emmons, H. W. The numerical solution of the turbulence problem, in *Proc. First Symp. Applied Mathematics* (E. Reissner, W. Prager, and J. J. Stoker, eds). Am. Math. Soc., Providence, R.I., **67**, 1949.

169. Thoman, D. C. and Szewczyk, A. A. *Proceedings IUTAM Symposium on High Speed Computing in Fluid Dynamics* (F. N. Frankiel and K. Stewartson, eds.), *Physics of Fluids*, Supplement II, **12**, 76, 1969.

170. Harlow, F. H. Stability of difference schemes; selected topics, *Rept. 2452*. Los Alamos Scientific Laboratory, Los Alamos, N.M., 1960.

171. Fromm, J. E. Finite difference methods of solution of non-linear flow processes with application to the Bénard problem, *Rept. LA-3522*. Los Alamos Scientific Laboratory, Los Alamos, N.M., 1967.

172. Hung, T. K. and Macagno, E. O. *LaHouille Blanche*, **21**, 391, 1966.

173. Torrance, K. E. *J. Res. NBS*, **72B**, 281, 1968.

174. Harlow, F. H. and Fromm, J. E. *Phys Fluids*, **7**, 1147, 1964.

175. Wilkes, J. O. The finite difference computation of natural convection in an enclosed rectangular cavity. Ph.D. Dissertion, University of Michigan, Ann Arbor, Michigan, 1963.

175a. Douglas, J., Jr. *SIAM J. Appl. Math.*, **3**, 42, 1955.

175b. Douglas, J., Jr. *Num. Math.*, **4**, 41, 1962.

176. Alonso, C. V. Time-dependent confined rotating flow of an incompressible viscous fluid. Ph.D. Dissertation, University of Iowa, Iowa City, Iowa, 1971.

177. Pearson, C. E. *J. Fluid Mech.*, **21**, 611, 1965.

178. Holt, M. *Proc. Second Int. Conf. on Num. Methods in Fluid Dyn.*, Lecture Notes in Physics 8. Springer-Verlag, New York–Berlin, 1971.

179. Hamielec, A. E., Hoffman, T. W., and Ross, L. L. *A.I. Ch.E. J.*, **13**, 220, 1967.

180. Rimon, Y. and Cheng, S. I. *Phys. Fluids*, **12**, 949, 1969.

181. Harlow, F. H. and Welch, J. E. *Phys. Fluids*, **8**, 2182, 1965.

182. Welch, J. E., Harlow, F. H., Shannon, J. P., and Daly, B. J. The MAC method, *Report LA 3425*. Los Alamos Scientific Laboratories, Los Alamos, N.M., 1966.

183. Hirt, C. W. and Harlow, F. H. *J. comp. Phys.*, **2**, 114, 1967.

184. Donovan, L. F. *AIAA J.*, **8**, 524, 1970.

185. Putre, H. A. *NASA-TN-D-5682*. NASA Lewis Research Center, Cleveland, Ohio, 1970.

186. Chan, R. K.-C., Street, R. L., and Fromm, J. E. See [178].

187. Amsden, A. A. and Harlow, F. H. *J. comp. Phys.*, **6**, 322, 1970.

188. Chorin, A. J. *Bull. Am. Math. Soc.*, **73**, 928, 1967.

189. Chorin, A. J. *J. comp. Phys.*, **2**, 12, 1967.

190. Aziz, K. A numerical study of cellular convection. Ph.D. Dissertation, Rice University, Houston, Texas, 1965.

191. Moon, P. and Spencer, D. E. *Field Theory Handbook*. Springer-Verlag, New York–Berlin, 1961.

192. Hirasaki, G. J. A General Formulation of the Boundary Conditions on the Vector Potential in Three Dimensional Hydrodynamics. Ph.D. Dissertation, Rice University, Houston, Texas, 1967.

193. Taylor, T. D. Numerical methods for predicting subsonic, transonic and supersonic flow, *AGARDograph No. 187*. NASA Langley, Langley Field, Va. 23365; 1973.

194. Orszag, S. A. and Israeli, M. *Annual Review of Fluid Mech.*, **6**. Annual Rev. Inc., Palo Alto, Calif., 1974.

195. Kreiss, H. and Oliger, J. *Methods for the approximate solution of time dependent problems*. UNIPUB Inc., P.O. Box 433, New York, 1973.

196. Roache, P. J. Recent developments and problem areas in computational fluid dynamics, in *Computational Mechanics*. Lecture Notes in Mathematics, 461. Springer, New York–Berlin, 1975.

197. Vichnevetsky, R. (ed.). *Proc. 1975 AICA International Symposium on Computer Methods for Partial Differential Equations*. AICA, Computer Science, Rutgers Univ., New Brunswick, N.J., 1976.

198. Brown, G. W. Monte Carlo methods, in *Modern Mathematics for the Engineer* (E. F. Beckenbach, ed.). McGraw-Hill, N.Y., 1956.

199. Barber, M. N. and Ninham, B. W. *Random and Restricted Walks*. Gordon & Breach, New York–London–Paris, 1970.

200. Meyer, H. A. (ed.) *Symposium on Monte Carlo Methods*. Wiley, New York, 1956.

201. Haviland, J. K. The solution of two molecular flow problems by the Monte Carlo method, in *Methods in Computational Physics* **4**, p. 109, 1965.

201a. Gentry, R. A., Harlow, F. H., and Martin, R. E. Computer experiments for molecular dynamics problems, in *Methods in Computational Physics* **4**, p. 211, 1965.

202. Orcutt, G. H., Greenberger, M., Rivlin, A. M., and Kerbel, J. *Micro-Analysis of Socioeconomic Systems—Assimilation Study*. Harper & Row, New York, 1961.

203. Liittschwager, J. and Ames, W. F. On transient queues: practice and pedagogy, in *Proc. Eighth Annual Sym. on Interface of Comp. Sci. and Statistics*, Feb. 1975.

204. Gear, C. W. *Numerical Initial Value Problems in Ordinary Differential Equations*. Prentice-Hall, Englewood Cliffs, N.J., 1971.

205. Hindmarsh, A. C. GEAR, Ordinary differential equations system solver, *UCID-30001 Rev. 2*. Lawrence Livermore Laboratory, 1972.

206. Liskovets, O. A. The method of lines, (English Translation) *J. Diff. Eqs.*, **1**, 1308, 1965.

207. Chang, J. S. and Madsen, N. K., Global transport and kinetics models, *UCRL-75062*. Lawrence Livermore Laboratory, 1973.

208. Madsen, N. K. and Sincovec, R. F. *Computational Methods in Nonlinear Mech.* (J. T. Oden, *et al.*, eds.) pp. 371–380. Texas Inst. for Comp. Mechanics, Austin, Texas, 1974.

209. Houghton, D. D. and Kashara, A. *Commun. Pure App. Math.*, **21**, 1, 1968.

210. Kashara, A. and Houghton, D. D. *Jour. Comp. Phys.*, **4**, 377, 1969.

211. Mikhlin, S. G. and Smolitskiy, K. L. *Approximate Methods for Solution of Differential and Integral Equations*. Elsevier, New York, 1967.

212. Walter, W. *Differential and Integral Inequalities*. Springer-Verlag, New York–Berlin, 1970.

213. Blackman, R. and Tukey, J. *The Measurement of Power Spectra*. Dover, New York, 1958.

214. Cooley, J. W. and Tukey, J. *Math. Comp.*, **19**, 297, 1965.

215. Hockney, R. W. *J. Assn. Comp. Mchry.*, **12**, 95, 1965.

216. Hockney, R. W. The potential calculation and some applications, in *Methods Computational Physics*, **9**. Academic Press, New York, 1971.

217. Orszag, S. *Studies in App. Math.*, **50**, 293, 1971.

218. Orszag, S. *J. Fluid Mech.*, **49**, 75, 1971.

219. Marchuk, G. On the theory of the splitting method, in *Numerical Solution of Partial Differential Equations II* B. Hubbard (ed.). Academic Press, N.Y., 469, 1971.

6
Weighted residuals and finite elements

6-0 Introduction

The *methods of weighted residuals* are general techniques for developing approximate solutions of operator equations. In all of these the unknown solution is approximated by a set of local basis functions containing adjustable constants or functions. These constants or functions are chosen by various criteria to give the 'best' approximation for the selected family. A general discussion of weighted residual methods is found in Ames [1] and Finlayson [2].

In treating the continuum it was quite natural for the practicing scientist and engineer to attempt to reduce it, in an essentially physical manner, to an assembly of discrete elements. The first attempts employed a decomposition of elastic continua into an assembly of rods or beams which are standard, well-understood, engineering structures. The early works of Hrenikoff [3], in which elasticity problems were solved by a 'framework' method, and McHenry [4], who introduced a lattice analogy for plane stress problems, bear special mention. Courant [5] formulated the essence of what is now called a *triangular finite element*. The paper of Newmark [6] should also be noted as well as the works of Prager and Synge [7] which evolved into the hypercircle method (cf., e.g. Synge [8]).

Credit for approximating directly to a continuum by an element with multiple connecting points and the use of the term *finite element* probably should go to Turner *et al.* [9] and Argyris [10]. These early efforts developed the basic properties by physical arguments related to the stress or displacement distribution within the subregion. The finite element method, arising as it did in structural mechanics, is now widely used in that field. A rapid expansion into other areas has been promoted by Zienkiewicz [11] (see references therein), Oden [12], and a variety of specialty conferences such as the Maryland Symposium of 1972 [13] and the International Conference on Computational Methods in Nonlinear Mechanics of 1974 [14]. From the mathematical viewpoint one of the weighted residual methods is a particularly convenient way to introduce the finite element procedure.

6-1 Weighted residual methods (WRM)

Consider the initial boundary value problem

$$\begin{aligned}
u_t &= Lu, & \mathbf{x} &\in V,\ t > 0 \\
u(\mathbf{x}, 0) &= u_0(x), & \mathbf{x} &\in V \\
u(\mathbf{x}, t) &= f_s(\mathbf{x}, t), & \mathbf{x} &\in s
\end{aligned} \tag{6-1}$$

for $u = u(\mathbf{x}, t)$, where L denotes a differential operator in the space derivatives of u, $\mathbf{x}$ is a vector of space variables, and V is a space domain with boundary s. There are several variations of any of the assorted weighted residual methods—the *interior*, *boundary* and *mixed* procedures. To develop these and also to list the various methods select a trial solution u_T in the form†

$$u_T(\mathbf{x}, t) = u_s(\mathbf{x}, t) + \sum_{i=1}^{N} c_i(t)u_i(\mathbf{x}) \qquad (6\text{-}2)$$

where the $u_i(\mathbf{x})$ are known basis functions (e.g. trigonometric, Legendre, etc.), selected from a set of complete (perhaps, orthonormal) basis functions, which satisfy

$$u_s = f_s, \qquad u_i = 0 \qquad \text{for } \mathbf{x} \text{ on } s.$$

Therefore, u_T satisfies the boundary conditions, but not the initial condition or the equation, for all functions $c_i(t)$. This is essentially the *interior method*, perhaps the easiest variation to apply for most problems. Of course, it is not necessary that the trial solution be linear in the c_i.

If trial solutions are selected that satisfy the differential equation but not the boundary conditions, the variant is called the *boundary method*. An intermediate situation exists in the so-called *mixed methods* where the trial solution does not satisfy either the equations or boundary conditions.

To select the optimal set of functions (constants) $c_i(t)$ the *equation residual*

$$R_E(u_T) = Lu_T - (u_T)_t \qquad (6\text{-}3)$$

and *initial residual*

$$R_I(u_T) = u_0(\mathbf{x}) - u_s(\mathbf{x}, 0) - \sum_{i=1}^{N} c_i(0)u_i(\mathbf{x}) \qquad (6\text{-}4)$$

are formed, for continuation of the interior method discussion. These are measures of how well the trial function u_T satisfies the equation and the initial conditions, respectively. If the trial function is the exact solution, then both residuals are zero. With increasing N the analyst 'hopes' that both residuals become smaller. In the WRM the functions (constants) $c_j(t)$ are chosen in such a way that the residuals are zero in some average sense. To develop those ideas we select N weighting functions w_j, $j = 1, 2, \ldots, N$, introduce the spatial average (inner product or weighted integral)

$$(w, v) = \int_V wv \, dV \qquad (6\text{-}5)$$

---

† Alternative forms are possible. For example $u_i = u_i(\mathbf{x}, t)$ with $c_i(t) = constants$ to be determined; $u_T = $ function of the right-hand side of (6-2) (see Ames [1], Finlayson [2]).

and set the N weighted integrals of the equation residual R_E equal to zero†,

$$(w_j, R_E(u_T)) = 0, \quad j = 1, 2, \ldots, N. \tag{6-6}$$

Equations (6-6) represent N simultaneous operator or algebraic equations for the c_j. If $c_j = c_j(t)$, the equations are ordinary differential equations; and if the c_j are constants, the N equations are algebraic. In a similar way, when the N weighted integrals of the initial residual R_I are set equal to zero, the initial conditions of the c_j, i.e. $c_j(0)$, are determined.

The particular WRMs differ from one another because of the choice of the weighting function w_j. The best known of these are described herein together with a little of their historical development.

(a) Subdomain (Biezeno and Koch [15, 1923])
Let the equation domain V be divided into N smaller subdomains V_j with

$$w_j(\mathbf{x}) = \begin{cases} 1, & \mathbf{x} \in V_j \\ 0, & \mathbf{x} \notin V_j \end{cases} \tag{6-7}$$

A modern extension is called the method of *integral relations* (Finlayson [2, p. 78]).

(b) Collocation (Frazer *et al.* [16])
Select N points $P_j = P_j(\mathbf{x}_j), j = 1, 2, \ldots, N$ in V with

$$w_j(\mathbf{x}) = \delta(\mathbf{x} - \mathbf{x}_j)$$

where δ is the Dirac delta generalized function.‡ Thus

$$(w_j, R_E) = \int_V \delta(\mathbf{x} - \mathbf{x}_j) R_E \, dV$$
$$= R_E(u_T(\mathbf{x}_j, t)) \equiv 0 \tag{6-8}$$

specifies that the residual is zero at N specified points P_j. As N increases the residual vanishes at more and more points. Lanczos [17] took the basis functions $u_i(\mathbf{x})$ as Chebyshev polynomials and used the roots to a Chebyshev polynomial as the collocation points. This procedure is sometimes referred to as *orthogonal collocation* in the recent literature (see Section 6-2).

(c) Least squares (Gauss–Legendre; cf. Hall [18], Sorenson [19])
Let the c_j be constants for all $j = 1, \ldots, N$. In the least squares method the functional

$$I(\mathbf{c}) = \int_V R_E^2 \, dV, \quad \mathbf{c} = (c_1, \ldots, c_N)$$

---

† The residual error is required to be *orthogonal* to each of the weighting functions.
‡ For a test function $\phi(\mathbf{x})$ that vanishes outside the compact set V, $\int_V \phi(\mathbf{x}) \, \delta(\mathbf{x} - \mathbf{x}_j) \, d\mathbf{x} = \phi(\mathbf{x}_j)$.

is to be made stationary. Thus

$$\frac{\partial I}{\partial c_j} = 2 \int_V R_{\mathrm{E}} \frac{\partial R_{\mathrm{E}}}{\partial c_j} \, dV = 0, \quad j = 1, \ldots, N \tag{6-9}$$

provides N algebraic equations for the c_j. The weighting functions are

$$w_j(\mathbf{x}) = \partial R_{\mathrm{E}}/\partial c_j.$$

This method often leads to cumbersome equations but has been applied to many complicated problems (cf. Becker [20]) in engineering. The (mean) square residual has some theoretical significance since error bounds can be obtained in terms of it. Questions of this type will be discussed in Section 6-4.

(d) Bubnov–Galerkin method (Bubnov [21], Galerkin [22])

Perhaps the best known of these approximate methods is the Bubnov–Galerkin procedure. In this method the weighting functions are chosen to be the basis functions of the trial solution, i.e.

$$w_j(\mathbf{x}) = u_j(\mathbf{x}).$$

Now, the basis functions were chosen as members of a complete set of functions, over some function space, so that any function in that space can be expanded as $f = \sum a_i u_i$. If the solution of (6-1) lies in that function space, then the trial solution, in the sense of Bubnov–Galerkin, is capable of representing the exact solution as $N \to \infty$.

(e) Moments (Yamada [23])

In this method, originally developed to study nonlinear diffusion and laminar boundary layer problems, successively higher moments of the residual are required to be zero. For the operator L in one variable x, the weighting functions are chosen as

$$w_j = x^j, \quad j = 0, 1, 2, \ldots.$$

For the *first approximation* the method of moments is identical to the sub-domain method, with the whole domain the subdomain, and this is sometimes called the *integral method* of von Kármán [24] and Pohlhausen [25].

(f) General WRM

With the weighting functions chosen from a complete set not the same as those used for the trial functions the procedure is called a *general* WRM.

The several criteria were unified under the title *weighted residual methods* by Crandall [26], while Collatz [27] used the term *error distribution principles*. Biezeno [15] and Courant [28] early recognized the similarity of the several methods.

(g) Stationary functional method (Rayleigh [29], Ritz [30])

Let Φ be a functional, e.g. variational integral, equivalent to the original problem in some sense. The *stationary functional method* consists of inserting the trial function (6-2) with constants c_j into Φ and setting

$$\frac{\partial \Phi}{\partial c_j} = 0, \quad j = 1, \ldots, N. \tag{6-10}$$

These N algebraic equations are solved for the c_j and the corresponding u_T represents an approximate solution. The problem here is that not every initial–boundary value problem has a convenient functional Φ or for that matter a variational principle. That there is no variational principle for the Navier–Stokes equations is well known (Finlayson [2, p. 285]). Of course this method is not a WRM, but its close relation to those procedures motivated its inclusion here.

PROBLEMS

6-1 The state of torsion of a uniform elastic cylindrical prism is characterized by a dimensionless stress function $\psi(x, y)$ ($\psi_x = -\tau_{yz}, \psi_y = \tau_{xz}$). Geometric compatibility requires ψ to satisfy $\psi_{xx} + \psi_{yy} = -2$, $\psi = 0$ on the boundaries $x = \pm 1$ and $y = \pm 1$. (See, e.g., Timoshenko and Goodier [31].) With $\psi_T = c_1(1 - x^2)(1 - y^2)$ evaluate the constant c_1 by *Galerkin's method* and by collocating at the origin.

6-2 For the equation of Problem 6-1 the equivalent variational problem is

$$\Phi = \tfrac{1}{2} \int_{-1}^{1} \int_{-1}^{1} \{(\psi_x)^2 + (\psi_y)^2 - 4\psi\} \, dx \, dy.$$

Using ψ_T of Problem 6-1 find the value of c_1 that renders Φ stationary.

Ans: $c_1 = \tfrac{5}{8}$.

6-3 Consider the boundary value problem

$$\psi_{xx} + \psi_{yy} = 0, \quad 0 < x < 1, \ 0 < y < \infty$$

with boundary conditions $\psi(0, y) = \psi(1, y) = 0$ for $y > 0$, $\psi(x, 0) = x(1 - x)$, $\psi(x, y \to \infty) = 0$ for $0 \le x \le 1$. Use the trial solution $\psi_T = c_1(y)x(1 - x)$. Find appropriate boundary conditions on c_1 and the equation residual R_E.

(a) Collocate along $x = \tfrac{1}{3}$ and find $c_1(y)$.   *Ans:* $c_1 = \exp(-3y)$.

(b) Use the method of moments.

6-4 Use the Galerkin method, with a trial function

$$\psi_T = c_1(y)x(1 - x) + c_2(y)x^2(1 - x)^2$$

having two undetermined functions, for the problem of 6-3.

$$Ans:\ c_1(y) = 0.8035 \exp(-3.1416y) + 0.1965 \exp(-10.1059y)$$
$$c_2(y) = 0.9105[\exp(-3.1416y) - \exp(-10.1059y)].$$

6-5 Consider the dimensionless equation $\psi_{xx} = \psi_{tt}$, $\psi = 0$, at $x = 0$ and $x = 1$; $\psi(x, 0) = x(1 - x)$, $\psi_t(x, 0) = 0$, governing the transverse oscillations of a string. Using the trial solution $\psi = x(1 - x)c_1(t)$ find $c_1(t)$ by collocation at $x = \frac{1}{3}$, subdomain, and Galerkin's method.

$$Ans:\ \cos wt;\ w^2 = 9;\ w^2 = 12;\ w^2 = 10.$$
The actual value is π^2.

6-2 Orthogonal collocation

If the researcher wishes only a first approximation to the answer of his problem, the methods of Section 6-1 are especially applicable. But more precise answers are often desired and they can be obtained by WRM also. Any method selected should be convenient to use and easy to generalize in the sense that improved accuracy is obtainable at the expense of more computation without reformulation or restructuring. One such method is the orthogonal collocation method of Lanczos [17]. The collocation method is the simplest WRM to apply. In higher approximations the choice of collocation points, while not crucial, can be done in ways that make the calculations convenient and accurate. If the collocation points are selected as the roots to orthogonal polynomials, the procedure is called orthogonal collocation.

Lanczos [17, 32], Clenshaw and Norton [33], Norton [34], and Wright [35], using Chebyshev series, were concerned primarily with initial value problems for ordinary differential equations. Villadsen and Stewart [36] and Villadsen [37] have developed the procedure for boundary value problems. The trial functions were chosen from sets of orthogonal polynomials that satisfied the boundary conditions, and the roots to the polynomials were selected as the collocation points. The *major advantages* are: the choice of collocation points is no longer arbitrary; the low order results are more accurate; and the results can be obtained as values of the solution at the collocation points rather than the unknown coefficients of the trial functions. In what follows the general procedure will be introduced including several examples.

Let $P_n(x)$ be a polynomial of nth order

$$P_n(x) = \sum_{j=0}^{n} c_j x^j \tag{6-11}$$

whose coefficients c_j are determined by the requirement that P_n is orthogonal on $a \leq x \leq b$ to all polynomials of order less than n, relative to the weighting function $w(x) \geq 0$,

$$\int_a^b w(x)P_n(x)P_m(x)\,dx = 0, \quad n = 0, 1, \ldots, m - 1. \tag{6-12}$$

This specifies each polynomial up to a multiplicative constant. The value assigned to that constant is usually determined by some such requirement as $P_n(1) = 1$. In the table below are tabulated some of the more frequently used orthogonal polynomials.

Legendre: $-1 \leq x \leq 1$,† $w(x) = 1$;

$$P_0 = 1, \qquad P_1(x) = x, \qquad P_2(x) = \tfrac{1}{2}(3x^2 - 1)$$

$$(r + 1)P_{r+1} = (2r + 1)xP_r(x) - rP_{r-1}(x)$$

Chebyshev: $-1 \leq x \leq 1$, $w(x) = (1 - x^2)^{-1/2}$;

$$P_0 = 1, \qquad P_1 = x, \qquad P_2 = 2x^2 - 1$$

$$P_{r+1}(x) = 2xP_r(x) - P_{r-1}(x)$$

Laguerre: $0 \leq x < \infty$, $w(x) = e^{-x}$;

$$P_0 = 1, \qquad P_1 = 1 - x, \qquad P_2 = 2 - 4x + x^2$$

$$P_{r+1}(x) = (1 + 2r - x)P_r(x) - r^2P_{r-1}(x)$$

Hermite: $-\infty < x < \infty$, $w(x) = e^{-x^2}$;

$$P_0 = 1, \qquad P_1 = 2x, \qquad P_2 = 4x^2 - 2$$

$$P_{r+1}(x) = 2xP_r(x) - 2rP_{r-1}(x)$$

Each polynomial of order $n > 0$ has n roots. Detailed properties are derived in such basic references as Jackson [38], Szegö [39], or Hildebrand [40]. Stroud and Secrest [41] give some of the zeros of the polynomials.

Additional properties for the trial function may be suggested by the problem under investigation. For example, suppose a problem solution is sought on the domain $-1 \leq x \leq 1$ which is symmetric about zero, and the value of the solution is specified on the boundary, say $u(1) = u(-1)$. Then a possible choice of a trial function for $u = u(x)$ might be

$$u(x) = u(1) + (1 - x^2) \sum_{i=1}^{N} a_i P_{i-1}(x^2) \tag{6-13}$$

where N is the number of interior collocation points to be used. The orthogonal polynomials of (6-13) are constructed using the orthogonality condition

$$\int_0^1 w(x^2)P_n(x^2)P_m(x^2)x^{a-1}\,dx = c_{(n)}\delta_{nm}, \quad m = 1, \ldots, n - 1$$

where $a = 1, 2, 3$ for planar, cylindrical, or spherical geometry. The orthogonal polynomials are then determined up to constants which may be evaluated

---

† The finite interval $a \leq x' \leq b$ can be transformed to $-1 \leq x \leq 1$ by a linear change in variables.

by a variety of assumptions—e.g. with the first coefficient equal to one. The trial solution (6-13) is then substituted into the differential equation to form the residual that contains the N undetermined coefficients a_i, $i = 1, \ldots, N$. The residual is set to zero at the N collocation points x_j, which are the roots to the polynomial $P_N(s) = 0$. This provides N algebraic equations to solve for the a_i.

The computer programs are often simpler if they are written in the values of the solutions at the collocation points $u(x_j)$ rather than the a_i. Since $P_{N-1}(x^2)$ is a polynomial of degree $N - 1$ in x^2, the trial function (6-13) is a polynomial of degree N in x^2, rewritten as

$$u(x) = \sum_{i=1}^{N+1} d_i x^{2i-2} \tag{6-14}$$

where the $(N + 1)$st collocation point is at $x = 1$. Thus the vectors and matrices below are of dimension $N + 1$. Upon calculating the first and second derivatives of (6-14) and evaluating them at x_j there results

$$u(x_j) = \sum_{i=1}^{N+1} x_j^{2i-2} d_i, \qquad u'(x_j) = \sum_{i=1}^{N+1} \frac{d}{dx}(x^{2i-2})\Big|_{x_j} d_i$$

$$u''(x_j) = \sum_{i=1}^{N+1} \frac{d^2}{dx^2}(x^{2i-2})\Big|_{x_j} d_i. \tag{6-15}$$

Using bold-faced lowercase letters for vectors and bold-faced uppercase letters for matrices, (6-15) is expressible in matrix form as

$$\mathbf{u} = \mathbf{Xd}, \qquad \mathbf{u}' = \mathbf{X'd}, \qquad \mathbf{u}'' = \mathbf{X''d} \tag{6-16}$$

where $\mathbf{X}$, $\mathbf{X'}$ and $\mathbf{X''}$ are $N + 1$ by $N + 1$ matrices formed from the function, derivative, and second derivative values—i.e.

$$X'_{ji} = \frac{d}{dx} x^{2i-2}\Big|_{x_j}.$$

The quantities $\mathbf{u}$, $\mathbf{u}'$, $\mathbf{u}''$, and $\mathbf{d}$ are $N + 1$ component vectors.

Upon solving $\mathbf{u} = \mathbf{Xd}$ for $\mathbf{d}$, the first and second derivatives can be rewritten as

$$\mathbf{u}' = \mathbf{X'X^{-1}u} = \mathbf{Au}, \qquad \mathbf{u}'' = \mathbf{X''X^{-1}u} = \mathbf{Bu}. \tag{6-17}$$

The derivatives are expressed in terms of the values of the function at the collocation points.

The accurate evaluation of integrals requires the use of quadrature formulas. For example,

$$\int_0^1 f(x^2) x^{a-1} \, dx = \sum_{j=1}^{N+1} w_j f(x_j) \tag{6-18}$$

where the w_j are determined by evaluating (6-18) for $f = x^{2i-2}, i = 1, 2, \ldots, N$,

$$\int_0^1 x^{2i-2} x^{a-1} \, dx = \frac{1}{2i - 2 + a} = \sum_{j=1}^{N+1} w_j x_j^{2i-2}.$$

Therefore, $\mathbf{Xw} = \mathbf{f}$ or $\mathbf{w} = \mathbf{X}^{-1}\mathbf{f}$ where $f_i = (2i - 2 + a)^{-1}$. Kopal [42, p. 390] shows that this integration is exact for polynomials of degree $2N$ in x^2 provided the interior collocation points are the roots of $P_N(s)$ where the P_N are those polynomials with $w(x) = 1 - x^2$.

As an illustration of these results consider the problem

$$u_{xx} + \lambda u_{yy} = 1, \quad -1 < x < 1, \quad -1 < y < 1$$
$$u = 0 \quad \text{on boundary} \tag{6-19}$$

which is symmetric about $x = 0$, $y = 0$, so that polynomials in x^2 and y^2 can be used. Let the polynomials be those with $w = 1 - x^2$ and the trial solution be

$$u(x, y) = (1 - x^2)(1 - y^2) \sum_{i,j=1}^{N} a_{ij} P_{i-1}(x^2) P_{j-1}(y^2). \tag{6-20}$$

Using matrix $\mathbf{B}$ to represent the second derivative, the collocation equations are

$$\sum_{i=1}^{N+1} B_{ji} u_{ik} + \lambda \sum_{i=1}^{N+1} B_{ki} u_{ji} = 1 \tag{6-21}$$

where $u_{i,j} = u(x_i, y_j)$ are the values to be determined at the collocation points, $u_{N+1,j} = u_{i,N+1} \equiv 0$, and the $\mathbf{B}$ matrices are as in (6-17). These linear equations are solved by any of the direct or iterative methods of Chapter 3.

For second-order equations on bounded domains which have no special symmetry properties, polynomials are needed that are orthogonal on $(0, 1)$ and that possess both even and odd powers of x. With these a typical trial function is

$$u(x) = \alpha + \beta x + x(1 - x) \sum_{i=1}^{N} a_i P_{i-1}(x) \tag{6-22}$$

which contains $N + 2$ constants. N of these are obtained by evaluating the residuals at the N collocation points, the N roots of $P_N(s) = 0$, and the other two are provided by the boundary conditions at $x = 0$ and $x = 1$. The polynomials are shifted Legendre polynomials, and the matrix expressions, analogous to (6-17), are the same except

$$X_{ji} = x_j^{i-1}, \qquad X'_{ji} = (i - 1)x_j^{i-2}, \qquad \text{etc.}$$

Many examples illustrating these methods are given in Finlayson [2].

PROBLEMS

6-6 Find the first three *Legendre* polynomials by using (6-11) and (6-12) with $P_n(1) = 1$. This technique is essentially the classical Gram–Schmidt method.

6-7 Find the first three polynomials orthogonal on $0 \le x \le 1$ with $w(x) = 1 - x^2$ using (6-12). What are their roots?

6-8 Apply orthogonal collocation to $y' = g(x, y)$, $y(0) = y_0$ with a trial function of the form

$$y(x) = y_0 + x \sum_{i=1}^{N+1} a_i P_{i-1}(x).$$

Calculate the matrix operators (6-17) and set up the computational form.

6-9 Unsteady diffusion in a sphere is governed by $c_t = r^{-2}(r^2 c_r)_r$, $c = 0$ at $t = 0$, $c = 1$ at $r = 1$, $t > 0$, $c_r = 0$ at $r = 0$, $t > 0$. Apply orthogonal collocation *in the r variable* thereby obtaining a set of ordinary differential equations. Use $w = 1$. The resulting equations can be solved numerically or by eigenvalue techniques.

$$Ans: \frac{dc_j}{dt} = \sum_{i=1}^{N+1} B_{ji} c_i,$$

$$c_j(0) = 0, \quad c_{N+1} = 1, \quad j = 1, \ldots, N.$$

6-10 Develop the first three Chebyshev polynomials by using (6-11) and (6-12). Using the resulting set $\{P_k\}$ as a basis, develop the equations analogous to (6-17). Notice here again that one should not use the basis functions $\{x^k\}$ with equally spaced collocation points.

6-3 Bubnov–Galerkin (B-G) method

Many discussions and applications of the present method exist in the cited literature. To illustrate the details in a simple problem we consider the initial boundary value problem in which

$$\psi_t = \psi_{xx}, \quad 0 < x < 1, \; 0 < t$$

$$\psi(x, 0) = 1, \quad 0 \le x \le 1$$

$$\psi(0, t) - \psi_x(0, t) = 0, \quad t > 0 \tag{6-23}$$

$$\psi_x(1, t) = 0, \quad t > 0.$$

The solution of this problem is approximated by a trial family

$$\psi_T = \sum_{j=1}^{N} c_j(t) \phi_j(x) \tag{6-24}$$

where the $\phi_j(x)$ are known basis functions. Using the *interior* method, we select the ϕ_j to satisfy the boundary conditions at $x = 0$ and $x = 1$, without placing any restrictions on the $c_j(t)$. Thus we require that each ϕ_j satisfy the conditions

$$\phi_j(0) - \phi_j'(0) = 0$$
$$\phi_j'(1) = 0, \quad j = 1, 2, \ldots, N. \tag{6-25}$$

A simple family of polynomials meeting this requirement is

$$\phi_j(x) = (1 + x) - x^{j+1}/(j + 1). \tag{6-26}$$

The trial family (6-24) with (6-26) satisfies the boundary conditions but not the initial conditions or the equation. By applying the B-G method the unknowns $c_j(t)$ are determined so the latter conditions are approximately met. For definiteness in what follows, only two terms will be considered.

When $t = 0$, $\psi = 1$ for $0 < x < 1$ so the initial residual [see (6-4)]

$$R_I(\psi_T) = R_I(c_1(0), c_2(0), x)$$
$$= 1 - \left(1 + x - \frac{x^2}{2}\right)c_1(0) - \left(1 + x - \frac{x^3}{3}\right)c_2(0). \tag{6-27}$$

$R_I(\psi_T)$ is a measure of the amount by which the condition $\psi = 1$ at $t = 0$ is not satisfied.

For $t > 0$, $\psi_t = \psi_{xx}$ must hold. By forming the equation residual [see (6-3)]

$$R_E(\psi_T) = R_E(c_1(t), c_2(t), x)$$
$$= \left(1 + x - \frac{x^2}{2}\right)c_1'(t) + \left(1 + x - \frac{x^3}{3}\right)c_2'(t) + c_1 + 2c_2 x. \tag{6-28}$$

Any combination of criteria can be applied to the two residuals. The B-G method gives the same results as the least squares method when applied to (6-27), so in that sense it appears to be an optimal method for fitting initial conditions (see Problem 6-11). Here the B-G method is applied to both residuals.

For the initial residual, the B-G method requires

$$\int_0^1 R_I \phi_j(x) \, dx = 0, \quad j = 1, 2$$

or

$$\frac{9}{5}c_1(0) + \frac{691}{360}c_2(0) = \frac{4}{3}$$
$$\frac{691}{360}c_1(0) + \frac{1291}{630}c_2(0) = \frac{17}{12}.$$

Therefore

$$c_1(0) = 3.0346, \quad c_2(0) = -2.1510. \tag{6-29}$$

For the equation residual, the B-G method requires

$$\int_0^1 R_E\phi_j(x)\,dx = 0, \quad j = 1, 2$$

or

$$\tfrac{9}{5}c_1' + \tfrac{691}{360}c_2' + \tfrac{4}{3}c_1 + \tfrac{17}{12}c_2 = 0$$

$$\tfrac{691}{360}c_1' + \tfrac{1291}{630}c_2' + \tfrac{17}{12}c_1 + \tfrac{23}{15}c_2 = 0. \tag{6-30}$$

The solution of these equations subject to the initial data (6-29) is

$$c_1 = 0.5862 \exp\left[-0.7402t\right] + 2.4484 \exp\left[-11.770t\right]$$

$$c_2 = 0.1444 \exp\left[-0.7402t\right] - 2.2954 \exp\left[-11.770t\right]$$

which, when substituted in (6-24), provides the B-G approximation for ψ.

In more complicated and nonlinear problems the equations corresponding to (6-30) may have to be solved numerically or by a reapplication of a WRM.

A relationship of the B-G method with finite difference techniques can be established through the use of continuous, piecewise linear, basis functions. These triangular functions are defined as

$$\phi_j(x) = \begin{cases} 1 - |x - x_j|/\Delta x & \text{for } |x - x_j| \le \Delta x \\ 0 & \text{elsewhere} \end{cases} \tag{6-31}$$

and $x_j - x_{j-1} = \Delta x$; or more generally, for a nonuniform mesh,

$$\phi_j(x) = \begin{cases} \dfrac{x - x_{j-1}}{x_j - x_{j-1}}, & x_{j-1} \le x \le x_j \\[2mm] \dfrac{x - x_{j+1}}{x_j - x_{j+1}}, & x_j \le x \le x_{j+1} \\[2mm] 0, & x \le x_{j-1} \text{ or } x \ge x_{j+1}. \end{cases} \tag{6-32}$$

Harrington [43] showed that the B-G method using these trial functions is related to an implicit finite difference method when applied to the *linear* diffusion equation

$$u_t = u_{xx}. \tag{6-33}$$

Using (6-31) the trial function is selected as

$$u(x, t) = \sum_{j=1}^{n} a_j(t)\phi_j(x). \tag{6-34}$$

Some of the implications of this choice are

$$u(x_j, t) = a_j, \qquad \frac{\partial u}{\partial t}(x_j, t) = \frac{da_j}{dt}.$$

Using the trial function (6-34), the B-G method for (6-33) yields

$$\sum_{j=1}^{n} \frac{da_j}{dt} (\phi_i, \phi_j) = \sum_{j=1}^{n} a_j(\phi_i, \phi_j''). \tag{6-35}$$

Of course, the term ϕ_j'' on the right-hand side of (6-35) is not defined. In order to calculate the right-hand side integrate by parts to obtain

$$(\phi_i, \phi_j'') = \int \phi_i \phi_j'' \, dx = \phi_i \phi_j' - \int \phi_j' \phi_j' \, dx. \tag{6-36}$$

Using (6-36) the right-hand side of (6-35) can now be evaluated (Problem 6-14) with the result

$$\frac{1}{6} \frac{da_{j-1}}{dt} + \frac{2}{3} \frac{da_j}{dt} + \frac{1}{6} \frac{da_{j+1}}{dt} = \frac{a_{j+1} - 2a_j + a_{j-1}}{(\Delta x)^2} \tag{6-37}$$

which is similar to an implicit method and also suggests an implicit method of lines. A reduction of (6-37) to finite differences is accomplished by approximating the time derivatives, whereupon

$$\frac{1}{\Delta t} \{\tfrac{1}{6}(a_{j-1}^{n+1} - a_{j-1}^n) + \tfrac{2}{3}(a_j^{n+1} - a_j^n) + \tfrac{1}{6}(a_{j+1}^{n+1} - a_{j+1}^n)\}$$

$$= \theta\left[\frac{a_{j+1}^{n+1} - 2a_j^{n+1} + a_{j-1}^{n+1}}{(\Delta x)^2}\right] + (1 - \theta)\left[\frac{a_{j-1}^n - 2a_j^n + a_{j+1}^n}{(\Delta x)^2}\right]. \tag{6-38}$$

where $a_j^n = a_j(n \, \Delta t)$ and an implicit method (see the Crank–Nicolson form of Section 2-3) with parameter θ has been used for the right-hand side. Upon rearrangement Eqn (6-38) becomes

$$\frac{a_j^{n+1} - a_j^n}{\Delta t} = \left[\theta - \frac{(\Delta x)^2}{6 \, \Delta t}\right]\left[\frac{a_{j+1}^{n+1} - 2a_j^{n+1} + a_{j-1}^{n+1}}{(\Delta x)^2}\right]$$

$$+ \left[1 - \theta + \frac{(\Delta x)^2}{6 \, \Delta t}\right]\left[\frac{a_{j+1}^n - 2a_j^n + a_{j-1}^n}{(\Delta x)^2}\right]. \tag{6-39}$$

When the choice $\lambda = \theta - (\Delta x)^2/6 \, \Delta t$ is made, (6-39) becomes (2-35), the general implicit form, when the proper identifications are made.

The preceding argument verifies that the finite difference equations for solving (6-33) can be regarded as a Galerkin method using the triangular basis functions (6-31) to construct the trial function (6-34). A straight-line interpolation is used to interpolate for values of u at $x \neq x_j$. For nonlinear problems, the correspondence does not hold. Of course, for a given problem, there may be other basis functions that play a similar role to that of the triangular functions used herein. Swartz and Wendroff [44] have discussed other generalized finite difference schemes using related arguments.

PROBLEMS

6-11 A function $f(x)$, $0 < x < 1$, is to be approximated by a linear combination, $\sum_{j=1}^{N} c_j \phi_j(x)$, of known functions $\phi_j(x)$. Define a residual $R = f(x) - \sum c_j \phi_j(x)$ and let the coefficients c_j be determined by an application of the B-G method to R. Show that the integral of R^2 over $0 < x < 1$ is minimized.

6-12 Consider Problem 6-3. Select a trial family, with two terms, using appropriate orthogonal polynomials from Section 6-2. Carry out the B-G method with that trial family.

6-13 The Burgers' equation (Section 2-12) $u_t + u u_x = u_{xx}$, with $u(0, t) = u(1, t) = 0$, $u(x, 0) = \sin \pi x$ is to be solved approximately by the B-G method using a trial family with one undetermined function. Can a trial family be selected satisfying both the initial and boundary data? Repeat with two undetermined functions.

6-14 Carry out the calculations to convert (6-35) into (6-37).

6-15 Repeat the computation from (6-33) to (6-37) using the step functions

$$\phi_j(x) = \begin{cases} 1 & x_j - (\Delta x)/2 \leq x \leq x_j + (\Delta x)/2 \\ 0, & \text{otherwise} \end{cases}$$

instead of (6-31). *Ans:* The left hand side is $\frac{1}{8}a'_{j-1} + \frac{3}{4}a'_j + \frac{1}{8}a'_{j+1}$.

6-4 Remarks on completeness, convergence, and error bounds

Many of the questions of this section require a special mathematical analysis called *functional analysis*. We have used some of the ideas previously. Here we sketch some further concepts, describe some of the useful complete sets of basis functions, and give some typical convergence theorems and error bounds. A substantial portion of the results are restricted to linear problems. Finlayson [2, Chapter 11] presents an excellent summary, although it is lacking in details (see also Mikhlin [45] and Collatz [46]).

Let E_n be n-dimensional Euclidean space, $x = (x_1, \ldots, x_n)$ a point in E_n, V a bounded domain in E_n, and ∂V the boundary of V. The union of V and ∂V, $\bar{V} = V \cup \partial V$ is the closure of V. For $T \geq 0$, $Q = V \times (0, T)$. The class of functions that are n times continuously differentiable on V is denoted by $C^n(V)$. The class of functions that are square integrable over Q is denoted by $L_2(Q)$; the *scalar product* [compare (6-5)] by

$$(u, v) = \int_Q uv \, dx \, dt \tag{6-40}$$

and the *norm* by

$$\|u\| = (u, u)^{1/2}. \tag{6-41}$$

A sequence of functions f_k is said to be *orthonormal* if $(f_k, f_j) = \delta_{kj}$, $k, j = 1, 2, \ldots$, where δ_{kj} is the generalized function, $\delta_{kj} = 1$ if $k = j$ and 0 otherwise.

The (Hilbert) space of functions $u(x)$ that are such that $u(x)$ and u_{x_i} are in $L_2(V)$ with scalar product

$$(u, v)_1 = \int_V uv \, dx + \int_V \sum_{i=1}^n u_{x_i} v_{x_i} \, dx \qquad (6\text{-}42)$$

is denoted by $H^1(V)$. The (Hilbert) space of functions $u(x, t)$ for which u, u_{x_i}, u_t are in $L_2(Q)$ with inner product

$$(u, v)_{1,1} = \int_Q uv \, dx \, dt + \int_Q \left(\sum_{i=1}^n u_{x_i} v_{x_i} + u_t v_t \right) dx \, dt \qquad (6\text{-}43)$$

is denoted by $H^{1,1}(Q)$. The class of functions that vanish on ∂V are called $H_0(V)$, and $H_0(Q)$ represents those that vanish on $\partial Qx(0, T)$.

Let L be a linear operator defined for all $u \in D_L$, the *domain* of L. The operator L is *symmetric* if for $u, v \in D_L$,

$$(u, Lv) = (v, Lu)$$

L is *positive definite* if for any $u \in D_L$, $u \not\equiv 0$,

$$(u, Lu) \geq 0.$$

L is *positive* and *bounded below* if for any $u \in D_L$ there exists $\gamma > 0$ such that

$$(u, Lu) \geq \gamma(u, u).$$

For a functional $F(u)$, the sequence $\{u_n\}$ forms a *minimizing sequence* for F if

$$\lim_{n \to \infty} F(u_n) = \inf F(u)$$

where the *inf* (infinum) is the 'greatest lower bound.' Even though the functional converges to inf $F(u)$, the minimizing sequence may not converge to a function in the space. Indeed, there may be no function in the space at which the minimum is attained.

A *classical* solution of

$$Lu = f \qquad (6\text{-}44)$$

obeys (6-44) everywhere in V. A *generalized (weak) solution* of (6-44) is one for which

$$(\phi, Lu - f) = 0 \qquad (6\text{-}45)$$

for all ϕ in a given class of functions.

In addition to the classical concepts of convergence and uniform convergence other ideas are useful. In particular, $\{u_n(x)\}$ *converges in the mean* if to any $\varepsilon > 0$ there corresponds N such that

$$\|u - u_n\| < \varepsilon \quad \text{whenever } n > N. \tag{6-46}$$

Convergence in energy requires

$$|u - u_n| < \varepsilon \quad \text{whenever } n > N \tag{6-47}$$

where the 'energy,' $|u| = (u, Lu)$, involves derivatives of u. In mean convergence the converging sequence may not approach the limit function at every point of the domain, but the regions in which they differ goes to zero as $n \to \infty$. Convergence in energy permits similar things to happen to the derivatives. Error bounds that correspond to these convergence concepts are *pointwise bounds*, *mean square error*, and *energy error bounds*. A sequence u_n is said to *weakly converge* to an element u of a function space if

$$\lim_{n \to \infty} (u_n, \phi) = (u, \phi) \tag{6-48}$$

holds for all ϕ in the space. The B-G method sometimes yields sequences that are weakly convergent to generalized functions.

A set of functions f_n is said to be *linearly independent* if the only solution to

$$\sum_{i=1}^{N} \alpha_i f_i = 0$$

is $\alpha_i = 0$ for all i. A set of functions f_n is *complete in the mean* in a space if any function of the space can be expanded in terms of that set,

$$\left\| u - \sum_{i=1}^{n} a_i f_i \right\| < \varepsilon, \quad n > N. \tag{6-49}$$

Of course, a set of functions complete for one space of functions need not be complete for another space—it is clearly necessary to specify the space considered. A set of functions is *complete in energy* if

$$\left| u - \sum_{i=1}^{n} a_i f_i \right| < \varepsilon, \quad n > N. \tag{6-50}$$

Some of the more useful results are listed below together with references for further study.

Theorem 6-4.1 (Mikhlin [45, p. 66])

Let the orthonormal set $\{f_n\}$ be complete, in the mean convergence sense, for some class of functions. Then the 'Fourier series' for any function u of the class

$$\sum_{i=1}^{\infty} (u, f_i) f_i \tag{6-51}$$

converges in the mean to u.

Theorem 6-4.2 (Mikhlin and Smolitsky [47, p. 237])

Let $\{f_n\} \in D_L$, L a positive definite operator, and let $\{Lf_n\}$ be complete in a Hilbert space H. Then $\{f_n\}$ is complete in the energy space H_L. Theorem 6-4.2 means that trial functions must be capable of representing functions ϕ and derivatives $L\phi$. *Thus completeness in energy is required of trial functions.*

As an example, consider the two-dimensional problem

$$Lu = -\sum_{i,j=1}^{2} (A_{ij}u_{x_j})_{x_i} + cu = f(x) \quad \text{in } V$$

$$u = 0 \quad \text{on } \partial V$$

(6-52)

with†

$$\sum_{i,j=1}^{2} A_{ij}\xi_i\xi_j \geq \mu(\xi_1^2 + \xi_2^2), \quad \mu > 0.$$

Any set of functions that is complete in the energy of the operator $-\nabla^2 u$ is also complete in the energy of L. Two such systems are

$$f_{i,j} = \sin(i\pi x/a)\sin(j\pi y/b)$$

(6-53)

in rectangular geometry, $0 < x < a, 0 < y < b$; and

$$f_{i,j} = c_{i,j}J_j(\gamma_{j,i}r)\cos j\theta$$

in circular geometry, where $\gamma_{j,i}$ is the ith positive root of $J_j(x)$ and $c_{i,j}$ are chosen so $\|f_{i,j}\| = 1$. Babuska *et al.* [48] discuss the choice of trial functions to give optimal approximations. The eigenfunctions constructed in eigenvalue problems provide a convenient source of complete sets.

Error bounds are usually developed using variational and reciprocal techniques, maximum and minimum principles, or differential–integral inequalities. For the development of bounds by variational methods, there is an elementary discussion in Crandall [26], more details in Kantorovich and Krylov [49] and Collatz [46], and a study of approximate solution error for diffusion problems by Yasinsky and Kaplan [50]. An introductory treatment for maximum–minimum principles is contained in Protter and Weinberger [51], while Walter [52] is an excellent source for differential–integral inequalities.

As a first example of error bound development consider the semilinear problem

$$Lu = f(u, x)$$

(6-54)

where the linear operator L has an inverse‡ L^{-1} and $f(u, x)$ satisfies a

---

† This is the uniformly elliptic condition.
‡ For example, L^{-1} may be expressed as the integral of a Green's function.

Lipschitz condition

$$\|f(u, x) - f(v, x)\| \leq K \|u - v\| \tag{6-55}$$

with Lipschitz constant $K > 0$. Let the trial (approximate) solution be denoted by u_n, whereupon the equation residual R_n becomes

$$Lu_n - f(u_n, x) = R_n. \tag{6-56}$$

Formal inversion yields

$$u_n = L^{-1}f(u_n, x) + L^{-1}R_n \tag{6-57}$$

from (6-56) and

$$u = L^{-1}f(u, x) \tag{6-58}$$

from (6-54). Subtracting (6-57) from (6-58) and taking the norm gives

$$\|u - u_n\| \leq \|L^{-1}\| \|f(u, x) - f(u_n, x)\| + \|L^{-1}\| \|R_n\|$$
$$\leq \|L^{-1}\| K \|u - u_n\| + \|L^{-1}\| \|R_n\|.$$

Upon solving for $\|u - u_n\|$ we have

$$\|u - u_n\| \leq \frac{\|L^{-1}\| \|R_n\|}{1 - K\|L^{-1}\|} \tag{6-59}$$

as a formal error bound valid when $K\|L^{-1}\| < 1$. A generalization of this result, used by Ferguson [53] to study diffusion and reaction in a spherical pellet, is given in Problem 6-16.

Many elliptic, parabolic, and hyperbolic equations obey a maximum principle. In the simplest form they assert that the maximum of the solution occurs on the boundary. A typical result from Protter and Weinberger [51, p. 187] is for nonlinear parabolic equations.

Let $F(\mathbf{x}, t, u, u_{x_i}, u_{x_i x_j})$ be elliptic in a domain E in (x, t) space; that is,

$$\sum_{i,j=1}^{n} \frac{\partial F}{\partial u_{x_i x_j}} \xi_i \xi_j > 0$$

for all real vectors $\xi = (\xi_1, \ldots, \xi_n)$ holds for each point in E and for any u. The nonlinear operator

$$Lu = F(\mathbf{x}, t, u, u_{x_i}, u_{x_i x_j}) - u_t \tag{6-60}$$

is said to be parabolic whenever F is elliptic. Let u be a solution of $Lu = f(\mathbf{x}, t)$ with L specified by (6-60) in E. Suppose $w = w(\mathbf{x}, t)$ satisfies $Lw \leq f$ in E. Form $v(\mathbf{x}, t) = u(\mathbf{x}, t) - w(\mathbf{x}, t)$ and consider the inequality that results:

$$F(\mathbf{x}, t, u, u_{x_i}, u_{x_i x_j}) - F(\mathbf{x}, t, w, w_{x_i}, w_{x_i x_j}) - \frac{\partial v}{\partial t} \geq 0. \tag{6-61}$$

Using the mean value theorem of the multidimensional calculus and evaluating the derivatives of F at the arguments

$$\theta u + (1 - \theta)w, \qquad \theta u_{x_i} + (1 - \theta)w_{x_i}, \qquad \theta u_{x_i x_j} + (1 - \theta)w_{x_i x_j}$$

$$\text{with } 0 \leq \theta \leq 1$$

there results

$$\sum_{i,j=1}^{n} \frac{\partial F}{\partial r_{ij}} v_{x_i x_j} + \sum_{i=1}^{n} \frac{\partial F}{\partial p_i} v_{x_i} + \frac{\partial F}{\partial u} v - \frac{\partial v}{\partial t} \geq 0. \tag{6-62}$$

Here $p_i = u_{x_i}$, $r_{ij} = u_{x_i x_j}$ are used as convenient notations, and we assume F is elliptic in E for all functions of the form $\theta u + (1 - \theta)w$ and their first and second derivatives. The left-hand side of (6-62) is a linear parabolic operator for v. Results from the linear theory can then be used to establish

Theorem 6-4.3

Let D be a bounded domain in E_n with $E = Dx(0, T)$. Suppose u is a solution of $Lu = f(\mathbf{x}, t)$ in E with L given by (6-60) subject to the initial condition $u(\mathbf{x}, 0) = g_1(\mathbf{x})$ in D, and boundary conditions $u(\mathbf{x}, t) = g_2(\mathbf{x}, t)$ on $\partial Dx(0, T)$. Let h and H satisfy the inequalities

$$LH \leq f(\mathbf{x}, t) \leq Lh \quad \text{in } E$$

$$h(\mathbf{x}, 0) \leq g_1(\mathbf{x}) \leq H(\mathbf{x}, 0) \quad \text{in } D$$

$$h(\mathbf{x}, t) \leq g_2(\mathbf{x}, t) \leq H(\mathbf{x}, t) \quad \text{on } \partial Dx(0, T)$$

and L be parabolic with respect to the functions $\theta u + (1 - \theta)h$, $\theta u + (1 - \theta)H$ for $0 \leq \theta \leq 1$. Then

$$h(\mathbf{x}, t) \leq u(\mathbf{x}, t) \leq H(\mathbf{x}, t) \quad \text{in } E.$$

This theorem says the approximate solution h is a lower bound on the exact solution u if it satisfies the initial and boundary conditions and the residual is everywhere positive in E. Some elementary applications are given in Problem 6-17 and 6-18.

PROBLEMS

6-16   Consider $\nabla^2 u_i = f_i(u_1, u_2, \ldots, u_N)$, $i = 1, 2, \ldots, N$ with

$$\| f_i(u_1, \ldots, u_N) - f_i(v_1, \ldots, v_N) \| \leq \sum_{j=1}^{N} K_{ij} \| u_j - v_j \|$$

$K_i = \sum_{j=1}^{N} K_{ij}^2$. Find the error bound for each component u_i using a trial solution $\bar{u}_i$. When is the bound valid?

6-17   One-dimensional diffusion is governed by the nonlinear equation $u_t = [k(u)u_x]_x$, $k > 0$, $|\partial k/\partial u|$ bounded. Using the fact that any constant satisfies this equation, apply Theorem 6-4.3 to conclude that for any solution u, the maximum or minimum values occur either at the initial time or on the boundary.

6-18 (Boley [54]) consider $Au_t = [k(u)u_x]_x$ with A (constant) > 0 and $k(u) = k_0(1 + au)$, $a > 0$, subject to $u(x, 0) = 0$, $u(0, t) = 1$. Show that the solution

$$u_1 = \text{erfc} \left[\frac{x}{(4k_0t/A)^{1/2}} \right]$$

of the linearized problem $k_0(u_1)_{xx} = A(u_1)t$, with the same initial and boundary data, is a lower bound for u. Suggest a candidate for an upper bound. (*Hint:* Try an invariant (similar) solution.)

6-5 Nagumo's lemma and application

The application of differential and integral inequalities has much in common with maximum and minimum principles, although it is of greater generality. The literature in this area is blessed by several fine volumes. In addition to the work of Walter [52], already mentioned, there is a volume by Szarski [55] and two by Lakshmikantham and Leela [56]. Two volumes on inequalities by Beckenbach and Bellman [57] and Mitrinovic [58] are also of assistance in this area of study. Here we shall be content to state a special case of the Nagumo–Westphal lemma† (see Walter [52, p. 187]) and then apply it to a semilinear diffusion problem.

Nagumo–Westphal lemma

Let $G = (0, T) \times (0, l)$, $\bar{G} = [0, T] \times [0, l]$ and $\Gamma = [0, T] \times \{0, l\} \cup \{0\} \times [0, l]$, which is the parabolic boundary of G. Consider the initial–boundary value problem (Walter [61])

$$u_t = u_{xx} + f(u) \quad t > 0, \, 0 < x < l \tag{6-63}$$

$$u(t, 0) = u(t, l) = 0, \quad t \geq 0 \tag{6-64}$$

$$u(0, x) = 0 \quad 0 \leq x \leq l \tag{6-65}$$

where f is locally Lipschitz [(6-55)], $f(0) \geq 0$, $f \to \infty$ as $u \to 1 - 0$. Let v and w be continuous in $\bar{G}$ with derivatives v_t, w_t, v_{xx}, w_{xx} continuous in G. If

$$\left. \begin{array}{l} v_t \leq v_{xx} + f(v) \\ w_t \geq w_{xx} + f(w) \end{array} \right\} \quad \text{in } G \tag{6-66}$$

$$\left. \begin{array}{l} v \leq 0 \\ w \geq 0 \end{array} \right\} \quad \text{on } \Gamma \tag{6-67}$$

† The lemma goes back to Nagumo [59] in 1939, but being written in Japanese was largely unknown. It was rediscovered by Westphal [60] in 1949 and used widely in parabolic problems.

then

$$v \le u \le w \quad \text{in } \overline{G} \tag{6-68}$$

where u is the solution of (6-63)–(6-65) which is assumed to exist in $\overline{G}$.

The specific example considered here was discussed by Kawarada [62] in 'quenching' studies and solved by Walter [61] using the preceding lemma. Kawarada's problem involved $f(u) = (1 - u)^{-1}$, which becomes singular as $u \to 1$. With this $f(u)$ the solution of (6-63)–(6-65) is said to *quench* if there exists a number $T > 0$ such that

$$\sup \{|u_t(t, x)| : 0 \le x \le l\} \to \infty \quad \text{as } t \to T - 0. \tag{6-69}$$

A necessary condition for (6-69) is

$$\max \{u(t, x) : 0 \le x \le l\} \to 1 \quad \text{as } t \to T - 0 \tag{6-70}$$

and Kawarada [62] shows that (6-70) implies (6-69) so the conditions are equivalent. Condition (6-70) is more easily accessible and hence will be used.

Global existence questions, that is existence for all $t > 0$, versus unboundedness for $t \to T - 0$, have received considerable attention. Quenching is a related phenomena when the *solution remains bounded but a derivative becomes unbounded as $t \to T - 0$*. Such phenomena are usually connected with singularities of $f(u)$. Kawarada [62] found that quenching occurs for $L > 2\sqrt{2}$. Here it is shown that $L > \pi/2$ implies quenching, while for $L < 1.5303$ there is none. Both *upper* and *lower* bounds are constructed.

Upper bounds

(a) The function $w = x(L - x)$ is ≥ 0 on Γ and (6-66) becomes $-2 + 1/[1 - x(L - x)] \le 0$. This inequality is clearly true as long as $L \le \sqrt{2}$, whereupon $0 < L \le \sqrt{2}$ implies global existence. This result is weaker than (b) but very little effort was required to obtain it!

(b) An upper bound $w = w(x)$ independent of t is established as a solution of the boundary value problem

$$w'' + \frac{1}{1 - w} = 0, \qquad w(0) = w(L) = 0.$$

Since w is symmetric with respect to $x = L/2$, set $y(x) = 1 - w(L/2 + x)$ which satisfies

$$y'' = y^{-1}, \qquad y(0) = y_0, \qquad y'(0) = 0 \tag{6-71}$$

where $y_0 = 1 - w(L/2)$. Our interest lies in the value of L such that $y(L/2) = 1$.

Integrating (6-71) by elementary techniques gives

$$\int_{y_0}^{y(x)} \frac{dy}{\sqrt{[\ln (y/y_0)]}} = \sqrt{2}\, x \qquad (6\text{-}72)$$

and with $x = L/2$,

$$\int_{y_0}^{1} \frac{dy}{\sqrt{[\ln (y/y_0)]}} = \frac{\sqrt{2}\, L}{2}. \qquad (6\text{-}73)$$

For any number $y_0 \in (0, 1)$, L is calculated from (6-73), $y(x)$ from (6-72), and a corresponding solution of w from

$$w(L/2 + x) = w(L/2 - x) = 1 - y(x), \quad 0 \le x \le L/2.$$

For all those L, w is an upper bound, independent of t, for the solution of the original problem. Thus global existence occurs.

To simplify the calculation set $t = \sqrt{[\ln (y/y_0)]}$ and $z = \sqrt{[\ln (1/y_0)]}$ in (6-73), whereupon $L = 2\sqrt{2}\, F(z)$ with

$$F(z) = e^{-z^2} \int_0^z e^{t^2}\, dt, \quad 0 < z < \infty.$$

Clearly $F(0) = 0$, $F(\infty) = 0$, F has one positive maximum at z_0 and $2F(z_0) = 1/z_0$. By iteration $z_0 \approx 0.92414$, $F(z_0) \approx 0.54104$, and the corresponding L_0 is $L \approx 1.53030$. Thus for $0 < L \le L_0$, the solution u of (6-63)–(6-65) exists globally.

Lower bounds

Lower bounds are sought of the form

$$v(t, x) = \alpha(t) s(x)$$

with $s(x) = \sin \lambda x$, $\lambda = \pi/L$, and $\alpha(t)$ to be determined. With this function v, inequality (6-66) becomes

$$\frac{\alpha'}{\alpha} + \lambda^2 \le \frac{1}{\alpha s (1 - \alpha s)}. \qquad (6\text{-}74)$$

Since the left-hand side of (6-74) does not depend upon x, the inequality still holds in the infimum of the right-hand side with respect to x. Thus

$$\frac{\alpha'}{\alpha} + \lambda^2 \le \begin{cases} 1/\alpha(1 - \alpha) & \text{if } 0 < \alpha \le \frac{1}{2} \\ 4 & \text{if } \frac{1}{2} < \alpha < 1. \end{cases} \qquad (6\text{-}75)$$

With $\alpha(t) = t$ and $L \ge \pi$, (6-75) is satisfied; i.e. for $L \ge \pi$ the solution quenches and the quenching time $T(L) \le 1$.

Now let $\alpha = \varepsilon t$, $0 < \varepsilon < 1$. Then (6-75) becomes

$$
\left(\frac{\pi}{L}\right)^2 \leq
\begin{cases}
\dfrac{1}{\alpha(1-\alpha)} - \dfrac{\varepsilon}{\alpha}, & 0 < \alpha \leq \tfrac{1}{2} \\[4mm]
4 - \dfrac{\varepsilon}{\alpha}, & \tfrac{1}{2} < \alpha < 1.
\end{cases}
$$

For $0 < \alpha \leq \tfrac{1}{2}$,

$$
\frac{1}{\alpha(1-\alpha)} - \frac{\varepsilon}{\alpha} \geq \frac{1-\varepsilon}{\alpha(1-\alpha)} \geq 4(1-\varepsilon)
$$

and for $\tfrac{1}{2} < \alpha < 1$,

$$
4 - \frac{\varepsilon}{\alpha} \geq 4 - 2\varepsilon.
$$

Consequently, for any $L > \pi/2$ there exists an $\varepsilon > 0$ such that (6-75) is satisfied. Hence the solution quenches for $L > \pi/2$ and the time to quench

$$
T(L) \leq \frac{4L^2}{4L^2 - \pi^2} \quad \text{for } L > \pi/2.
$$

A better estimate for $T(L)$ is achieved if $\alpha(t)$ is found from (6-75) using the equality sign. Then α satisfies

$$
\alpha' = \frac{1}{1-\alpha} - \lambda^2\alpha, \quad \alpha(0) = 0.
$$

The resulting upper bound (Problem 6-19) for $T(L)$ is

$$
T(L) \leq t_0 + \frac{1}{4 - \lambda^2} \log 2 \tag{6-76}
$$

where t_0 is such that $\alpha(t_0) = \tfrac{1}{2}$.

This technique is easily generalized to problems of the form $u_t = Lu + f(u)$ where Lu is a linear parabolic operator.

PROBLEM

6-19 Carry out the analysis leading to the improved estimate (6-76) for $T(L)$ and calculate $T(\pi)$, $T(\infty)$.   *Ans:* $T(\pi) < 0.6772$.

6-6 Introduction to finite elements

Acceptance and application of approximate methods that bear the title 'finite element' has been staggering. One bibliography by Norrie and de

Vries [63], covering the period 1956–1974, has over 3800 citations! This bibliography is highly recommended for all embarking on a new problem in this well-explored field. In the historical development of the method the criterion used for determining the unknown coefficients was a variational principle. This was acceptable for those problems in structural mechanics that possess variational principles. More recently the Galerkin method† has been used as the criterion, thus permitting application of the idea to problems for which no variational principle exists. The Bubnov–Galerkin method is employed throughout our discussion. Alternatives and additional examples are available in Zienkiewicz [11], Oden [12], Aziz [13], and Strang and Fix [64].

A two-dimensional diffusion problem in Cartesian coordinates (x, y)

$$\frac{\partial}{\partial x}\left(D\frac{\partial u}{\partial x}\right) + \frac{\partial}{\partial y}\left(D\frac{\partial u}{\partial y}\right) = f(x, y) \quad \text{in } A$$

$$u = u_0 \quad \text{on } \partial A$$

(6-77)

will be used to introduce finite element concepts.

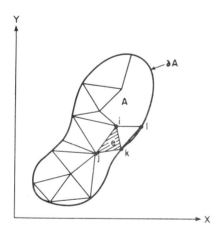

Fig. 6-1   Finite element geometry

First, the domain A is separated by imaginary lines (surfaces) into a number of finite elements (triangles, rectangles, cubes, tetrahedrons, etc.) as in Fig. 6-1. Second, the elements are assumed to be interconnected at a discrete number of nodal points on their boundaries. The values of u at these nodal points will be the basic unknown parameters. Zienkiewicz [11] discusses the introduction of parameters other than the nodal ones. Other nodeless parameters can also be employed, but we do not do so here (see

† Other WRM can also be used.

e.g. Pian [65] in elasticity calculations). Third, a trial function (or functions) is chosen to uniquely define the state of concentration (other dependent variable) within each finite element in terms of its nodal concentrations. Fourth, the Bubnov–Galerkin method is applied to find the 'optimal' values of the unknown parameters.

Consider now the *triangular* finite element e of Fig. 6-1. Within this triangle, with nodes denoted by i, j, k, the concentration u is taken as a linear function of x and y with parameters α_1, α_2, and α_3,

$$\bar{u} = \alpha_1 + \alpha_2 x + \alpha_3 y. \tag{6-78}$$

Equation (6-78), when evaluated at each of the three nodes i, j, k of e, e.g.

$$\bar{u}_i = \alpha_1 + \alpha_2 x_i + \alpha_3 y_i \tag{6-79}$$

provides a set of three equations for the constants (α_1, α_2, α_3) of (6-78). The solutions, written in terms of the nodal concentrations $\bar{u}_i$, $\bar{u}_j$, $\bar{u}_k$, are

$$\bar{u} = N_i(x, y)\bar{u}_i + N_j(x, y)\bar{u}_j + N_k(x, y)\bar{u}_k \tag{6-80}$$

where

$$N_i = (a_i + b_i x + c_i y)/2\Delta \tag{6-80a}$$

$$a_i = x_j y_k - x_k y_j, \qquad b_i = y_j - y_k, \qquad c_i = x_k - x_j \tag{6-80b}$$

plus permutations on i, j, k,

$$2\Delta = 2 \text{ (area of element } e)$$

$$= \begin{vmatrix} 1 & x_i & y_i \\ 1 & x_j & y_j \\ 1 & x_k & y_k \end{vmatrix} = (x_j y_k - x_k y_j) + (x_k y_i - x_i y_k) + (x_i y_j - x_j y_i).$$
$$\tag{6-80c}$$

Relations among the coefficients are often useful. From elementary arithmetic we find

$$a_i + a_j + a_k = 2\Delta$$
$$b_i + b_j + b_k = c_i + c_j + c_k = 0 \tag{6-80d}$$

and

$$N_i + N_j + N_k = 1. \tag{6-80e}$$

Equations (6-80) provide the form of the trial function within the finite element e.

When the point (x, y) does not fall within a triangle possessing a vertex at (x_i, y_i), then $N_i(x, y) = 0$. *Thus the trial solution consists of a sum of these trial functions, each of which is a pyramid on some triangle (finite element) and zero elsewhere.* The Bubnov–Galerkin method, with weighting function N_n and trial solution (6-80), is now applied to (6-77) with the result

$$\int_A N_n(x, y)[-(D\bar{u}_x)_x - (D\bar{u}_y)_y + f]\, dx\, dy = 0. \tag{6-81}$$

Within any triangular element possessing a vertex at (x_n, y_n), N_n is given by (6-80a). Clearly N_n is different in different elements because the a, b, c are different.

Upon examining (6-80) it is apparent that $\bar{u}$ does not have a (classical) second derivative. To evaluate those terms we are motivated by integration by parts to define

$$\int_e N_n[\nabla \cdot D\nabla u]\, dA = [N_n\, D\boldsymbol{n} \cdot \nabla \bar{u}]_{\partial e} - \int_e D\nabla N_n \cdot \nabla \bar{u}\, dA \tag{6-82}$$

where† ∂e is the boundary of e and $\boldsymbol{n}$ is the outward normal vector. Since both N_n and $\bar{u}$ have first derivatives, the right-hand side of (6-82) can be evaluated. Equation (6-82) is correct within the element e, but another difficulty is encountered in the evaluation of (6-81). To compute (6-81) the discontinuity in the first derivative of $\bar{u}$ at the boundary, and hence the infinite value of the second derivative there must be accounted for! This problem is usually resolved (in one dimension) by *defining the integral along the discontinuity* to be

$$\int_{a-\varepsilon}^{a+\varepsilon} p(x)q''(x)\, dx = p(a)[q'(a+) - q'(a-)] \tag{6-83}$$

where p is continuous at $x = a$.

A generalization of (6-83) to two dimensions is used to resolve the aforementioned problem. Consider the boundary i-k which is common to element ijk and ikl. For element ijk, the weighting functions are denoted by N_i, N_j, N_k and for element ikl by N_i', N_k', and N_l'. Each of these six functions contribute to (6-81) from the discontinuity along i-k. However, the value of the flux is constant in elements ijk and ikl. Thus the discontinuity in the derivative gives rise to the contribution

$$\int (N_i + N_j + N_k)\, \nabla^2 \bar{u}\, dA + \int (N_i' + N_k' + N_l')\, \nabla^2 \bar{u}\, dA$$

$$= D\left\{\boldsymbol{n} \cdot \nabla\bar{u}\Big|_{\substack{\text{along } ik \\ \text{inside } ikl}} - \boldsymbol{n} \cdot \nabla\bar{u}\Big|_{\substack{\text{along } ik \\ \text{inside } ijk}}\right\} \tag{6-84}$$

† It is convenient to use $\nabla = \boldsymbol{i}\, \partial/\partial x + \boldsymbol{j}\, \partial/\partial y$ where $\boldsymbol{i}$ and $\boldsymbol{j}$ are unit vectors.

where $\mathbf{n}$ is the normal pointing outward from the element ijk. The boundary term in (6-84) cancels the sum of the boundary terms in (6-82) at each interior boundary. Thus the interior boundary terms give no net contribution and can be ignored here.

Next let us examine the situation on the boundary of the region. Suppose k-l is an external boundary on which the concentration $u = u_0$ is specified. Along the boundary x and y are related by means of

$$\frac{y - y_l}{y_k - y_l} = \frac{x - x_l}{x_k - x_l}. \tag{6-85}$$

No weighting functions N_l' or N_k' are used in (6-81), and it is easy to verify that N_i' is equal to zero because of (6-85) (Problem 6-20). Thus the external boundary term in (6-81) also vanishes.

Next, the weighted residual must be evaluated for the finite element e with nodes ijk using (6-82). Thus

$$\int_e N_n \, \nabla \cdot D \, \nabla \bar{u} \, dA = -\int_e D \, \nabla N_n \cdot \nabla \bar{u} \, dA$$

$$= -\frac{1}{(2\Delta)^2} \int_e D[(b_i \bar{u}_i + b_j \bar{u}_j + b_k \bar{u}_k)b_n$$

$$+ (c_i \bar{u}_i + c_j \bar{u}_j + c_k \bar{u}_k)c_n] \, dx \, dy. \tag{6-86}$$

If D is constant, some simplifications occur. The terms in the integral are then constant, and the value of the integral is their value times the area of the element. If f is also constant, then the last integral becomes

$$f \int_e N_n \, dx \, dy = \frac{f}{2\Delta} (a_n + b_n \bar{x} + c_n \bar{y}) \, \Delta$$

where

$$\bar{x} = \frac{1}{\Delta} \int_e x \, dx \, dy, \qquad \bar{y} = \frac{1}{\Delta} \int_e y \, dx \, dy.$$

The centroid calculation gives

$$\bar{x} = (x_i + x_j + x_k)/3, \qquad \bar{y} = (y_i + y_j + y_k)/3$$

$$a_n + b_n \bar{x} + c_n \bar{y} = 2\Delta/3.$$

Assembling the final equations we have the total contribution due to N_n is then

$$\sum_{q=i,j,k} h_{nq} \bar{u}_q + F_n \tag{6-87}$$

$$h_{nq} = D[b_n b_q + c_n c_q]/4\Delta, \qquad F_n = f\Delta/3$$

and the final equations are found by the sum of (6-87) over all of the nodes

$$\sum_{\text{all nodes}} \left[\sum_{q=i,j,k} h_{nq}\bar{u}_q + F_n \right] = 0. \tag{6-88}$$

Equation (6-88) is a set of linear equations for the nodal concentrations $\{\bar{u}_q\}$. Note that the nodal concentration appears once for each element that has a vertex at that node. The foregoing derivation is simpler if a variational principle exists (Problem 6-21).

When a time derivative is included, the term

$$\int_e N_n \frac{\partial u}{\partial t} \, dx \, dy = \frac{d}{dt} \int_e N_n [N_i\bar{u}_i + N_j\bar{u}_j + N_k\bar{u}_k] \, dx \, dy$$

must be added to (6-81).

In its essentials this is the finite element method. Of course other bases can be used and many modifications are possible. The rich and varied literature should be consulted for these details.

PROBLEMS

6-20 Show that N_i' and N_k' are not used in the external boundary calculation and that $N_i' = 0$ when the boundary is at $k\text{-}l$ of Fig. 6-1.

6-21 The variational principle for (6-77) with D and f constant, is to minimize

$$\phi(u) = \int_A [\tfrac{1}{2} D \, \nabla u \cdot \nabla u + fu] \, dA + \int_{\partial A} (\tfrac{1}{2}u^2 - u_0 u) \, ds.$$

Using triangular elements show that Eqns (6-88) are again obtained. Here it is not necessary to define generalized derivatives.

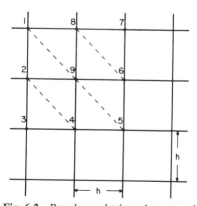

Fig. 6-2 Regular and triangular network

6-22   Consider the regular network shown in Fig. 6-2. With D and f constant, a five-point finite difference molecule applied at point 9 would yield

$$\frac{D}{h^2}(\bar{u}_2 + \bar{u}_4 + \bar{u}_6 + \bar{u}_8 - 4\bar{u}_9) - f = 0.$$

Compare this result with that using the finite element method where the triangular elements are those shown with dotted lines.

Ans: The results are equivalent.

Note: Other arrays do not necessarily yield equivalent results.

6-23   Describe a finite element calculation for an axisymmetric heat flow problem in which the equation is

$$\frac{\partial}{\partial r}\left[rk\,\frac{\partial T}{\partial r}\right] + \frac{\partial}{\partial z}\left[rk\,\frac{\partial T}{\partial z}\right] + Q = 0$$

with

$$\frac{\partial T}{\partial n} + q + \alpha T = 0$$

on the boundary.

REFERENCES

1. Ames, W. F. *Nonlinear Partial Differential Equations in Engineering*, Vol. I, 1965, Vol. II, 1972. Academic Press, New York.
2. Finlayson, B. A. *The Method of Weighted Residuals and Variational Principles.* Academic Press, New York, 1972.
3. Hrenikoff, A. *J. Appl. Mech.*, **8**, 169, 1941.
4. McHenry, D. *J. Inst. Civil Eng.*, **21**, 59, 1943.
5. Courant, R. *Bull. Am. Math. Soc.*, **49**, 1, 1943.
6. Newmark, N. M. Numerical methods of analysis in bars, plates and elastic bodies, in *Numerical Methods of Analysis in Engineering* L. E. Grinter (ed.). Macmillan, New York, 1949.
7. Prager, W. and Synge, J. L. *Quart. Appl. Math.*, **5**, 241, 1947.
8. Synge, J. L. *The Hypercircle in Mathematical Physics.* Cambridge Univ. Press, London and New York, 1957.
9. Turner, M. J., Clough, R. W., Martin, H. C., and Topp, L. J. *J. Aeronaut. Sci.*, **23**, 805, 1956.
10. Argyris, J. H. *Energy Theorems and Structural Analysis.* Butterworth, London 1960 (reprinted from Aircraft Eng. 1954–1955).
11. Zienkiewicz, O. C. *The Finite Element Method in Engineering Science.* McGraw-Hill, London, 1971.
12. Oden, J. T. *Finite Elements of Nonlinear Continua.* McGraw-Hill, New York, 1972.
13. Aziz, A. K. (ed.) *The Mathematical Foundations of the Finite Element Method with Applications to Partial Differential Equations.* Academic Press, New York, 1972.

14. Oden, J. T. (ed.) *Computational Mechanics*, Lecture Notes in Mathematics 41. Springer-Verlag, Berlin–New York, 1975. (See especially the articles by G. Fix, R. H. Gallagher, J. T. Oden and L. C. Wellford, and P. J. Roache.)
15. Biezeno, C. B. and Koch, J. J. *Ing. Grav.*, **38**, 25, 1923.
16. Frazer, R. A., Jones, W. P., and Skan, S. W. *Approximations to functions and to the solutions of differential equations*, *Gt. Britain Air Ministry Aero. Res. Comm. Tech. Rept.*, **1**, 517, 1937.
17. Lanczos, C. *J. Math. Phys.*, **17**, 123, 1938.
18. Hall, T. *Carl Friedrich Gauss*. MIT Press, Cambridge, Mass., 1970.
19. Sorenson, H. W. *IEEE Spectrum*, **7**, 63, 1970.
20. Becker, M. *The Principles and Applications of Variational Methods*. MIT Press, Cambridge, Mass., 1964.
21. Bubnov, I. G. *Sborn. inta inzh. putei soobshch*, **81**, USSR All Union Special Planning Office (SPB), 1913.
22. Galerkin, B. G. *Vestn. Inzh. Tech.* (USSR) **19**, 897, 1915; (translation 63-18924, Clearinghouse, Fed. Sci. Tech. Info., Springfield, Va.).
23. Yamada, H. *Rept. Res. Inst. Fluid Eng.*, *Kyushu Univ.*, *Kyushu, Japan*, **3**, 29, 1947.
24. von Kármán, T. *Z. Angew. Math. Mech.*, **1**, 233, 1921.
25. Pohlhausen, K. *Z. Angew. Math. Mech.*, **1**, 252, 1921.
26. Crandall, S. H. *Engineering Analysis*. McGraw-Hill, N.Y., 1956.
27. Collatz, L. *The Numerical Treatment of Differential Equations*. Springer, Berlin–New York, 1960.
28. Courant, R. *Proc. First Int. Cong. Appl. Mech. Delft.*, **17**, 1924.
29. Rayleigh, 3rd Baron (Strutt, J. W.), *Trans. Roy. Soc. London*, **A161**, 77, 1870.
30. Ritz, W. *J. Reine u. Angew. Math.*, **135**, 1, 1909.
31. Timoshenko, S. and Goodier, J. N. *Theory of Elasticity*, 2nd. edit., p. 259. McGraw-Hill, New York, 1951.
32. Lanczos, C. *Applied Analysis*. Prentice-Hall, Englewood Cliffs, N.J., 1956.
33. Clenshaw, C. W. and Norton, H. J. *Comp. J.*, **6**, 88, 1963.
34. Norton, H. J. *Comp. J.*, **7**, 76, 1964.
35. Wright, K. *Comp. J.*, **6**, 358, 1964.
36. Villadsen, J. V. and Stewart, J. P. *Chem. Eng. Sci.*, **22**, 1483, 1967.
37. Villadsen, J. V. *Selected Approximation Methods for Chemical Engineering Problems*. Inst. for Kemiteknik, Numer. Inst., Danmarks Tekniske Højskole, 1970.
38. Jackson, D. *Theory of Approximation*. Amer. Math. Soc. Colloq. Pub. Vol. 11, New York, 1930.
39. Szegö, G. *Orthogonal Polynomials*. Amer. Math. Soc. Colloq. Pub. Vol. 23, New York, 1959.
40. Hildebrand, F. B. *Introduction to Numerical Analysis*, Chapter 7. McGraw-Hill, New York, 1956.
41. Stroud, A. H. and Secrest, D. *Gaussian Quadrature Formulas*. Prentice-Hall, Englewood Cliffs, N.J., 1966.
42. Kopal, Z. *Numerical Analysis*. Wiley, New York, 1955.
43. Harrington, R. F. *Field Computation by Moment Methods*. Macmillan, New York, 1968.
44. Swartz, B. and Wendroff, B. *Math. Comp.*, **23**, 37, 1969.
45. Mikhlin S. G. *Variational Methods in Mathematical Physics*. Pergamon, Oxford Univ. Press, London, 1964.

46. Collatz, L. *Functional Analysis and Numerical Mathematics*. Academic Press, New York, 1966.
47. Mikhlin, S. G. and Smolitsky, K. L. *Approximate Methods for Solution of Differential and Integral Equations*. American Elsevier, New York, 1967.
48. Babuska, I., Prager, M., and Vitasek, E. *Numerical Processes in Differential Equations*. Wiley, New York, 1966.
49. Kantorovich, L. V. and Krylov, V. I. *Approximate Methods in Higher Analysis*. Wiley, New York, 1958.
50. Yasinsky, J. B. and Kaplan, S. *Nucl. Sci. Eng.*, **31**, 80, 1968.
51. Protter, M. H. and Weinberger, H. F. *Maximum Principles in Differential Equations*. Prentice-Hall, Englewood Cliffs, N.J., 1967.
52. Walter, W. *Differential and Integral Inequalities*. Springer, Berlin, 1970.
53. Ferguson, N. B. Orthogonal collocation as a method of analysis in chemical reaction engineering. Ph.D. Thesis, Univ. of Washington, Seattle, Wash., 1971.
54. Boley, B. A. *Proc. Symp. Nav. Struct. Mech.*, *3rd, Columbia Univ.*, pp. 260. Pergamon Press, Oxford, 1964.
55. Szarski, J. *Differential Inequalities*. Monografie Matematyczne, Vol. 43, Warsaw, Poland, 1965.
56. Lakshmikantham, S. and Leela, S. *Differential and Integral Inequalities*. Academic Press, New York, 1971.
57. Beckenbach, E. F. and Bellman, R. *Inequalities*. Springer, New York, 1965.
58. Mitrinovic, D. S. *Analytic Inequalities*. Springer, New York–Berlin, 1970.
59. Nagumo, M. See Walter [52, p. 335].
60. Westphal, H. Zur Abschätzung der Lösungen nichtlinearer parabolischer Differentialgleichungen, *Math. Zeit.*, **51**, 690, 1949.
61. Walter, W. Parabolic differential equations with a singular nonlinear term. Lecture at Oberwolfach, Germany, November, 1975.
62. Kawarada, H. On solutions of initial-boundary problems for $u_t = u_{xx} + (1 - u)^{-1}$, *Publ. RIMS, Kyoto Univ., Japan*, **10**, 729, 1975.
63. Norrie, D. H. and de Vries, G. *A Finite Element Bibliography* (I. Author Listing; II. Keyword Listing; III. Citation Listing) Mech. Eng. Dept. Repts. 57, 58, 59. University of Calgary, Canada, (May) 1975.
64. Strang, G. and Fix, G. *An Analysis of the Finite Element Method*. Prentice-Hall, Englewood Cliffs, N.J., 1973.
65. Pian, T. H. H. *AIAA J.*, **2**, 576, 1964.

Author Index

Numbers in parentheses are reference numbers and indicate that an author's work is referred to although his name is not cited in the text. Numbers in italics show the page on which the complete reference is listed.

Subject Index